W9-DFS-468

ENERGY FOR MAN

*the text of this book is printed
on 100% recycled paper*

ENERGY FOR MAN

FROM WINDMILLS TO NUCLEAR POWER

HANS THIRRING

With a New Introduction by
Murray Bookchin

Carnegie Library
Livingstone College
Salisbury, North Carolina 28144

HARPER COLOPHON BOOKS

HARPER & ROW, PUBLISHERS

NEW YORK, HAGERSTOWN, SAN FRANCISCO, LONDON

This book was originally published in Great Britain in 1956 by
George G. Harrap & Co. Ltd. A second edition was published in the
United States in 1958 by Indiana University Press and is available in
hardcover from Indiana University Press. It is here reprinted by
arrangement.

ENERGY FOR MAN. Copyright © 1958 by Indiana University Press.
Introduction to the Harper Colophon edition copyright © 1976 by
Harper & Row, Publishers, Inc. All rights reserved. Printed in the
United States of America. No part of this book may be used or reproduced in any manner without written permission except in the case
of brief quotations embodied in critical articles and reviews. For information address Harper & Row, Publishers, Inc., 10 East 53d Street,
New York, N.Y. 10022. Published simultaneously in Canada by Fitz-
henry & Whiteside Limited, Toronto.

First HARPER TORCHBOOK edition published 1962.

First HARPER COLOPHON edition published 1976.

STANDARD BOOK NUMBER: 06-090527-1

76 77 78 79 80 10 9 8 7 6 5 4 3 2 1

621.4
T447

Contents

94689

caveat

MURRAY BOOKCHIN

Introduction to the Harper Colophon Edition

"ADDING solar heat to the energy system of a country helps to increase the wealth of the nation, and if all houses in areas with favourable conditions were equipped with solar heating systems, fuel saving worth millions of pounds yearly could be achieved. The work of Telkes, Hottel, Löf, Bliss, and other scientists who are paving the way for solar heating is real pioneer work, the full significance of which will emerge more clearly in the future."

Written in the late fifties, when much of the Western world had rushed into a heady commitment to oil-heating systems, more and larger automobiles, sweeping highways, and staggering increases in the consumption of petroleum, this vision by a sober physicist must have seemed bizarrely utopian. Today it is compelling fact, acknowledged even by the staid U.S. Energy Research and Development Administration. At present we would have to add such new names to individuals working in solar space-heating as Steve Baer, Eugene Eccli, Harold Hay, Robert Reines, and Harry Thomason among many others. We would have to emphasize the importance of whole-systems approaches to the energy problems that include horticulture in combination with a wide variety of energy sources—work that has been pioneered by John and Nancy Todd of the New Alchemy Institute of Massachusetts, the Institute for Social Ecology at Goddard College in Vermont, and the Farallones Institute in California. The people and institutions involved in alternative energy systems would easily fill a sizeable directory, and new ones, including new designs, reach us at an almost dizzying tempo.

Thirring, to be sure, forewarns his readers that, although he is not a "science-fiction writer, but a professional scientist, part of this book deals with the possibilities and prospects of future power production." This is a caveat to let his readers and colleagues know that they are dealing with an imaginative man. To provide evidence of his prophetic insights, Thirring retrospectively adduces a "good many" of his predictions that time "fully verified later on." He cites two of his earliest beliefs that "aeroplanes" rather than lighter-than-air craft held the future of aviation and that silent films would be replaced by sound films, an opinion that, according to Thirring, even Charlie Chaplin regarded as "quite impossible both for technical and artistic reasons." Perhaps more ominously, he foresaw the construction of a hydrogen bomb four years before Truman revealed the project of making such a weapon. Thirring does not "boast of any visionary gifts," but the fact is that he has them and his credentials on this score are as impressive as his technical background. And above all, he is a humanist. "What is the use of all the progress in science and technology, of the increase in industrial output," he asks, "if its greater part is consumed in the production of destructive weapons? As long as the great national efforts are opposed to one another all progress on one side will be counteracted by increased efforts on the other side, thus causing more economic strain on each nation, more labour for the toiling population, without alleviating the lack of food, housing, and clothing of all classes except a small minority of well-to-do people."

It is this rare mix of seer, technician, and humanist that makes a book written nearly two decades ago so valuable today. In a field that has developed with such incredible rapidity, Thirring's *Energy for Man* still ranks as one of the most solid introductions to technical fields as disparate as steam engines and electric generators, internal combustion engines and windmills, gas turbines and solar collectors, heat pumps and hydraulic generators, plant-originated fuels and radioactive elements. One would reasonably think that a book of less than four hundred pages could not possibly cover so wide a range of material without leaving frustrating gaps in basic concepts. Yet what makes *Energy for Man* so remarkable a work is that Thirring, with unmatched artfulness and selectivity, manages to glean from the highly specialized literature on energy precisely the appropriate technical information that one would need to have an excellent insight into each particular subject. The reader in the field does not have to wade through scores of volumes in physics and engineering to learn the essentials of thermodynamics or electrochemistry. They appear, here, in Thirring's book—neither too little to frustrate us nor too much to burden us.

Thirring's material is marked neither by a superficial eclecticism nor by ponderous specialization. Stylistically, it tends to be clearest at exactly those points where one would expect a numbing use of definitions, tables, and formulas. Here and there, the totally uninformed reader may stumble if he or she does not take the pains to read the text with care. Happily, the book is so remarkably comprehensive as to make no scientific demands other than a knowledge of rudimentary mathematics and the use of a reasonably good dictionary. Even when Thirring discusses chemical formulas in connection with the constituents of petroleum (and he does go into that kind of detail, however synoptically, when he deals with a subject), he is at pains to be lucid in distinguishing aromatic from chain-like structural formulas. At its best, the book has a richly imaginative quality, prophetic insight, and a deep sensitivity to human needs. It can be read from cover to cover as comfortably as it can be used as a reference work. Finally, Thirring's *Energy for Man* contains that indispensable bedrock information which teachers, students, and above all, the citizen who is concerned with energy policy must know if he or she is not to be victimized by "experts" who are in the service as much of special interests as of their own specialties. Once read, digested, and used properly, it becomes a technical declaration of independence for the citizen-amateur to utilize in exercising social control over a vital area of life—the energy that moves our technology and a basic stratum of technology that shapes our everyday lives. As a teacher, I cannot conceive of a satisfactory course on energy and power technology—conventional or alternative—that does not use this work. As a libertarian, I cannot think of a society whose members can afford to ignore its wealth of information in managing technology, unless, of course, they are prepared to surrender their autonomy to specialists and professionals.

What have the passing years altered in this book? What would be required to update it? These questions are difficult to answer—not for want of sufficient technical material but precisely because there is such an abundance of data, indeed so many new designs flooding in upon us from such a wide variety of disciplines, that what is probably the most advanced information today may well be rendered obsolete a year or two from now. Even energy consumption patterns seem to be changing as a result of soaring inflation, so that the patterns cited by Thirring for the 1950s may well hold true for the future, however different they may appear today. Consider the relative consumption of coal, petroleum, natural gas, water power, and wood for the United States in Figure 6, page 40. The pattern suggests that the largest single source of energy was coal (about 40 per cent), followed by petroleum (about 35 per cent), and natural gas (about 25 per cent). By 1970 coal had moved to the bottom, occupying third place; natural gas ranked second; and petroleum was the leading source of power. Projections tended to increase the expected proportionate consumption of petroleum and natural gas with respect to coal and assigned an increasing role to nuclear power.

Ironically, the mid-seventies suggest that we may well return to the very pattern that Thirring described for his own day in the 1950s. France and West Germany, to be sure, have veered toward greater use of nuclear power; the United Kingdom, with immense oil reserves in the North Sea, may have a heavily oil-based energy system and the same may be true for other northern European countries. The Soviet Union, in turn, with its recently discovered Siberian oil fields, may well rival the Arab countries as a petroleum producer. So far as the United States is concerned, however, I could find no satisfactory substitute for the patterns charted by Thirring in the 1950s. If fossil fuels remain the principal source of energy, Thirring's data for 1950 may not be so far-fetched as they seemed a decade ago. Generally, coal may prove to be the most plentiful fossil fuel available to American and ultimately all high-technology societies in the future.

Thirring's chapter on steam is still as valid and comprehensive as it was when he wrote it. To his chapter on internal combustion engines one might add a section on the rotary engine and hydrogen-powered engines. Since much of the chapter deals with motor vehicles as such, observations on steam and electric engines might well be appropriate. The chapter on gas turbines would require some brush strokes to bring it up to date, but it is still a solid and highly reliable work for what it tries to achieve. Similarly, Thirring's chapters on heat pumps and electricity require little emendation. His chapter on coal (his prognosis that coal will be used very extensively in the future is truly prescient) could benefit from additional material on coal gasification and synthetic fuels. Pilot research projects on petroleum derived from oil shale are being established listlessly after much press bombast. More serious efforts seem to be underway by public utilities—with ample governmental assistance—to establish pilot projects for the gasification of coal into a highly combustible fuel. At this writing, on the Navajo reservation in New Mexico, the El Paso Natural Gas Company is converting some 26,000 tons of coal per day into 250 million cubic feet of pipeline gas by using the Lurgi process. Methanation, which could more than double the heating value of the gas, forms the final step in a basic coal gasification process in which oxygen and steam are added under conditions of high pressure and heat to form a gas of carbon oxides, hydrogen, methane, and sulfur compounds. Another process, the so-called "Hygas" technique, involves mixing coal with light oil and hydrogenating the mixture to produce methane. This technique is being developed by several large Chicago corporations. Still another process is Donath's "bi-gas" technique, a two-stage process in which coal is turned into methane-rich gas or, if desirable, into synthetic oil and gasoline.[1] It may be appropriate to add, here, that Germany, during World War II, constructed twenty-four coal-hydrogenation plants, ten of which furnished the Luftwaffe with most of its aviation gasoline.

On electric power generation, it would be necessary to add some remarks on magnetohydrodynamic (MHD) generators, which directly produce electricity from hot gas, thereby avoiding the need for steam generation. MHD generators operate at temperatures substantially higher than those of conventional steam plants and at efficiencies that exceed these plants by nearly 50 per cent. Although the technique (no new sources of energy are involved) at this writing has not advanced much beyond the research stage in the United States, it is being used as a power source on a limited though increasing scale in the Soviet Union. A discussion of improved fuel cells, geothermal energy, hydrogen fuels, and the use of solid wastes would also have to be included in the book to update it. Geothermal energy and solid wastes from cities already afford appreciable sources of energy in various parts of the world. They have been so widely popularized in newspapers and pulp magazines that they hardly require more than mere mention here. Newly developed fuel cells that, in principle, are no different from our ordinary batteries could probably best be used as energy sources on a decentralized scale for individual buildings, where they could be installed and provided with methane to generate electric power. Wilson Clark, in a fascinating review of the subject, notes that the newer fuel cells could be most efficiently employed when they are "dispersed and serve the ultimate customer directly, because this eliminates the energy-wasting step of transmitting electricity." But the home fuel cell, although being investigated in pilot projects here and there, is apparently viewed in Clark's opinion "as a distinct threat to [the public utilities'] growing control over home energy use. The reason electric utilities are ordering fuel cells is that they would like to control the future development of this technology and prevent the wide use of fuel cells in the best location—the homes and buildings, the sites of power consumption."[2]

[1] For a comprehensive appraisal of five practical processes for coal gasification that are now under study, see Harry Perry, "The Gasification of Coal," *Scientific American,* March 1974, pp. 19–25.

[2] Wilson Clark, *Energy for Survival* (New York: Doubleday & Company, Anchor Press, 1974), p. 228. This book is a marvel of research and forms a valuable companion work to Thirring's book.

Thirring's discussion of fuels from vegetation would intrigue those interested in alternative energy sources today. They would welcome his emphasis on the use of wood as a fuel in highly forested areas—and his concern for preserving our woodlands for conservation as well as for economic reasons. Here, too, we would have to take note of improvements that have been made in wood-burning stoves and in the design of fireplaces.[3] In his discussion on farm wastes, Thirring was apparently not aware of the methane digesters that were to become so fashionable a decade and a half after his book was published.[4] His extended discussion on chlorella as a fuel source, while quite understandable in the fifties, when algae and seaweed seemed to hold considerable promise as a combustible material, is no longer satisfactory. These bioconversion techniques may prove valuable at some future time, but today they are marginal at best when compared with other sources of alternative energy.

The most exciting chapter in Thirring's book deals with solar, wind, and tidal power. Thirring's choice of solar home-heating to open his discussion is typical of his prescience: today, space-heating by solar radiation is one of the most promising techniques for diminishing and, in many areas, eliminating the use of fossil fuels for home-heating purposes. Nearly two decades ago, Thirring was in no position to deal with the elaborate and costly plumbing technology used in the famous Dover House in Massachusetts. Encumbered by valves and a network of pipes, the experiments that approximated a high-technology vision of solar space-heating may well have served to set back rather than advance the use of this technology. A number of designers were driven to despair over the prospects of solar energy as a competitively significant source of home heating. Solar cells, even more costly as devices for directly converting solar energy into electricity, seemed to hold promise

[3] See Ken Kern, "Heating and Cooking with Wood," in *Producing Your Own Power*, edited by C. H. Stoner with the technical guidance of Eugene and Sandra Fulton Eccli (New York: Random House, Vintage Books, 1975). Any work that has involved Eugene Eccli—and most certainly his new compilation, *Low-Cost, Energy-Efficient Shelter* (Emmaus, Pa.: Rodale Press, 1976)—can be singled out as a source of the most authoritative material on alternative sources of energy.

[4] For the most comprehensive, readable, and practical treatment of methane digesters, the reader should examine *Methane Digesters for Fuel and Fertilizer, New Alchemy Newsletter* No. 3 (Spring 1973), edited by Richard and Yedida Merrill (available from the New Alchemy Institute, Box 432, Woods Hole, Mass. 02543).

only for the distant future. Despite the sophistication of the cells, their increased efficiency, and their use in space programs, the cost of producing them on a large scale seemed prohibitive.[5]

Although a dark cloud of pessimism began to gather around the tested space-heating designs of the late fifties and early sixties, a decisive breakthrough had already occurred at exactly this time. Working largely on his own, Harry Thomason, a patent attorney for the Army Signal Corps and a former refrigeration engineer, developed a remarkably simplified collector design and storage system that appreciably reduced the cost and maintenance problems associated with earlier efforts. By stripping down his pitched roof collector to corrugated metal sheets in which the troughs replaced pipes, and by using a water tank embedded in rocks as his principal storage system, Thomason removed the need for an elaborate system of collector tubing and a fairly vulnerable heat-retention technique involving eutectic salts. The first Thomason house had been constructed by the late fifties in Washington, D.C. On the basis of price levels during this period, Thomason reported that the cost of the collector averaged about $1 per square foot and the entire installation had involved an expenditure of approximately $2,500. About 95 per cent of the house's heating requirements during the winter of 1959–60 had been supplied by solar energy. The cost of operating Thomason's conventional oil-heated backup system had totalled less than $5. During the following winter, the coldest Washington had experienced in more than forty years, the bill had increased merely another $1.50. The collector, circulating water during Washington's summer nights, served as a cooler, diminishing house temperatures by eight to fifteen degrees in the hot, humid days that occur so frequently in the area.[6]

The essential feature of Thomason's structure was not so much its cheapness as its almost beguiling simplicity. The Dover House had been plagued by mechanical problems and difficulties in dealing with the cans of Glauber's salts that were used in its storage system. The Thomason design is now very much in vogue for a host of uses, al-

---

[5] Which is not to question the future of mass-produced solar cells as a major source of domestic electric power. There is a strong likelihood that sheets of cadmium sulfide cells will be available in the next few years for newly constructed homes, but whether they will prove to be fully compatible with conventional electric systems remains to be seen.

[6] Accounts and house plans of the Thomason houses are obtainable at a very modest price from the Edmund Scientific Co., 150 Edscorp Building, Barrington, N.J. 08007.

though in highly modified form. Water may be sprayed over flat plates instead of corrugated ones; the storage system may consist of a concrete tank instead of a metal one. More recently, the Thomason-type roof collector has been reared over a well-insulated wall or a greenhouse, often using plexiglass-type sheets instead of glass, a combination that has been developed at the Goddard College Institute for Social Ecology to serve as a greenhouse for both seedlings and edible food.

In New Mexico, Steve Baer has stacked blackened, water-filled 55-gallon drums to form a wall for collecting solar energy during the day. In the evening, the heat in the drums is prevented from dissipating into the surrounding air by movable walls. Baer's work has been so strikingly innovative that full descriptions of his bead wall, night wall, sky lid, and drum wall, among other items, would require an entire section in their own right.[7] In California, Harry Hay has developed a remarkably clever "water bed" collector system for flat-roofed structures that combines practicability with inexpensiveness. Day Charhoudi, currently associated with the Massachusetts Institute of Technology, has designed a 950-square-foot "biosphere" that integrates a home and a solar heater based largely on a greenhouse in which earth is used as a heat-storage resource. At Coos Bay, Oregon, Henry Mathew uses flat vertical plates to collect solar energy for a fairly conventional house, but his installation involved the use of an enormous 8,000-gallon tank to store the heat acquired by the collector system. Robert Reines has worked out an integrated laboratory complex of his own involving the use of wind generators, solar collectors, and domes.[8]

One could list an ever-growing variety of small-scale solar space-heating techniques. The number of designs is certain to increase in the years ahead. Increasingly, the emphasis in small-scale solar technology has shifted from broad principles, many of which have been known for years, to clever adaptations of these principles to uniquely different

[7] The interested reader is advised to communicate with Baer's Zomeworks Corporation, P.O. Box 712, Albuquerque, N.M. 87103, for his flyer listing plans and explanations of his designs.

[8] The most detailed account I have seen of Hay's house appears in *Alternate Sources of Energy,* a compilation edited by the Ecclis and available from Alternate Sources of Energy, Route 2, Box 90A, Milaca, Minn. 56353. Charhoudi's work appears in a large, fairly detailed poster that, to my knowledge, is not readily available. The Mathew house plans are available for $10 from Henry Mathew, Route 3, Box 768, Coos Bay, Ore. 97420. Reines has not published his material, but the interested reader can communicate with Integrated Living Systems, Star Route 103, Tijeras, N.M. 87059.

regions. Moreover, in sharp contrast to the large-scale, high-technology installations that initially glamourized solar energy as a topic for public discussion—projects such as Peter Glaser's space station, which would use solar cells to produce microwaves and beam them 22,000 miles back to earth, or Aden and Marjorie Meinel's proposal for converting some 5,000 square miles of desert land into large "solar farms"—attention has focused largely on decentralized technologies scaled to human dimensions, technologies that are comprehensible to the public, simple to produce, and subject to direct control by moderate-sized or small companies.

Much the same emphasis appears in the studies that are being currently made in wind power. Thirring's discussion of windmills is almost exclusively confined to wind as a source of electric power and deals in large part with the massive Smith-Putnam turbine reared in the early forties on Grandpa's Knob, a hill twelve miles west of Rutland, Vermont. This structure, although dramatic in appearance (it could be seen from distances of twenty-five miles) and pioneering in its goals, would now be obsolete. Eight-ton blades, such as those used on the turbine, are not likely candidates for future designs unless large-scale installations based on recent aeronautic designs prove to be practical.[9] William E. Heronemus of the University of Massachusetts at Amherst has proposed stringing a modified Danish wind-generator design on soaring 850-foot towers across flat country or on marine installations rather than mountains. The plan is not dissimilar in its scope and gigantism from the Meinels' "solar farms," although it is obviously more practical than Glaser's science-fiction satellite conception. High technology does not merely focus on size and immensity of scale; more laudibly, it has shown a keen and progressive concern with the application of aeronautic rotor designs. Indeed, if any impressive breakthroughs occur in developing high-technology wind-generator installations, they will probably be in this area of research rather than in high towers and complex transmission systems.

Thirring's section on wind power lacks any discussion of small-scale windmills. This is understandable if one considers the time during which it was written. By the late fifties, rural electrification had virtually eliminated the inexpensive "prairie" windmills that were prevalent

---

[9] A delightful personal account of the Smith-Putnam turbine appears in Palmer C. Putnam's *Power from the Wind* (New York: Van Nostrand Reinhold Co., 1948), a book that despite the passing years is well worth reading today.

in central and western regions of the United States forty years ago. Like many newer designs in use today—and these vary so greatly in number that I can only single out the most popular, such as the vertical Savonius rotor, the vertical axis ("egg beater") wind turbine, the Windworks "12-footer," the exotic-looking Zephyr, and the "three-wheel" RD 7000—the old "prairie" windmills were designed to pump water. Both the old and the new designs are eminently serviceable for collecting water for domestic purposes, irrigation, and recirculating the water in aquacultural installations such as those at New Alchemy Institute. Perhaps the simplest, least expensive, and most serviceable windmill for this purpose is a very traditional design, the sail-wing windmill, which involves the use of sails as rotors. The array is then connected to a crankshaft that translates the rotary motion of the sails into the vertical motion of a connecting rod for operating a pump. The New Alchemy Institute's design has proved so highly effective that, with various modifications, it is being widely emulated.[10]

Almost every windmill I have discussed could also be used to generate electric power, although the results vary enormously from one design to another. If one accepts a kilowatt as the theoretical threshold for a viable domestic unit (and the reader must clearly bear in mind that such a figure is meaningless except for comparative purposes, because a fairly windy area, even at low velocities, may provide more electric power than one with high winds that are merely episodic); if, moreover, we follow the most commonly used or recommended elevations for specific wind generators, the following results are likely to obtain: a typical three-stage Savonius is not likely to reach our theoretical threshold until wind velocities reach 32 miles per hour; the "egg-beater" design requires 14–16 miles per hour; the Windworks, 20 miles per hour; the Zephyr, 10–12; the RD 7000, 14. These are very rough figures; they merely hint at the different efficiencies at which the units operate. On the other hand, a Savonius rotor will begin to rotate, however low its r.p.m., at the most trifling wind speeds, while a Zephyr hardly begins to get underway until the wind exceeds 8 miles per hour. Clearly the most effective use of small-scale wind power requires a combination of several different designs at a single site. Either singly or in combination, they can also be used to supplement electric power acquired from conventional sources. One of

---

[10] A detailed account and the appropriate schematics for constructing the windmill appear in the 1974 issue of *The Journal of the New Alchemists* (The New Alchemy Institute, P.O. Box 432, Woods Hole, Mass. 02543; $4.00).

the most attractive wind generators for electric power is still a relatively old design, the famous Jacobs 1800 and 2500 watt plants. These superbly wired generators and their European modifications begin recharging batteries at fairly low wind speeds and reach our threshold in 12-mile-per-hour winds.

Finally, as to Thirring's very extensive discussion of nuclear energy, one may fairly say that it is immensely informative for the space allotted by the author, but I would add little more in praise of his account. The issue of reactor technology is beclouded by such conflicting claims that many nuclear technologists vary anywhere from ten to fifty years on when we will enter the "fusion age." It would certainly be very hazardous to take a responsible stand on the basis of current estimates. Nor is it entirely clear that fusion is as "clean" a source of energy as many of its proponents claim it to be. Tritium, if produced in any appreciable quantity, is a fairly long-lived radionuclide. Indeed, any neutron activity opens the risk of radioactive pollution. As to fission, I cannot emphasize too strongly my firm conviction that an energy program based on this technology is carrying us on a headlong course to ecological disaster. The growing costliness of this technology, its wastefulness of precious resources, the multitude of mechanical defects that have closed down one reactor after another for lengthy periods, and the mindless irresponsibility of the Atomic Energy Commission in disposing of highly radioactive wastes—all this wanes beside the prospect of an energy technology based on millennia of nerve-wracking monitoring of some of the most dangerous toxicants known to science. We are accumulating, leaking into complex food webs, and haphazardly dealing with toxic materials that at present levels alone will be the objects of grim concern for tens of thousands of years. The entire history of civilization, with its series of devastations and upheavals, would form only a trifling fraction of the time span this technology has "invested" in the storage and monitoring of already accumulated, highly lethal, and indisputably carcinogenic wastes from reactor operations. It is numbing to think that the present generation can seriously discuss, much less implement, nuclear programs that will increase the immense burden we and future generations will be obliged to bear in the form of long-lived radioactive wastes.

Whatever humanity's hopes may be for a world in which human needs, technology, and the natural world will be reconciled rest in great part on the "ecotechnologies" (to use the much-borrowed term I coined years ago) based on solar, wind, methane, geothermal, and related sources discussed above. Each of these technological components is only marginally significant when taken by itself, except in uniquely favorable areas and biomes. It is doubtful if the downtown districts of large cities could be lighted by solar energy or wind power. It is also doubtful if society's needs for combustible fuels could be satisfied by methane digesters. In a highly centralized society based on densely populated areas, we would require energy in such massive and concentrated quantities that the techniques developed by contemporary pioneers in alternative technology would seem irrelevant, if not utterly utopian. Yet we know only too well that our sources of fossil fuels are limited and that nuclear alternatives open the long-range prospect of ecological disaster. The change in our energy practices involves not only far-reaching technological decisions but also social ones. Increasingly, we must think of energy in terms not merely of rational, indeed ecological, techniques for producing it, but also of the social changes needed to utilize these techniques in a truly ecological fashion.

Primary on the social agenda that would render alternative technology relevant to contemporary human problems is the decentralization of the giant concentrated cities and industries on which our society is based. "Small *is* beautiful," to use E. F. Schumacher's phrase, not merely because the technology that smallness yields is "soft" or "intermediate." However felicitous these features may be, the terms "soft" and "intermediate" deal with external properties of a phenomenon, just as color, shape, and weight may help to identify a thing but tell us very little about its intrinsic character. Small really becomes beautiful when it is ecologically different—rather than dimensionally different—from conventional forms of urban, industrial, and one could justifiably add, political life. Accordingly, the small-scale technology and whole systems pioneered by groups such as The New Alchemy Institute are exciting because they are also human-scaled and gently tailored to the ecosystems in which they are located. Not only do they allow for an artistic integration of technology with the surrounding biotic environment, but they are controllable, comprehensible, and usable to people without the burdening, often obstructive expertise of technicians, who tend to preempt the individual's control over his or her own life. Solar houses, windmills, and methane digesters need not be remote technologies, removed from popular control by an inordinate professionalism that denies people any sense of relationship with technology and with the natural world. Such alternative technologies are inherently libertarian. They tend to give people a new recognition of their ability to control the instrumentalities not only of social life but of day-to-day material life.

From a technological viewpoint, decentralized alternative energy systems allow for the use and interplay of an immense variety of new components as part of a complex energy system, indeed, an energy ecosystem. Power derived from the sun can be integrated with power derived from the wind; both can be further used to produce an artful integration of solar and wind power with methane, hydroelectric, geothermal, hydrogen, and like sources of energy. Decentralized communities need not pose the formidable energy problems raised by giant cities and metropolitan areas. Allowing for varying shifts in emphasis if our communities exist in areas of considerable direct sunlight or fairly constant winds, we could emphasize one new technological component over another. But we would not want to sacrifice variety as such and the many compensating factors that would serve to move alternative technology from the margin of society to its center.

Yet even the most far-reaching measures toward decentralization would be inadequate if we failed to alter our sensibility toward each other and toward nature. That we can no longer think in terms of the "domination of nature" has become so widely acknowledged that it is now a truism. That this concept of domination stems from the domination of human by human is less easy to see. Ultimately, the manner in which we deal with each other—whether as objects of exploitation or as subjects in a relationship of mutuality and understanding—will either sustain or subvert the manner in which we deal with the natural world. The tendency, today, is to cope with our ecological problems largely as though they were technological ones. Our hopes are placed not in new ecocommunities but in huge solar satellites in space and great wind generators lined across our plains and oceans. If we are prepared to accept an alternate technology, it is still within a social framework that will permit the encroachment of concrete on our diminishing fertile soils, an insensate system that fosters an appalling waste of materials, an exploitative economy, and utterly dehumanizing social relations.

I can hardly emphasize too strongly that such an approach to energy problems is "ecotechnocratic," not ecological. In the long run humanity would gain little from technological advances in alternative energy systems that "plugged" society into the grid system of multinational corporate enterprises—a system that could irreparably damage the planet merely by simplifying it, turning varied organic ecosystems into homogenized inorganic industrial systems—no matter to what extent homes were removed from the grid system of conventional power utilities. Thirring, in the course of questioning the value of scientific and technological progress that fails to meet human needs, could well ask the same question in terms of alternate energy. What would be the value of progress even in new energy systems if, without changing our sensibilities and social relations, we devised less-polluting, seemingly "inexhaustible" sources of energy? This question, I submit, lies at the heart of any serious discussion of the future of technology. No technological issue can be defined on its own terms. It always has a social dimension, however much this dimension may be masked by technical jargon. A discussion of solar energy, wind power, methane, and other popular sources of alternative energy that fails to deal with the way these sources will be used inherently presupposes the way energy is being used at present—in an exploitative manner toward humanity and toward nature. The reader of Thirring's book must ultimately search beyond the promise of technology to its roots in human needs and social relations. Without that search, not only do our energy difficulties remain unsolved; worse, we unknowingly incorporate the social roots of our difficulties into the problem itself and are necessarily captive to false "solutions."

MURRAY BOOKCHIN
November, 1975

School of Metropolitan &
    Community Studies
Ramapo College of New Jersey
Mahwah, New Jersey

Institute for Social Ecology
Goddard College
Plainfield, Vermont

The Three Great Tasks

HUMANITY to-day is faced by three great tasks which are mutually interconnected.

1. *Food Production*

The number of inhabitants of our globe is approximately 2400 millions, and is increasing by about 70,000 every day, so that by the end of the century world population will reach between 3000 and 4000 millions. Food production, insufficient as it is now, is decreasing rather than increasing, because of wars and civil strife in some of the most important agricultural areas, and because of the dangerous progress of soil erosion in many parts of the world. The present irresistible advance of the southern border of the Sahara may serve as a warning signal of similar dangers elsewhere.

Two-thirds of the whole population of the world is suffering to-day from squalor, hunger, and deficiency diseases. Unless the disproportion between increasing birth-rates and stagnating food production is changed radically within the next generations 90 per cent of humanity will live a miserable life in the midst of hunger and disease. And it would be a fatally short-sighted policy of the haves towards the have-nots not to be concerned about their troubles. For the awakening nationalism of the former colonial peoples, strengthened by their overwhelming population figures, will make them refuse to tolerate the export of food from regions of starvation, and therefore the supplying of food to the haves who, like Britain and others, are dependent on imports of food will become ever more difficult. In a world which, thanks to scientific and technological progress, is tending more and more to become one great society differences in the level of standards of life, though unavoidable, of course, cannot be maintained any longer in such extreme proportions as they have been from the beginnings of civilization till to-day. In order to save human society from collapse we need a world-wide increase of food production in the second half of this century, and a stabilization of world population at a ceiling of, say, between six and ten thousand millions in a further future.

2. *Power Production*

One of the most important auxiliary problems connected with food production is industrial power. Individual farmers, as well as co-operative societies, need all kinds of agricultural tools, ranging from plough and scythe to tractors and harvesters, to machinery and building material for drainage and irrigation, flood-control, etc. And the complementary task to that of producing food—namely, its distribution—cannot be fulfilled unless there are sufficient transportation means available—railways, wagons, engines, motor-cars, trucks, ships, canals, locks, and so on. In order to build and manufacture all this we need iron, coal, timber, metals, and other materials, as well as energy in all its forms—heat, mechanical power, and electricity. In contrast to the demands for other commodities and goods, such as textiles, which in periods of depression remain unsold, the demand for power is nearly always and everywhere greater than the supply. The provision of sufficient amounts of power at cheap prices will add directly to a nation's welfare. By exploiting all natural sources of energy we can save manual labour, promote food production, and raise the general standard of living.

3. *Education and Re-education*

The successful carrying out of world-wide plans depends on the possibility of fighting illiteracy and raising the educational level in backward countries. Teaching and education in undeveloped areas are, therefore, among the tasks with which UNESCO (United Nations Educational, Scientific, and Cultural Organization) has been vitally concerned since its foundation in 1946.

All this is, however, a necessary, but not the only, requirement for improving human welfare. The most serious troubles of our time are caused not by the illiterates in the backward countries, not even by the great masses of the civilized nations, but by the educated people, and particularly by the reigning classes of the Great Powers themselves. What is the use of all the progress in science and technology, of the increase in industrial output, if its greater part is consumed in the production of destructive weapons? As long as the great national efforts are opposed to one another all progress on one side will be counteracted by increased efforts on the other side, thus causing more economic strain on each nation, more labour for the toiling population, without alleviating the lack of food, housing, and clothing of all classes except a small minority of well-to-do people.

And, quite apart from international conflicts, the private life of innumerable people is spoilt by futile quarrels caused by lack of mutual understanding, by an ill-tempered nervousness, and by the stubborn struggle for aims which in sober consideration hardly deserve being fought for. As a result of this state of mind, a good many of the haves, although enjoying every kind of modern comfort, are continually worried, upsetting each other and being concerned with matters of petty interest. These people are in many respects worse off than some of the have-nots, who, in spite of poverty and want, may be enjoying the happiness of love and human understanding within their immediate surroundings, and who are able to laugh and joke, detached from the imaginary obligations and vanities of the rich.

What we need, therefore, is a thoroughly new system of teaching and education capable of changing the next generation. We need a reform in the teaching of history and an education inspired by the great ideals of humanity, responsibility, and co-operative helpfulness. We must do away with the prejudices, vanities, superstitions, taboos, and idols implanted by certain parts of our traditional education, which have caused the wrong attitudes of former and present generations—namely, the narrow-mindedness of chauvinistic nationalism, the struggle for personal and collective power, religious fanaticism in some places and periods, fanatic class hatred in others—and, in addition, the lack of understanding for one's neighbours and the inability to judge the relative value of personal interests and human issues.

It is only a society led by men educated in the way described here that will be able to carry out global welfare projects, and will use the progress achieved in technological and economic development in a wiser manner, to open the doors to a better, brighter, and happier world.

The Aim of This Book

Among the three tasks outlined here the first seems to be that of the most immediate urgency, and the third the most noble one —the one which I personally feel deserves priority. Still, this book will deal exclusively with the second task: Power Production.

Here is a well-defined problem of incontestable international importance, and a full knowledge of possible solutions and their implications will be useful not only to technicians, but to all people concerned with economic and industrial planning. The investment of large capital and labour in long-term power projects, like TVA (Tennessee Valley Authority) and others, is a matter of great responsibility, and the men who have to decide finally will feel safer when, instead of relying only on the opinions of their experts, they can form a judgment of their own on possible future developments which might bear on the profitableness of the project in hand. A question which has been put frequently since 1945 is: Should we take the responsibility of investing hundreds of millions in large-scale hydro-electric plants if in the near future we can get electricity cheaper from atomic energy?

Hence, although I am not a science-fiction writer, but a professional scientist, part of this book deals with the possibilities and prospects of future power production. We physicists are by no means the captains of the ship of human fate, but we are in the position of the man on the foretop whose outlook is ahead of that of the rest of the crew. Looking back on former writings of mine, I find a good many predictions which were criticized and contested at the time, but which were fully verified later on. When I wrote my first newspaper article on aviation in 1913, maintaining definitely that the then greatly favoured Zeppelins would later play no rôle at all in civil and military aviation comparable with that of aero-planes, I was strongly attacked by the supporters of the lighter-than-air system. A dozen years later, at a time when the sound film was believed by many experts, including Charlie Chaplin, to be quite impossible both for technical and artistic reasons, I predicted clearly the radical replacement of the silent film by sound films. Later on, like all other physicists who read the current literature, I foresaw the advent of the atomic age from the time when Hahn published his discovery of uranium fission in 1939, and in 1946, four years before President Truman revealed the project of making the hydrogen bomb, I published hints as to the construction of such a bomb, adding numerical data on its possible energy output.

I do not mention these facts to boast of any visionary gifts, but only to give the reader the reassurance that long-term prognoses on the general trend of technical developments can be made with a higher degree of correctness and reliability than weather forecasts for the morrow. There are hundreds of physicists who are equally well informed—and dozens having a much better insight than my-self into possible technological developments—all of them able to

ıake correct predictions on matters concerned with industrial
lanning and capital investment if only they are prepared to descend
·om the high chair of pure science and discuss matters with lay-
ıen in plain language.

The aim of this book is, therefore, to give a comprehensive
urvey of possible means of present and future power production,
ogether with relevant numerical data, including information on
ıvailable sources of energy and on the relative merits and draw-
ɔacks of the different methods. Plain language and suitable illus-
:rations are intended to make the volume readable and interesting
enough for a public ranging from secondary-school boys to industrial
managers.

ACKNOWLEDGMENTS

I wish to express my heartiest thanks to all the people who either helped me in collecting data and material for the book or gave me valuable advice and hints by answering questions and criticizing my drafts.

For their assistance in obtaining numerical data and illustration material I am indebted to:

Peter Aufschnaiter, of New Delhi, India.

Lieutenant-Colonel F. M. Bailey, C.I.E., of Stiffkey, England.

Dr Paul Boschan, Ph.D., New York, U.S.A.

Robert Chorherr, Librarian at the University of Vienna.

C. H. Gray, Secretary of the World Power Conference, London.

Dr G. Schubert, Ph.D., Professor of Physiology in the University of Vienna.

Dr R. Stahl, Director-General of the Österreichischen Verburdgesellschaft, Vienna.

Dr Maria Telkes, of the Solar Energy Project, College of Engineering, New York University.

F. Wachs, of Vienna.

Gerald Wendt, of UNESCO, Paris.

The Escher-Wyss Engineering Works, Ltd, Zürich.

The Gebrüder Sulzer A.G. Winterthur, Switzerland.

The M.A.N. Maschinenfabrik Augsburg-Nürnberg A.G.

The Österreichischer Rundfunk, Vienna.

The Shell Austria A.G., Vienna.

The Standard Oil Company (New Jersey), New York, U.S.A.

The Steyer-Daimler-Puch A.G., Steyr, Austria.

The Vereinigte Aluminium Werke A.G., Ranshofen, Austria.

The Wiener Stadtwerke, Elektrizitätswerke, Vienna.

I am grateful for valuable advice and hints given by:

Dr K. Baumann, of the University of Vienna.

Dr Karl Beck, of Vienna.

Dr H. Bertele, of Purley, England.

Dr H. Melan, of the Technischen Hochschule, Vienna.

Dr K. Peters, of the Technischen Hochschule, Vienna.

Dr H. Robl, of Duke University, Durham, U.S.A.

Permission for quotations and the reproduction of tables and figures was kindly granted by the publishers of the following works:

AYRES, E., and SCARLOTT, C.: *Energy Sources: the Wealth of the World* (McGraw-Hill, New York, Toronto, London, 1952).

DAVIES, S. J.: *Heat Pumps and Thermal Compressors* (Constable, London, 1950).

Energy in the Service of Man (mimeographed papers prepared by UNESCO, Paris).

Energy Resources of the World (Department of State publication, 3428, U.S. Government Printing Office, Washington, D.C., 1949).

PUTNAM, P. C.: *Energy in the Future* (Van Nostrand, New York, 1953).

RUHEMANN, MARTIN: *Power* (Sigma Books, London, 1946).

My friend Wolfgang Foges in London suggested the idea of writing this book, and introduced me to Messrs George G. Harrap and Co., Ltd. I wish to thank the publishers particularly for the kind understanding and the patience with which they treated their slow author. My brother Ernst helped me efficiently in correcting the proofs, and my wife Toni deserves my gratitude, because not only did she type the manuscript, but also succeeded in keeping visitors at bay while I was working.

ENERGY FOR MAN

CHAPTER I

Power, Energy, and Heat

THE words ' power,' ' energy,' and ' work,' which will be used frequently throughout this book, have many different meanings in different spheres of human activity. Hence, in order to avoid confusion, it will be useful to give clear definitions of these words when they are used in physics and engineering. The reader who is familiar with elementary laws, terminology, and units of physics is advised to pass quickly over this chapter, being sure first, however, that he really understands the meaning of the Second Law of Thermodynamics.

Even within the realm of science and technology the word ' power ' can be used in two different ways, which will be treated separately here.

The correct meaning of power is energy per second; energy in its turn is the physical ability to do work; and work, again, is done when a body is moved by a force. The absolute unit of work in the C.G.S. (centimetre-gramme-second) system is one erg—that is, the work done when a force of one dyne acts through one centimetre in its own direction. It is a very small unit, being the work done when a body of 1.02 milligrammes ($= 0.0000359$ oz) weight is lifted vertically by one centimetre against the gravitational force at sea-level.

Ten million ergs are called a joule:

$$1 \text{ joule } = 10^7 \text{ ergs} \quad \dots\dots\dots\dots\dots\dots\dots\dots\dots(1)$$

and a joule per second is a power unit, called a watt:

$$1 \text{ watt } = 1 \text{ joule/sec } = 10^7 \text{ erg/sec} \quad \dots\dots\dots\dots(2)$$

A kilowatt (kW), being equal to one thousand watts, is the practical unit of power.[1] Although the unit watt is named after **James Watt,** the inventor of the steam-engine, the power unit introduced by

---

[1] The units of electric current and voltage have been so adopted that the units joule and watt have a very simple meaning in electricity. If a voltage of V volts produces a direct current of i amps the power consumed is iV watts.

Watt himself is a different one—namely, the horse-power (H.P.), which he used in selling his engines to farmers. By experimenting he found out the amount of work that an average horse could do in unit time. Multiplying it by a safety factor (1·5)—in order to avoid reclamations from farmers who believed that their horses were stronger—he defined a work of 550 foot-pounds per second as a horse-power. The corresponding metric unit, Pferdestärke (P.S.), is 75 kilogramme-metres per second. The ratio between the three units is:

$$1 \text{ H.P.} = 1·01387 \text{ P.S.} = 0·745476 \text{ kW}$$

A complete conversion table between metric and British units is given in Appendix I.

Since power is energy divided by time, the product of power and time is energy again. Hence

$$\text{joule/sec} = \text{watt} \dots\dots\dots\dots\dots\dots\dots\dots\dots(3)$$
$$\text{watt-second} = \text{joule} \dots\dots\dots\dots\dots\dots\dots(4)$$

A practical unit for energy is the kilowatt-hour (kWh)—that is, the work done by a 1-kW engine in 1 hour. 1 kWh is equal to $3·6 \times 10^{13}$ ergs (36 million million ergs), and it will lift a load of 1 metric ton over a difference in level of 367 metres (= 1119 ft).

Different Forms of Mechanical Energy

Mechanical energy exists in two different forms—as *potential energy* and *kinetic energy* (energy of motion). Heavy bodies on a higher level can do work by descending to a lower level and hoisting by means of rope and pulley a cage with a load from the depths of a shaft. In this case their potential energy is simply converted into potential energy of the load.

Neglecting energy losses by friction, any pulley system having a mechanical advantage of n will allow a load of n times the dropping weight (or pulling force) to be raised by it only through $\frac{1}{n}$ of the distance moved through by the dropping weight. The potential energy is therefore defined as the product of its weight and the height above a certain zero-level. Denoting the weight of a body by w, and the height of its position above zero-level by h, its potential energy E_p is given by

$$E_p = wh \dots\dots\dots\dots\dots\dots\dots\dots\dots\dots(5)$$

Potential energy is always related to a certain zero-level, because a body cannot do work by losing its potential energy unless it is allowed to descend to a lower level.

Kinetic energy is gained by a body when it is set in motion. If water rushes down the tubes of a power-plant its potential energy is turned first into kinetic energy, and this, again, is converted into mechanical work by driving the turbine-wheels, which in their turn drive the rotors of the dynamos, generating thereby electrical energy. A mass m moving with velocity v has the kinetic energy

$$E_k = \tfrac{1}{2}mv^2 \dots\dots\dots\dots\dots\dots\dots\dots\dots\dots\dots(6)$$

this expression giving a measure of its ability to do work. If a power-plant is fed by a water-flow of 10 cubic metres per second dropping down a vertical height of 1000 feet, the water will enter the turbines with a speed of 254 ft/sec, and, assuming an overall efficiency of the plant of 90 per cent, they will produce a power of 27,000 kW.

Dissipation of Energy

If, on the other hand, the water falls into a larger basin, as Niagara did in the old days, and as the great untapped African falls are still doing to-day, the potential energy is certainly also turned into kinetic energy first, but this is subsequently more or less lost in the calming waters below, so that there is apparently a loss of energy in this process. As a matter of fact, viewed from the practical standpoint of an engineer, the energy of the falling water is wasted in this case, but in the strict sense it cannot be lost. The physicist will explain to you that the kinetic energy which a falling body possessed as a bulk will not be lost utterly as soon as it comes to rest at ground-level, but will be *dissipated* into the invisible motion of its single molecules and atoms. Just as a complex elastic structure like a stringed musical instrument, when bumping on the floor, is set into vibrating motion, so the molecules of any moving body which is stopped short without doing work are set into trembling motion. In other words, when a falling body hits the ground its ordered, directed motion is converted into disordered motion of the atoms.

The energy of the invisible molecular motion of a body is, as we know to-day, nothing else but its heat content. *Heat is the energy of the disordered and invisible motion of the atoms and molecules of a body.* Energy cannot be destroyed or generated from nothing; it can only be converted from one form into another.

When a stone falls from a roof to the ground and lies there motionless, its original potential energy has been converted into kinetic energy during the fall, and afterwards partly converted into mechanical work by making a hole in the ground and partly dissipated into heat.

The Mechanical Equivalent of Heat

When this basic discovery was made by **Joule** and **Robert Mayer** about a century ago it was received with great scepticism, because up to then nobody had noticed any temperature rise in a fallen stone or at the bottom of a waterfall. (Falls of meteors, with their very spectacular exhibition of heat, are such rare events that their mere existence had been disputed up to the end of the eighteenth century.) The lack of observations of this kind is explained by the fact that the heat equivalent of mechanical energy is so very small. The water of Niagara at the bottom of its giant fall is only one-eighth of a degree centigrade warmer than on top, and in order to raise its temperature by one degree it would have to drop 427 metres = 1400 ft.

Taking as the heat unit the kilogramme-calorie (kcal) (which is the heat necessary to raise the temperature of one kilogramme of water by 1°C.), and the kilogramme-metre (kgm) as the unit of work, the mechanical equivalent of heat is:

$$1 \text{ kcal} = 427 \text{ kgm} \quad \dots\dots\dots\dots\dots\dots\dots(7)$$

Another exchange-rate worth remembering is

$$1 \text{ kWh} = 860 \text{ kcal} = 3412 \cdot 75 \text{ B.Th.U.} \quad \dots\dots\dots(8)$$

Similarly, as in the middle of the nineteenth century the smallness of the heating effect of mechanical work prevented for some years general acknowledgment of the principle of the conservation of energy, it was in the first three decades of this century that a conclusion from **Einstein's** Theory of Relativity was strongly contested both by experimental physicists and by philosophers. According to this theory, the mass of a body is increased with increasing velocity, and would even become infinite if the velocity of light were reached. The relation between a body's 'rest mass' m_0 (at velocity zero) and its mass m at a velocity v is given by

$$m = \frac{m_0}{\sqrt{1 - v^2/c^2}} \quad \dots\dots\dots\dots\dots\dots\dots(9)$$

where c is the velocity of light (3×10^{10} cm/sec). Of course, nobody had ever noticed the increase in mass of a moving body, for even an ocean liner of 50,000 tons, sailing at a speed of thirty knots, would raise its mass by only one-sixteenth of a milligramme, which is equal to two-millionths of an ounce. And yet to-day the validity of Einstein's equation has been proved beyond doubt by the experiences with the modern big accelerating machines producing particle velocities approaching closely the velocity of light. The grave difficulties occurring in the construction of cyclotrons, owing to the mass increase of very swift particles, can be overcome only by theoretical calculations based on this equation.

The First Law of Thermodynamics

Returning to the knowledge gained by Robert Mayer and Joule (a medical doctor and a brewer incidentally being the discoverers of one of the most fundamental principles of science), we may state the Law of Conservation of Energy, or the First Law of Thermodynamics:

The total sum of all kinds of energy in a closed system is constant.

This means that energy which by natural and artificial processes is permanently converted from one form into another is never lost or generated from nothing.[1] The apparent loss of energy in a body stopped dead is fully compensated by the heat produced, the exchange-rate being strongly in favour of the heat, so that only a small amount of heat is obtained in the conversion from a large sum of mechanical work. This circumstance, together with the fact that in the steam-engines of Watt a conversion of heat into mechanical work was actually performed, tended to raise new hopes as to possible cheap power production. For although the First Law of Thermodynamics turned down all chances of making ordinary perpetual-motion machines, it seemed to open promising prospects for thermal power production on account of the high value of heat assigned by the conversion factor 1 kcal = 427 kgm. Suppose we could succeed in constructing a device which would cool by one

---

[1] The theory of relativity has shown, again, that this is not strictly true. As explained more fully in Chapter XIV, energy can be converted into matter, and matter into energy. Hence the classical laws of the conservation of mass and the conservation of energy do not hold true each by itself; instead they fuse into the one universal law of the conservation of the sum of mass and energy. Of course, a proper conversion factor of mass and energy underlies this new principle; *cf* p. 309.

degree a layer only 4 inches thick of the water streaming along a ship's planks, and convert all the heat taken from the water into mechanical power driving the screws. Then we could travel with such an engine for all eternity without using any fuel, and have ample surplus energy for heating and lighting the ship.

Another still more striking example may be given by figuring out what could be achieved if the waters of the Mississippi could be cooled by $1°C$. at its mouth, and the heat thus taken away fully converted into electrical energy. The yearly delivery of the Mississippi river being approximately 6×10^{14} kg of water, the annual energy output of such work would be 6×10^{14} kcal, equivalent to approximately 7×10^{11} kWh, which is twice as much as the total consumption of electrical energy in the United States during 1952.

The Second Law of Thermodynamics

Unfortunately there is another basic principle of physics, called the Second Law of Thermodynamics, which quite explicitly states the impossibility of a "perpetual-motion machine of the second kind." This is defined as "*a periodically working machine which does nothing else but lift a load and cool a heat reservoir.*"

As everybody knows, the steam-engine is a periodically working machine which can lift a load and at the same time take heat from the burning fuel. Still, it is not a perpetual-motion machine of the second kind, because it does not fulfil the further condition *of not doing anything else.* As a matter of fact, any heat engine necessarily does something else than produce power: by heating its surroundings it transfers a part of the heat collected from the hot combustion chamber to the much cooler outer atmosphere or to the cooling water.

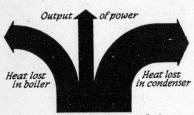

FIG. I. HEAT FROM FUEL COMBUSTION PARTLY CONVERTED INTO MECHANICAL WORK AND PARTLY GIVEN OFF TO THE COOLER SURROUNDINGS

Theory and experience teach us that the conversion of heat into mechanical energy is possible only when there are temperature differences at our disposal. Steam from the boiler expands and pushes the piston of the engine before entering the condenser or the outer atmosphere. In the same way, it is the very hot gases of an internal-combustion engine like the ordinary car-motor which do work in expanding before exhausting and giving off a part of the combustion heat to the surrounding air.

This might seem analogous to the fact that the enormous gravitational force pulling down all water on the earth cannot be used for power production unless there are differences in level (and, accordingly, potential differences) down which the water falls. The difference between the two cases consists, however, in the following circumstance. In the case of water-power any small difference in level can be used for fully converting the potential energy of water flowing down into useful work (provided that the frictional and other losses are avoided). In the case of the heat engine, however, only a part of the heat can be converted into mechanical work, the percentage of convertible energy depending on the difference in temperature, and becoming negligibly small for small temperature differences. To achieve economy, therefore, we must make our heat engines work between temperature limits spaced as far apart as possible.

The Efficiency

By the 'efficiency' of any device is meant the ratio between useful delivery and total amount of input. In the case of a heat engine the useful delivery is the mechanical work done by the shaft within a given time—for instance, during one rotation—and the input is the amount of heat consumed during the same interval of time. A conclusion drawn from the Second Law of Thermodynamics gives a formula for the maximum efficiency which could be achieved in a heat engine operating between a higher temperature T_1 and a lower temperature T_2 under the best possible conditions.[1] With T_1 and T_2 expressed in Kelvin degrees—that is, degrees centigrade plus 273°—the formula is simply

$$\text{efficiency} = \frac{T_1 - T_2}{T_1} \quad\text{......................................(10)}$$

---

[1] The following explanations are given for those readers who would like to know more about the subject: Processes performed at constant temperature are called *isothermal*; processes during which no heat is supplied or rejected are called *adiabatic*. A series of operations or processes performed with any substance, as, for instance, a series of compressions and expansions of a gas, is called a *cycle* if the substance returns to its original condition at the end of these processes. A theoretically interesting process is the so-called *Carnot cycle* of a gas, which consists of four steps performed in a cylinder with a moving piston:

(*a*) The cylinder is brought into contact with a source of heat of temperature T_1,

Thus, for instance, with $T_1 = 600°$ K. $= 620°$ F., and $T_2 = 360°$ K. $= 188·6°$ F., the efficiency of the Carnot process would be $240/600 = 40$ per cent. For each 100 kilo-calories consumed by burning fuel 40 kcal would be converted into 17,080 kgm of mechanical work, while the greater part—namely, 60 kcal—would be transferred to the lower temperature level of the surrounding air. With the existing heat engines the efficiency of the real process is, however, much lower, and it is only with greater differences in temperature that efficiencies of 40 per cent can be achieved. More details will be given in Chapters III and IV.

The reverse process, on the other hand—namely, the conversion of mechanical work or of electrical energy into heat—can be performed completely with a 100-per-cent output. If, for instance an electric voltage of 100 applied to a resistance of 10 ohms produces a current of 10 amps the power is

$$10 \times 100 = 1000 \text{ watts} = 1 \text{ kW}$$

and the heat generated in one hour will be the exact equivalent of 1 kWh, which is 860 kcal, according to equation (8). In the same way, the potential energy of a car or train running down a slope, and being subsequently stopped by applying the brakes, is nearly completely converted into heat, a very small residue being consumed for the mechanical work of attrition of the wheels and brakes.

---

and the gas expands isothermally, receiving thereby heat from the source and doing work by forcing the piston forward.

(b) The source of heat is removed, and the gas continues to expand and to do work, but during this step it expands adiabatically—that is, without receiving or rejecting heat. During the expansion it cools to a temperature T_2.

(c) A cold body at a temperature T_2 is brought into contact with the cylinder; the piston reverses, thus compressing the gas isothermally, while heat is rejected to the cold body. During this and the next step work is done on the piston, and hence on the gas, by an external force.

(d) The cold body is removed, and the remainder of the compression stroke is performed adiabatically, the temperature rising again to T_1, so that the cycle is completed.

As a result of these four steps more work is delivered from the gas to its surroundings than *vice versa*, and more heat is taken from the hot body than rejected to the cooler one. Hence altogether heat has been consumed and mechanical work produced. The efficiency of this process—that is, the ratio between the work done and the heat received from the hot body—is given by formula (10). It can be proved from the Second Law of Thermodynamics that no cyclic process operating between the same limits of temperature can have a greater efficiency than the Carnot cycle.

Unfortunately the Carnot process cannot be performed in practically existing heat engines; hence the practically attainable efficiencies of all engines working between two temperature limits T_1 and T_2 are distinctly lower than $\dfrac{T_1 - T_2}{T_1}$

It would be a mistake, however, to believe that on account of the full convertibility of electrical energy into heat the efficiency of all electric heating devices must be 100 per cent. Certainly an electric boiler turns all the power consumed into heat, but only a part of it is used for heating the water, while a certain percentage is wasted in heating the walls and the outer atmosphere. In talking of the efficiency of electric heating devices we do not mean, therefore, the ratio between the total heat generated and the electrical energy consumed (which is 100 per cent), but the heat transferred to the body to be heated divided by the electrical energy (both being expressed, of course, in the same units—for instance, kWh). Thus the efficiency of electric heaters and stoves may be sometimes less than 100 per cent. On the other hand, performance ratios of more than 100 per cent can be achieved in *heat pumps*. This is possible without violation of the principle of conservation of energy, because a heat pump, as its name suggests, is a device which in consuming electrical or mechanical energy pumps heat from a lower level of temperature to a higher one. The total heat transferred to the higher level—which may be a radiator used for heating a room—is the sum of the electrical energy consumed and the heat taken from the lower level, and is therefore greater than the electrical energy input. This very interesting device will be described in Chapter VI.

The efficiency in converting electrical energy into mechanical work by means of electric motors may reach values of well over 90 per cent, which compares very favourably with the efficiency of heat engines, varying between 5 per cent, or even less, in the older steam locomotives and about 35 per cent in modern large turbine plants or in Diesel engines. Hence, although all sorts of energy are equivalent according to the Law of the Conservation of Energy, they are of different practical value. And, therefore, our terminology makes a distinction between *free energy*, like electrical and mechanical energy, which can be freely converted into any desirable form, and *heat energy*, the conversion of which is restricted. Free energy and heat play a similar rôle in engineering to that of *hard* and *soft* currencies in the economic field.

In giving numerical data on power resources and available energy we have to state clearly whether heat or electrical energy is meant. A stock of 1000 metric tons of bituminous coal with a calorific value of 7200 kcal/kg, which is about 13,000 B.Th.U./lb, can supply an amount of 7200 million kilo-calories = 8·4 million kWh of *heat energy*. The use of the electrical unit kWh should not cause the reader, however, to believe that 8·4 million kWh of electrical energy could be obtained from the given stock of coal. For the conversion of heat into electrical energy is done with an efficiency considerably less than 100 per cent; hence only about 1·68 million kWh or so of electrical energy can be obtained from 1000 tons of coal.[1]

Although the thermal efficiency of a steam turbine itself may be in some cases 0·40 (40 per cent), or even more, the overall efficiency of the whole steam plant converting heat into electricity will be less on account of the other losses which are involved. The overall efficiency is the ratio between the electrical energy produced and the heat value of the fuel consumed. Losses are caused by heat escaping through the flues of the boiler system, heat delivered to the condensers of the turbines, heat produced by friction and electric currents in the process of converting mechanical work into electrical energy, and, lastly, in the transformers and transmission lines. Each step of the process has an efficiency smaller than unity, and the overall efficiency is the product of the efficiencies of the single steps. Thus, for instance, in older plants:

Boiler efficiency .... 0·68
Thermal efficiency of the turbine .... 0·32
Generator efficiency .... 0·96
Transmission efficiency .... 0·96
Overall efficiency = 0·68 × 0·32 × 0·96 × 0·96 = 0·20 = 20 per cent

As a consequence 1000 kWh of electrical energy are more valuable than 1000 kWh of heat for all applications of power, except heat as an end in itself.

Heat Value, Coal Equivalent, and Electricity Equivalent

In comparing the contribution made to the energy budget by the different sources it is necessary to use a common measure, because we cannot simply add tons of coal, wood, and petroleum. As long as only fuels are taken into account the *heat values* may be used. The calorific value of a fuel is the heat generated by the combustion of the unit weight of this substance, and the heat value of a given quantity of fuel is the product of its calorific value and its weight.

The contributions of different fuels can also be summed up by expressing them in *coal equivalents*. With a calorific value of

[1] Compare the work on 'electricity equivalent' on the opposite page.

7200 kcal/kg (=13,000 B.Th.U./lb) for average coal, and 10,600 kcal/kg for petroleum, one ton of petroleum is equivalent to 1·47 tons of coal, or, in other words, we obtain the coal equivalent of a given quantity of petroleum by multiplying its weight by 1·47. Assuming an average density of 0·85 for petroleum, one U.S. barrel (= 159 litres) of petroleum has a weight of 135 kg, and its coal equivalent is 135 x 1·47 = 199 kg of coal. Using round figures, we may keep in mind that the coal equivalent of one U.S. barrel of petroleum is approximately 0·2 tons. In all our tables metric tons are used. The conversion factors to British units are:

> 1 short ton or net ton (2000 lb) = 0·907185 metric tons.
> 1 long or gross ton (2240 lb) = 1·016047 metric tons.

If, in addition to different fuels, a considerable contribution to the energy income of a country is made by electricity gained from water-power it would be unfair to assess the contribution of the latter by simply taking the heat value of one kilowatt-hour, which amounts to 860 kcal. For, as mentioned already, different forms of energy, even if equivalent in the purely physical sense, are by no means economically equivalent—just as, for instance, a given food value offered in the form of potatoes has a lower market-price than the same food value in the form of beefsteaks or ham. The reason lies, as mentioned above, in the greater efficiency of producing mechanical work from electricity than from heat. Assuming the former efficiency to be nearly 100 per cent, and the latter on the average 20 per cent, we can say that from a given quantity of electrical energy five times as much mechanical power can be generated as from the same quantity of heat energy.

The notion of *electricity equivalent* has been introduced, therefore, and is used as a basis for comparing the energy drawn from different sources in the Guyol Report (*Energy Resources of the World*).[1] Assuming fuels converted to electricity at 20-per-cent efficiency, *their electricity equivalent is defined as one-fifth of their heat value.*

Table 1 below, taken from the Guyol Report, gives the heat values and electricity equivalents of a series of fuels. It may be noted that only approximate average figures are given, while different sorts of coal, oils, gases, and wood-fuels possess heat values which deviate to both sides of the values given in the table.

[1] All works and papers quoted are given in alphabetical order in the bibliography at the end of this book.

TABLE 1. HEAT VALUE AND ELECTRICITY EQUIVALENT OF DIFFERENT FUELS[1]

Fuel	Unit	Heat Value (1000 ton-calories[2] per unit)	Electricity Equivalent[3] (1000 kWh per unit)
Coal:			
German	Metric ton	7·0	1·63
Other	,, ,,	7·2	1·68
Brown coal and lignites:			
Czechoslovakian	,, ,,	4·9	1·14
German	,, ,,	2·2	·51
Other	,, ,,	2·8	·65
Peat	,, ,,	3·6	·84
Coke	,, ,,	6·0	1·40
Coal briquettes	,, ,,	7·2	1·68
Lignite briquettes:			
Czechoslovakian	,, ,,	7·0	1·63
Other	,, ,,	4·8	1·12
Wood-fuel	Cubic metres	1·8	·42
Mineral oil and derivative oil-fuels	Metric ton	10·6	2·47
Benzol	,, ,,	10·6	2·47
Alcohol	,, ,,	5·5	1·28
Natural gas	1000 cubic metres	9·6	2·24
Manufactured gas	,, ,,	4·3	1·00
Refinery gas	,, ,,	12·5	2·91
Blast-furnace gas	,, ,,	·8	·19
Electricity	1000 kilowatt-hours	·86	1·00

[1] The connexion between the figures in the third and fourth columns may be explained by reference to the first line:

Heat value of 1 ton of German coal $= 7$ ton-calories $= 7000$ kcal $= \dfrac{7000}{860}$ kWh $= 8\cdot15$ kWh.

Therefore electricity equivalent, $\dfrac{8\cdot15}{5}$, $= 1\cdot63$ kWh.

[2] 1 ton-calorie $= 1000$ kg-cal.

[3] At 20-per-cent efficiency, except electricity (100 per cent).

Remarks about Terminology: the Restricted Sense of the Word 'Power'

The meaning of the word ' power ' differs somewhat in scientific terminology from that in its colloquial use. As stated at the beginning of this chapter, the physical meaning of power is energy per unit time, energy comprising thereby all forms of energy, including heat. Hence in describing, for instance, the performance of a stove delivering, say, 17,000 B.Th.U. per hour we should not be altogether wrong in stating that its ' power ' amounts to 5 kilowatts. Still, an expression like this would be usual only in speaking of electric stoves, because in the ordinary use of the word power its meaning is restricted to motor-power—namely, mechanical work per second —and to electrical power. In speaking of the different items of the energy budget we shall, therefore, make a distinction between ' heat ' and ' power,' confining the latter term to engine-power.

The Use of Decimal Multiples

In order to avoid the use of very large or very small numbers, certain symbols for decimal multiples and sub-multiples have been introduced, and these are listed in Table 2 below. In the chapters on water-power frequent use will be made of the symbols MW, GWh, and TWh.

TABLE 2. SYMBOLS FOR DECIMAL MULTIPLES

Multiple	*Prefix*	*Symbol*	*Example*
10^{-6}	micro	μ	μs=microsecond= 1 millionth of a second
10^{-3}	milli	m	mm=millimetre
10^{-2}	centi	c	cm=centimetre
10^{-1}	deci	d	dm=decimetre=$\frac{1}{10}$metre
10^{2}	hecto	h	hl=hectolitre
10^{3}	kilo	k	kg=kilogramme km=kilometre kV=1000 volts kW=kilowatt
10^{6}	Mega	M	MW=1000 kilowatts
10^{9}	Giga	G	GWh=1 million kilo-watt-hours
10^{12}	Tera	T	TWh=1000 million kWh

In estimating the total power resources of the earth another, still larger unit has been introduced—Q, which means a trillion ($= 10^{18}$) British Thermal Units. Taking 1 B.Th.U. = 0.252 kcal we get 1 Q = 10^{18} B.Th.U. = 2.52×10^{17} kcal = 2.93×10^{14} kWh = 293,000 TWh.

The reader is also reminded that the words billion, trillion, etc., have different meanings in Europe and America:

	Europe	America
Billion	10^{12}	10^{9}
Trillion	10^{18}	10^{12}
Quadrillion	10^{24}	10^{15}
n-tillion	10^{6n}	10^{3n+3}

CHAPTER II

Survey of Sources and Uses of Power

A SURVEY of power economy and its sources cannot be confined to power in the restricted sense of the word—namely, motor-power produced by electricity or by steam or internal-combustion engines—but must be extended to all sorts of energy, including heat. For all the energy—in the physical meaning of the word—which we are using is, or can be, fed from the same set of sources—fuel, solar heat, water-power, etc.—and in times of local or universal shortage any increase in the demand for one form of energy will cause cuts in the others. From a given amount of fuel we can get more heat and less electricity and transport power, or *vice versa*.

Taking heat into account, we can say that mankind has been a consumer of energy ever since the day far back in the Upper Palæolithic Age, more than ten thousand years ago, when man acquired the ability to use fire.

In a still wider sense, every living organism is a consumer of energy, since animal life depends on metabolism fuelled by food, which supplies to the body the necessary amount of energy. The total energy income resulting from ' burning ' the food-fuel in the process of metabolism is used for different purposes, among which three play a dominant rôle:

(1) Chemical energy used for building up and regenerating the living tissues of the body.
(2) Energy needed for doing muscular and mental work (the latter, though very important, forming only quite a small item of the physiological energy budget).
(3) Heat for compensating the losses by conduction, radiation, and evaporation.

In the energy budget of any single warm-blooded creature, as well as in that of the whole of human society, the consumption of energy in the form of heat exceeds that of any other branch.

Muscular work, done by labourers, housewives, slaves, and draught animals, contributed the greatest part to the total needs of mechanical work among civilized nations in former times, and still does so among primitive peoples to-day.

Although, therefore, man-power and animal-power are an integral part of our energy budget, the survey below will make a sharp distinction between animate, or physiological, energy, maintained by food and fodder, and inanimate, or purely physical, energy, fed by using fuel, wind, and water-power, etc. For while the different sources of inanimate energy can replace each other, the whole set of sources of inanimate energy is not interchangeable with that of the food resources.[1]

It was the combination of the use of tools and the use of inanimate energy which caused the decisive step in the evolution of mankind. The application of fire and its heat led to the transition from the Stone Age to the Bronze Age, and subsequently to the Iron Age, and in addition it rendered greater, by the provision of warmth, the world's habitable area.

For thousands and thousands of years, through all the prehistoric ages, heat was the only form of inanimate energy used by men. To do the work necessary to supply food, clothing, and shelter man had to use his own muscles or those of his slaves or animals. It was only in historic ages, many millennia after the invention of fire, that a small part of human and animal labour began to be replaced by machines using the natural forces of water and wind. The Chinese were probably the first to use windmills, and water-wheels, originally devised by the Babylonians for irrigation, were used in the Roman Empire for driving mills—saw-mills and hammer-mills.

Owing to the large exchange rate between heat and mechanical work (1 kcal = 427 kgm or 1 B.Th.U. = 778 foot-pounds), the relative amount of power in the whole energy budget of mankind was but a tiny fraction of that of heat, certainly less than 1 per cent. It was only after the invention of the steam-engine that the percentage of mechanical energy rose gradually, so that in our day up to 30 or 40 per cent of all the fuel consumed serves for generating power for transport and industry.

With the small human population and with the modest *per capita* consumption of heat and power during antiquity and the Middle Ages, the global energy budget was fairly balanced. For the sources tapped at that time were wood as a fuel and wind and water for power, the latter being inexhaustible and the former

[1] This holds at least for the present state of matters. In future the possibility may arise of producing sugar from wood and synthesizing butter from coal.

recurrent by natural growth. With the exception of certain areas which were ruthlessly devastated, like some North African landscapes, or the cedar woods of the Lebanon, or, later on, the mountain-ranges of the east coast of the Adriatic, the balance between consumption and regrowth of wood-fuel was fairly well held. Towards the end of the Middle Ages, however, the first of the great fossil fuels, coal, revealed its advantage as a more concentrated source of energy, and since then up to the beginning of World War I coal reserves were exploited at an ever-increasing rate. Fig. 2 gives an approximate idea of the rise of coal production in Britain and the U.S.A. during the last two hundred years. The standstill reached at the beginning of this century, and continued so far, does not mean that power production has not increased since then. The reason is that engines with better efficiency have been employed, and part of the transport of persons and goods has shifted to motor vehicles driven by petroleum products. Although, therefore, the rate of coal consumption has remained almost constant for one or two generations, the level reached long ago is no longer compatible with a balanced budget. For humanity is consuming to-day more coal in a single year than had been generated in a hundred centuries or so during the process of carbonization some millions of years ago.

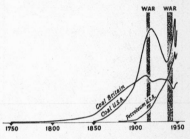

FIG. 2. THE RISE OF COAL PRODUCTION IN BRITAIN AND THE U.S.A.

Civilization is, therefore, in a transitory stage which cannot continue indefinitely. Anyhow, it seems rather fortunate that Nature has provided us with rich stores of fuels sufficient for several centuries, so that we may build up large industries and corresponding scientific development, until the time when humanity is mature enough to cope with its energy needs by using recurrent sources or by mastering nuclear fusion reactions.

Table 3 gives merely a rough idea how in the evolution of human civilization new sources and new contrivances of inanimate energy were added step by step to the already existing means in the main fields of energy demand—namely, domestic and industrial heat, industrial and transport power.

It should be noted that the table gives only the means of energy newly added at each stage of development, while actually the whole stock of previously known means is generally being used along with the new ones.

TABLE 3. SOURCES OF INANIMATE ENERGY AS THEY APPEARED FIRST AT DIFFERENT STAGES OF CIVILIZATION

Stage of Development	Domestic Heat	Industrial Heat	Industrial Power	Transport Power
Animal stage	—	—	—	—
Primitive cave-man	Wood-fire	—	—	—
Prehistoric civilization	„	Wood-fire	—	—
Antiquity	„	„	Wind, water	Wind
Since about 1250	„	Coal	„	„
„ „ 1400	Coal	„	„	„
„ „ 1710	„	„	Steam	„
„ „ 1820	„	„	„	Steam
„ „ 1890	„	„	Electricity	Electricity
„ „ 1900	Oil-prods.	Electricity	„	Petrol motors
„ „ 1955	„	„	Atomic power	Atomic power

A more quantitative representation of man's use of energy is given in Figs. 3–6, in which the *per capita* consumption of inanimate energy is contrasted with that of animate energy. While the former has been increased by civilization, the latter has not changed considerably. The global average energy content of human food is estimated at 2200 kcal per day, or 930 kWh annually. Deviations from the average amounting to plus 100 per cent for athletes or labourers doing heavy work and minus 50 per cent for underfed or even starving people may be taken into account.

In comparing the energy quota derived *from various sources* the conversion method used in the Guyol Report will be adopted, giving the numerical data in units of *electricity equivalents* (kWh or powers of ten × kWh). The reader is reminded that the electricity equivalent of a given stock of fuels is defined as one-fifth of its total energy content represented by its heat value.

We are making, however, an exception to this rule in the

representation of the apportioning of energy *among various uses*. In Figs. 3–5 the areas of the single sectors are drawn proportional to the *gross energy* consumed for heat, power, and chemical work.

Throughout this chapter the expression ' consumption' will be used for *per capita* consumption.

Animate Energy

Fig. 3 refers to animate energy consumed in the human body under different conditions. The area of the circles is drawn in proportion to the energy consumed per hour by the processes of metabolism, and the areas of the three sectors give the proportions of the main uses of this energy. The blank sector indicates the heat

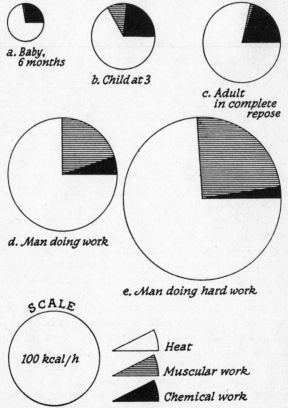

FIG. 3. ENERGY CONSUMPTION PER HOUR IN THE HUMAN BODY

produced, the hatched sector the muscular work done, and the black sector the energy spent on the chemical reactions of the diverse physiological processes.

Under normal conditions in moderate climates heat production will just suffice to compensate the losses to the cooler surroundings, and thus to keep the interior of the body at its normal temperature of about 98°F. (37°C.). Unfortunately the physiological regulating mechanisms are not such that in hot surroundings, when no heat loss takes place, the ' flame of the stove ' is turned down. Human metabolism is like a flame with a given minimum size; it can be made brighter, but cannot be turned down more than about 10 per cent below a certain level, given by the so-called ' basal metabolic rate ' (B.M.R.), represented by the circle of Fig. 3c. And, again, rather unfortunately, it does not become brighter automatically when more heat is needed in very cold surroundings. By merely supplying more fuel in the form of food we cannot prevent freezing at low temperatures without warm clothing. Turning up the flame to higher levels is done only by muscular exercise, quite irrespective of the real demands for heat. Hence in performing muscular work we not only, by increased oxygen intake, burn more fuel to keep the engine going, but, as shown in Figs. 3d and 3e, we also increase the heat production, sometimes even far in excess of the desirable amount. Nature is acting in this respect like a stoker overheating a building and at the same time opening all windows in order to keep the rooms at a tolerable temperature. Raising the body temperature above the normal level of 98°F. is avoided by three different measures. First of all there is a buffering reserve of cold in the subcutaneous outer layers of the body. It is only the inner kernel which is kept at 98°F. under normal conditions, while a layer nearly an inch in depth is some degrees cooler. When more heat is produced this outer shell of the body is warmed up, the skin temperature rises, and accordingly the heat loss to the surroundings is increased. About a hundred kilo-calories can be absorbed in this process, which by enhancing the blood circulation is quite a healthy occurrence. If, after warming up the outer shell, the excess heat production goes on, the worker or sportsman will feel too hot, and will start, therefore, to get rid of all unnecessary clothing, thus increasing the heat losses by radiation and convection. At the same time the automatic process of sweating is set going, and in evaporating sweat much more heat can be discharged than by convection and radiation. In this way even a heat production amounting to four times the basal rate, as shown in Fig. 3e, can be compensated without undue rise of the temperature of the inside of the body. The efficiency of the muscular engine—that is, the ratio between mechanical output and energy consumption—is not bad, after all. In the example given in Fig. 3*e* it is 24 per cent, and in trained persons efficiencies of well over 30 per cent, thus successfully competing with our best modern heat engines, have been observed.

In spite of its efficiency, the use, or rather misuse, of the human body for heavy work should be avoided for two reasons. First of all, physical efficiency is not equivalent to economic efficiency. The 3000 or 4000 kilo-calories which a worker needs daily have to be supplied in the form of food, which costs much more than a pound of coal giving as many calories. Apart from that, the human engine cannot run continuously, and heavy work in particular cannot be done for more than eight hours daily, whereas the fuel-consuming metabolism continues during the hours of rest. Secondly—and this is far more important—social considerations have to be taken into account. The optimum conditions for mental and physical health of men are achieved by doing regular muscular work of moderate amount without permanent overstrain. While, on the one hand, a life without any physical exercise is very unhealthy, the same holds for continued exhausting labour. Young people may in sport, for instance, strain their muscular forces to a maximum output for short periods without serious damage. But continued excessive strain leads to degenerative processes, mainly of the cardiac system, and hence to low life expectancies. The high death-rate of the rickshaw coolies in South-east Asia, with a life expectancy of thirty years or even less, may serve as a warning. Progressive civilization should, therefore, remove as far as possible all heavy work from human labour and replace it by engine work, at the same time, however, seeing to it that part of the free time gained is spent in light physical exercise like sport or gardening.

Alongside heat production and mechanical output, some physiological processes are running continuously, and these also consume energy. The building-up of all tissues, bones, muscles, etc., during the period of growth is one of these items; others are the processes of purification and regeneration ('hæmatopoiesis') of blood, or the production and secretion of hormones in the glands. While the basal metabolic rate averages 0·016 gramme-calories per gramme and minute in the whole body, the share of the kidneys is 0·170 cal/gmin and of the liver 0·057 cal/gmin. A part of the surplus consumption in the liver is spent in keeping this organ at a slightly higher temperature, while the rest, together with practically all the surplus energy consumed in the kidneys, serves for doing chemical or organic work. In growing children additional work is done to build up new tissues; hence the relative amount of organic work (black sector) is greatest in Figs. 3*a* and 3*b*, representing the metabolic rate of children. In contrast to the heat production, which increases when muscular work is done, the chemical work decreases during exercise—in the same way as in human society cuts are made in certain branches of industry when efforts are concentrated on production of war material. It can be seen from Figs. 3*d* and 3*e* that not only the relative share, but also the absolute amount, of chemical work decreases with increasing output of mechanical work.

This fact in itself, combined, moreover, with the specific fatigue phenomena, forbids the performance of very strenuous efforts over long periods. Hence the machine age (born late in the eighteenth century), in spite of some drawbacks, is, taken all in all, a blessing for mankind, not only because it increased the quantity and quality of consumer goods, reducing at the same time their prices and the distance between countries, but also because it relieved millions of toiling people from the heavy strain of very hard muscular work.

Inanimate Energy

The following figures referring to inanimate energy indicate the annual *per capita* consumption. The scale is given by the blank inner circle, representing 1000 kWh *per annum,* which is approximately equal to the global average of the energy content of yearly *per capita* food consumption.[1] The annular area round the inner circle is drawn proportional to the additional consumption of inanimate energy, the area of the whole circle thus indicating the total amount of energy—neglecting only the contribution of animate power exerted by draught animals—which is yearly at the disposal per head of population.

[1] 1000 kWh, or 860,000 kcal, yearly are as much as 2350 kcal daily, which might appear too little in view of the fact revealed in Figs. 3*d* and 3*e* that under working conditions the metabolic rate lies approximately between 150 and 300 kcal *hourly*. It must be considered, however, that this consumption rate is not continued through twenty-four hours. Hence the daily consumption of people doing muscular work is not 24 × 300 kcal, but approximately one-half of this figure. As a matter of fact, food consumption of labourers in countries of good social level amounts to between 3000 and 4000 kcal, or even more. Still, on the global average this high figure is compensated by the much lower ones of children and very old people and —unfortunately—by the hundreds of millions of underfed people, resulting in a world average as low as about 2200 kcal *per capita* daily, or 930 kWh annually.

The meanings of the different sectors are:

Blank. Heat for domestic, public, and agricultural uses, including particularly house-heating, cooking, boiling, etc.

Dotted. Heat in industry, mining, etc.

Horizontal Hatching. Mechanical energy—for sake of brevity called power—used for transport means, such as railways, motor-cars, aeroplanes, and ships.

Cross-hatching. Power used for all other purposes, particularly in industry and mining.

Black. Electrical energy used as such, without being converted into mechanical work or heat. The largest part of this sector is spent in electrolytic plants, like those for producing aluminium and copper. The energy used for lighting, X-raying, radio, TV, etc., belongs also in this sector, the latter items, however, being an almost negligibly small part of the whole consumption.

While the areas of the hatched sectors in Fig. 3 referring to animate energy were drawn proportional to the net output of muscular work, their meaning in Figs. 4 and 5 is somewhat different. They indicate the *energy consumed in power-generating engines.* On account of the low efficiency of thermal power production, a large part of this energy—namely, nearly 80 per cent—is lost as waste heat, and therefore the *useful output* of mechanical work and of electric energy is a considerably smaller fraction of the total energy than the fractions of the circles in Figs. 4 and 5 indicated by the hatched and black areas. Thus in the world's energy balance-sheet about two-thirds of the whole energy is spent in producing heat used for all kinds of purposes, and one-third is spent in feeding machinery to produce mechanical and electrical power. But owing to the small average efficiency of these devices about four-fifths of the latter third is again converted into heat—which is mostly wasted—and only about one-fifth of one-third—that is one-fifteenth—of the total input is obtained as useful power output.

Figs. 4a – 4e show the gradual rise of the use of inanimate energy supporting man's own physical efforts. While the power production rose from tiny beginnings in ancient times to its present level, which in industrialized countries amounts to several thousand kWh per person annually, the consumption of fuel for domestic purposes in a wider sense, such as heating, cooking, lighting, and protection

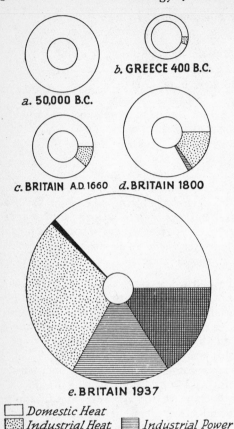

a. 50,000 B.C.

b. GREECE 400 B.C.

c. BRITAIN A.D. 1660 *d.* BRITAIN 1800

e. BRITAIN 1937

☐ *Domestic Heat*
▨ *Industrial Heat*   ▤ *Industrial Power*
▓ *Transport Power*   ■ *Chemical Energy*

FIG. 4. PER CAPITA CONSUMPTION OF INANI-
MATE ENERGY IN DIFFERENT EPOCHS

against wild beasts, may have been quite considerable even in prehistoric times. No statistics are available, of course, and the figures referring to those times cannot claim to be quantitatively correct. Still, inferences can be drawn from the vast consumption of wood-fuel among rather primitive tribes of present-day Africa. Thus, for instance in Nigeria and the Cameroons, as well as in Tanganyika Territory, the consumption of inanimate energy in the form of wood-fuel amounts to more than 5000 kWh annually, an astounding figure in view of the hot climate. The reason appears to be the habit of burning night-long large fires as a protection against assaults from wild animals.[1] It seems rather likely that these customs are very old, dating far back to the Palæolithic Ages, and it is even conceivable that the fateful deforestations, changing formerly fertile tropical landscapes into deserts, have been caused by men's original sin of wasting fuels. The basic difference between the inanimate energy budget of Palæolithic and modern man does, therefore, lie not in the amount of consumption as such, but in

[1] The energy consumed for this purpose is included in Fig. 4*a* in the blank area representing domestic heat.

the distribution between heat and power. In several countries of our present-day world the *per capita* consumption of all sorts of energy may be just a few per cent of that of certain African tribes about 50,000 years ago. But practically no inanimate power at all seems to have been used in pre-glacial times (Fig. 4*a*). In ancient Greece less fuel may have been consumed than in the earlier African stages of humanity, but part of it was used for industrial purposes

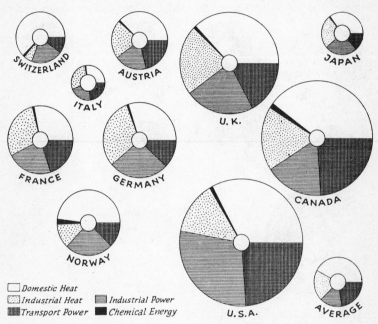

FIG. 5. PER CAPITA CONSUMPTION FOR VARIOUS USES OF
ENERGY IN 1950

in foundries, forges, and the like (Fig. 4*b*). Both the total amount of consumption and the percentage used for industrial heat were increased as soon as the use of coal became more and more widespread, and a stage like that shown in Fig. 4*c* was reached in England in the middle of the seventeenth century. Through all the periods, however, up to the triumph of Watt's improved steam-engine towards the end of the eighteenth century the contribution of inanimate mechanical power, fed by fuel, wind, or water, to the total energy budget of any nation was too small to be shown in graphic representation. It was only at the turn of the eighteenth

century that such a small sector of the total energy consumption
as that shown in Fig. 4*d* was spent for mechanical power in
industry, and later on in means of transport. The stage reached
just before World War II is shown in Fig. 4*e*.

The *per capita* consumption of various countries in 1950 is
represented in Fig. 5 by the annular area surrounding the inner
circle, which, again, is drawn proportional to the annual *per capita*
consumption of human animate energy. The existing statistics give

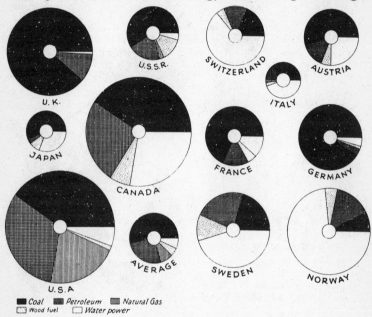

■ *Coal*   ▦ *Petroleum*   ▥ *Natural Gas*
▨ *Wood fuel*   ☐ *Water power*

FIG. 6. SOURCE OF THE ENERGY CONSUMED IN 1950
Areas proportional to *per capita* consumption in electricity equivalents
(fuels taken at 20-per-cent efficiency).

fairly correct data for the total consumption; but the apportionment
to the five main groups of uses represented by the five different
sectors of the figures is based on estimates, and cannot claim the
degree of accuracy of that of the whole consumption figures. In
other words, while the radii of the circles are correct within a
margin of error of a few per cent or so, the angles of the single
sectors may show deviations of many per cent from the true value.

The single sectors in Fig. 6 have quite a different meaning from
those of the foregoing figures, as they represent the percentage of

the *per capita* consumption in various countries drawn from the five main sources of energy—namely, coal + lignite, petrol, natural gas, wood, and water-power. The contribution of minor sources like peat, wind-power, etc., which totals less than a fraction of 1 per cent, is neglected. Conforming to the procedure adopted in the Guyol Report, the areas of the sectors are drawn proportional to the *electricity equivalent* of the respective energies. This means that the energy gained from water-power is taken at its full value, while—assuming an average efficiency of 20 per cent—only one-fifth of the energy content of all fuels is taken into account. The scale is chosen so that the inner circle, representing here 200 kWh of electricity equivalent yearly, has approximately the same size as in Figs. 4 and 5.

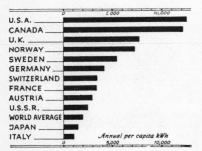

FIG. 7. ORDER OF RANK OF TWELVE COUNTRIES ACCORDING TO THEIR PER CAPITA CONSUMPTION OF ENERGY

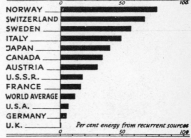

FIG. 8. ORDER OF RANK OF THE SAME COUNTRIES ACCORDING TO THE PERCENTAGE OF ENERGY DRAWN FROM RECURRENT SOURCES

With the blank and the dotted sectors representing the contribution of energy drawn from the recurrent sources (water-power and wood), a glance at the figures shows quite clearly how much our civilization is exploiting irreplaceable fuel reserves. It is only in Norway, Switzerland, and Sweden, countries with a total population of less than 1 per cent of the world's, that more than half of the electricity equivalent consumed is drawn from recurrent sources, while the vast majority of civilized men depend on the use of irreplaceable fossil fuels.

As shown in Figs. 7 and 8, quite a different hierarchy among the twelve countries considered here results when they are ranked in view of their *per capita* consumption of energy and in view of the percentage use of recurrent energy sources.

Our list containing an arbitrary selection of a dozen countries is, of course, incomplete. But even after fitting in all the others of the eighty-odd countries of the world the U.S.A. would remain No. 1 as regards the amount of energy consumption, and Norway would remain No. 1 as regards percentage of hydro-electricity. Re-

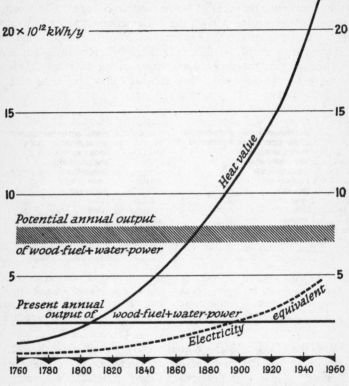

20×10^{12} *kWh/y*

Heat value

Potential annual output

of wood-fuel + water-power

Present annual output of *wood-fuel + water-power* *equivalent*

Electricity

1760 1780 1800 1820 1840 1860 1880 1900 1920 1940 1960

FIG. 9. THE RISE OF GLOBAL INANIMATE ENERGY CONSUMPTION
SINCE 1760

markably enough, three big and heavily industrialized countries—the U.S.A., Germany, and Great Britain—are far behind the average as regards the use of recurrent energy sources, and are likely to suffer first as soon as fuel reserves run short.

Fig. 9 indicates the rapid growth of the world's annual consumption of inanimate energy since the beginning of the Industrial Revolution, the unbroken line denoting the total energy, with fuel taken at its heat value, and the dotted line the electricity equivalent, with fuels taken at 20-per-cent efficiency. The figure gives merely a sketchy representation of the general development since 1760, neglecting minor fluctuations caused by wars or economic depressions.

In contrast to the foregoing figures representing the *per capita* consumption, the ordinates of the curves in Fig. 9 denote the total global consumption. The height of the horizontal straight line over the zero-line indicates the present global amount of $2\cdot1 \times 10^{12}$ kWh of energy gained yearly from the recurrent sources—wood and water-power. The horizontal shaded strip across the figure gives an approximate estimate of the energy which could be gained after full development of all potential water-power of the world, and by optimum utilization of all wood reserves. The exact value is not known yet; hence the amount of energy to be gained in future from the recurrent sources cannot be represented by a sharp line, but by the broader area of the shaded strip in Fig. 9, covering the range from 7 to 8 $\times$ 10^{12} kWh/y.

From the fact that the dotted curve does not yet reach the shaded strip it might be concluded that with fully developed water-power resources all over the world our present demand for pure power could still be satisfied—not, however, the demand for heat. As

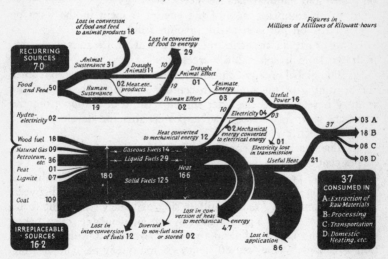

FIG. 10. ORIGIN AND UTILIZATION OF THE WORLD'S ENERGY IN 1937

Reproduced from " Energy Resources of the World "

can be seen from the unbroken line crossing the strip, it was in the second half of the nineteenth century that the total energy consumption crossed the margin of the possible yield of recurrent sources of the kind we are using to-day. And we have to expect that with the ever-increasing demand for energy the dotted curve will cross the margin at a date in the future earlier than that by which a full development of the water-power reserves can be accomplished. More detailed information about this will be given in Chapter X.

A very instructive illustration of the low efficiency of present power economy is given in Fig. 10, taken from the Guyol Report, which gives a complete survey of all the energy sources, including human and animal effort, showing what a small part of the total energy of 23·2 millions of millions of kWh consumed in 1937 was effectively used. Waste heat is by far the largest item of expenditure.

An analogous figure for 1956 would differ only by an approximately 25-per-cent increase of the total amount of consumption of inanimate energy, and by a decline in the preponderance of coal over petrol, natural gas, and hydro-electricity. The percentage of waste has not, however, changed considerably.

Summing up our survey, we can make the following statements on global and national balance-sheets of energy:

1. In all countries most of the energy is spent in the form of heat—partly on purpose, partly involuntarily as waste.

2. In nearly all countries more than half of the energy income is drawn from irreplaceable sources.

3. The global energy consumption has risen nearly tenfold since the beginning of the nineteenth century, and since the end of the century has exceeded the utter limit of the contribution from the recurrent sources (wood and water-power) after full development of all hydro-electric resources.

4. The world's present power economy is wasting more than 80 per cent of the total income.

With the present prices of coal, petrol, etc., the cash value of the percentage of fuel which is annually wasted amounts to between £7000 million and £14,000 million approximately. It ought to be added, however, that a part of the loss is an unavoidable consequence of the Second Law of Thermodynamics, while a considerable fraction is caused by the low efficiency of many engines and heating devices, which, in spite of their being old-fashioned, are still in use.

In view of these facts, the following questions will be dealt with in the next chapters:

A. What is the reason for the enormous losses, and how much of them can be avoided by better economy?

B. How great are the reserves of the most important fossil fuels—coal and petroleum—and how long will they last if the present trend of increasing rate of energy consumption continues?

C. What other sources of energy production can be mobilized in addition to those we are tapping to-day?

The description given in the next chapters of the gradual development of steam-engines, turbines, and internal combustion motors will help us to understand the problems of the efficiency of converting heat into power.

ENGINES USED IN POWER PRODUCTION

CHAPTER III

Steam

The Principle of the Piston Engine

ANY male readers interested in technical matters may have played as boys with a toy steam-engine, and will therefore be familiar with the simplest type of double-acting piston engine, as used in railway locomotives. The principle of the device can be easily understood by a glance at Fig. 11. A gliding valve operated from the main shaft alternately opens and closes the connexions from the inlet port *i* to both halves of the cylinder, at the same time closing and opening the connexions to the exhaust port *e*. A device like this can be kept working temporarily by any difference in pressure between inlet and exhaust pipes. It will do work when we connect the inlet pipe with the outer atmosphere and the exhaust with a big vacuum reservoir, and it will work as well when we connect the exhaust with the open air and the inlet with a container of air or gas under pressure. Certainly in both cases the device will not do work for a long period, because in the first instance the vacuum chamber will eventually be filled with air, until the pressure within is equal to that without. In the second case the container will be emptied after a while, its pressure dropping to that of the atmosphere.

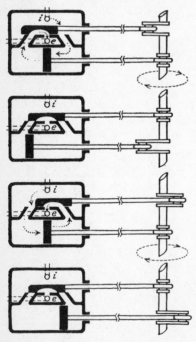

FIG. 11. DOUBLE-ACTING PISTON OF
A STEAM-ENGINE

In our toy engine the inlet was connected to the boiler, which provided steam under pressure as long as enough fuel and water were supplied. Without a proper water-injecting device, no continuous working was possible in our little steam-engine. When all the water had been evaporated we had to stop the engine, open the boiler, and refill it with water. But the real steam locomotives, which, except for some improvements explained later, work on the same principle, are equipped with appliances for feeding water into the boiler during operation, thus enabling the machine to run continuously. Injecting water against the pressure of the boiler is, of course, an operation which consumes energy, and it appears, therefore, on the debit side of the power account. This debit is, however, relatively quite small, because the power both yielded and consumed by the engine equals the product of pressure and volume (of steam on the asset side and water on the debit side). The specific volume of steam at normal pressure being about 1600 times greater than that of the water, the percentage of power consumed by injecting water into the boiler is insignificant.

The uninitiated might be tempted to suppose that this very simple model of a steam-engine was the prototype of the machines produced in the time of James Watt. Curiously enough, the actual development was on different lines.

The Story of the Steam-engine

The idea of the invention of the steam-engine is connected in the popular mind with the name of **James Watt** (1736–1819). The fact is, however, that Watt, whose creative genius should not be denied, was not the *inventor*, but, rather the most important *improver*, of the steam-engine, who by increasing its efficiency took the decisive step towards making it a machine in widespread use.

The first urgent demand for power engines did not arise in the field of transport or of operating contrivances like looms, spindles, or lathes in factories, but for the heavy work of pumping water from mines. To lift innumerable tons of water from the depths of a deep shaft requires an enormous amount of work, too much for manual labour in countries where, and at times when, no more slave labour is available.

In places where water-power was near at hand the pumps could be driven by water-wheels; in many cases, however, no usable water-power was available, and there the urgent need for

mechanical power was a strong incentive for the inventive spirit
of engineers.

The first fuel-driven engines for water-pumps seem to have been
inspired by a device for power transmission constructed as far
back as 1650 by **Otto von Guericke,** Mayor of Magdeburg, the
inventor of the air-pump. Fig. 12 shows schematically an air-
pump connected by a pipe-line to a cylinder closed by a piston.

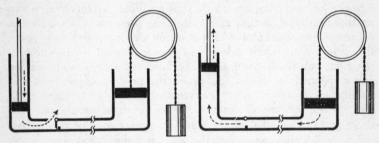

FIG. 12. GUERICKE'S DEVICE FOR TRANSMITTING POWER

If the lower part of the cylinder is evacuated the atmospheric
pressure (approximately one kilogramme per square centimetre, or
14·7 lb per square inch, or, more exactly, 1·033 ata = 14·696 psia[1])
pushes down the piston, which by means of a lever can lift a load
or do some other useful work. A contrivance like this—which in
our age has found its practical use in millions of vacuum brakes
for railway coaches—is, however, only a means of transmitting,
not of producing, mechanical work.

The next step was a fuel-operated power engine actually used
for pumping water from mines. The principle is explained in Fig.
13, which gives a diagrammatic view of **Savery's** engine of 1698.
The machine consists of a boiler B connected alternately to two

---

[1] The data concerning pressures will be given in metric units, British units
sometimes being added. The metric unit of pressure is 1 kg/cm², which is also called
'physical atmosphere' (abbreviation ' at '), differing slightly from the ' standard
atmosphere' (abbreviated ' At '), which is the atmospheric pressure corresponding
to a barometric reading of 760 mm Hg=29·921 in Hg. The letter ' a ' added in
' ata ' and ' psia ' (pounds per square inch) is meant to denote ' absolute ' pressure
—that is, the pressure above absolute zero pressure—in contrast to the *gauge*
pressure, which means the excess of the pressure in a given vessel (for instance,
steam pressure in a boiler) over the atmospheric pressure. The conversion factors
are:

$$1 \text{ Ata} = 1 \cdot 0333 \text{ ata}$$
$$1 \text{ ata} = 1 \text{ kg/cm}^2 = 14 \cdot 223 \text{ psia}$$
$$1 \text{ psia} = 0 \cdot 0703 \text{ ata}$$

oval vessels V and V', which in their turn are connected to a pipe leading down into the water to be pumped from the shaft. When the valve m is opened steam enters the vessel V, whereupon any air or water contained in V is displaced and driven out through valve n and exhaust pipe t. The valves n

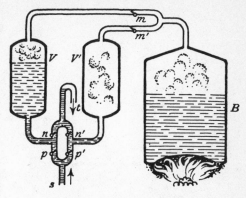

FIG. 13. SAVERY'S STEAM-ENGINE

and p, n' and p', open only upward. As soon as V is filled with steam, valve m is closed, and a jet of cold water from a cistern above (not shown in Fig. 13), streaming over the outer surface of V, causes the steam to condense. Thus a vacuum is created in V, and water is sucked in through pipe s and valve p until V is nearly filled. Then valve m is opened again, and the cycle starts afresh. The valves m', n', and p' are operated with a phase-shift of 180°, so that steam enters V' while water enters V, and *vice versa*.

This device has the obvious advantage of being a pump without pistons or other moving solid parts. There are, however, two decisive drawbacks. The lifting-power of the pump is strictly limited to depths of 10 metres = 32·8 ft—that is to say, as long as the water-pressure does not exceed the atmospheric pressure which causes the water to enter the evacuated vessel. A second, still graver disadvantage lies in the low efficiency (1 per cent or less) of the engine, which consumed an enormous amount of fuel for lifting relatively small quantities of water. Still, lacking better machines, a good many British coal-pits used Savery's engine, which was the first man-made device in history converting heat energy into mechanical work of many horse-power hours (H.P.h).

The idea of using a piston, originally introduced by the French inventor **Denis Papin** in his rather crude engine, was applied in 1705 by **Thomas Newcomen** and his assistant **John Cawley** to Savery's steam-pump.

In Newcomen's engine, which is shown in Fig. 14, the piston is connected by a rod and a chain to the beam C, which on the other side carries a pumping-rod P and a counterpoise W. Steam

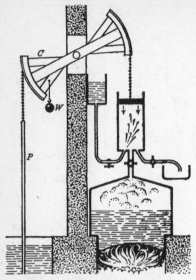

FIG. 14. NEWCOMEN AND CAWLEY
STEAM-ENGINE

admitted to the cylinder raises the piston to its upper dead centre, then the steam valve is closed, and a jet of cold water entering the cylinder causes the steam to condense, thus creating a vacuum. As a consequence, the piston is forced down by the atmospheric pressure pulling down the right arm of the beam and doing work on the pump attached to its left arm.

Newcomen's steam-engine was used for half a century for pumping water from many British coal-mines, until Watt entered the scene in 1763. He was at that time an instrument-maker in Glasgow, and while repairing a model of New-

comen's engine for the university he realized that condensing the steam within the cylinder involved at the same time an alternate chilling and heating of the far greater mass of the piston and the cylinder. To avoid the waste caused by this process it was necessary to keep the cylinder permanently hot. Hence Watt added to the engine a new organ, the separate condenser, into which the steam is allowed to escape during the back stroke of the piston, to be condensed there by cold water surrounding the condenser. As soon as the cylinder itself was no longer used as condenser it could be kept hot by being wrapped up with non-conducting bodies. By these improvements, patented in 1769, not only the waste of steam for alternately heating and cooling large masses could be avoided, but also an increase in the speed of the reciprocating cycle of steam inlet and exhaust could be attained.

It was only in his second patent (1781) that Watt introduced a mechanism for turning the reciprocating motion of the piston into the revolving motion of a shaft provided with a fly-wheel. In this way the use of the engine, originally restricted to pumping purposes, could be extended to all kinds of driving mechanisms familiar to every one to-day.

In his third patent (1782) Watt disclosed two further improvements. One was the double action of the piston, which is familiar to us from our toy engines. The cylinder being closed at both ends, the piston-rod goes through a stuffing-box which prevents leakage of steam. While one side of the cylinder is connected with the condenser, steam is admitted to the other side, and *vice versa*. The application of this device was delayed by technical difficulties of manufacturing really effective stuffing-boxes, and it was **Trevithick** who, about 1800, made perfect double-acting engines.

The other important improvement was the use of expanding steam. In the first steam-engines steam was admitted during the whole fore-

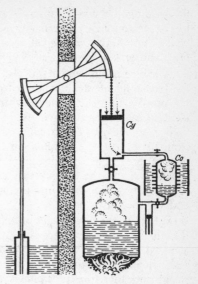

FIG. 15. WATT'S STEAM-ENGINE

Cy=cylinder; *Co*=condenser. The automatic control of the valves is not shown in the figure.

stroke, and a good deal of its energy was wasted when the steam, being still under pressure at the end of the stroke, puffed out into the condenser after the opening of the exhaust valve. Hence, in order to ensure better fuel economy, Watt changed the device so as to stop the admission of steam when the piston had made only a part of its stroke, allowing the rest of the stroke to be performed by the steam expanding from boiler-pressure to the low pressure of the condenser. Thus, for instance, with a 50-per-cent admission factor (the intake valve being closed at half the stroke), a 50-per-cent cut in steam consumption caused a reduction of only about 30 per cent of work at each stroke, improving in this way the efficiency by a factor of 7/5.

A further development led to the construction of the compound engine with two cylinders by **Jonathan Hornblower** in 1781. Steam is first admitted into a smaller cylinder, where it expands from boiler-high pressure to an intermediate pressure, and is then passed over into a larger cylinder, where it expands down to the condenser pressure doing work against a piston in each cylinder. This construction gives a better uniform motion, and at the same time,

Carnegie Library
Livingstone College
Salisbury, North Carolina

by dividing the whole range of expansion into two parts, reduces the losses caused by alternately heating and cooling the cylinder-walls. The high-pressure cylinder is kept permanently at a higher temperature, the other at a lower.

All these improvements helped to increase the efficiency and the fuel economy of steam-engines from the extremely low values of the pre-Watt times to a level which, although poor still, was sufficiently high to extend their use to more and more appliances, and, finally, to means of transport. The descent of the Watt engine from its forerunners, inspired in their turn by Guericke's vacuum devices, proved to be a decisive factor, because it led to the use of a condenser on the low-pressure side of the thermal cycle, instead of exhausting the steam into the open air, as the simple toy engines and the locomotives do.

What are the advantages of the condenser? One lies in the fact that the working substance is a given quantity of water performing a closed cycle consisting of evaporation, expansion, condensation, and, finally, feed-back into the boiler again. It can be kept clean, and therefore precipitates less fur in the boilers than the continually refilled water of the exhaust steam-engines. Far more important, however, is the other advantage, which lies in the fact that, with all other factors—as, for instance, boiler pressure and temperature—being kept constant, the efficiency of the condensing engine is considerably greater. The reason can be easily understood. With given dimensions of cylinder and piston, the force pushing the piston-rod will be proportional to the pressure difference on both sides of the piston. An engine A using a boiler pressure of, say, 2 ata $= 28.44$ psia (which is about twice the atmospheric pressure) and exhausting the steam into the atmosphere does approximately the same mechanical work per stroke as an engine B of equal geometrical dimensions using half the boiler pressure of A and exhausting the steam into a good vacuum. Both engines are consuming the same volume, but not the same weight, of steam, because the steam used in A has, on account of the double pressure, twice the density of that in B. Hence the steam consumption of A will be twice that of B, and, as both are doing the same work, we find that the efficiency of A will be only half that of B. We can improve A by using a condenser. If, instead of exhausting the steam of engine A into the atmosphere, we admit it to a second cylinder in which it expands to the low pressure of a well-cooled condenser, the steam and fuel consumption will remain unaltered, but the power will be increased.

The gains in efficiency secured by the use of the condenser paved the way for an increased demand for steam-engines, and thus helped to usher in the technical age. It was not possible, however—and is not even practically feasible to-day—to use condenser engines on locomotives. For, in order to obtain a vacuum with a pressure well below atmospheric pressure, the condenser must be properly cooled, a condition which can be fulfilled only if sufficient quantities of cooling water are available. It is not possible to carry in a train the amount of water (millions of gallons daily) used in stationary plants for cooling condenser engines of a power equal to that of a locomotive. Experiments have been made with closed circuits of cooling water using radiators like those in motorcars, but with the given area of cross-section of a train and the big engine-power of the locomotive the necessary degree of cooling cannot be achieved. Hence practically all the steam locomotives of the world are exhaust engines, and have accordingly rather poor efficiencies.

The Struggle for Higher Efficiencies

The improvements made since the days when Stephenson's *Rocket* was the champion of all locomotives were achieved by the gradual rise of boiler pressure and steam temperature. A very considerable part of the losses in all kinds of steam plants is due to the high value of the latent heat of vaporization of water. At a pressure of 2 kg/cm^2 absolute ($= 28.44$ psia) the boiling temperature of water is about 120°C. You need approximately 100 kilo-calories to heat 1 kg of water from the injection temperature to the boiling-point, and another 526 kilo-calories to evaporate it. Further raising of the steam temperature can be done with less expense of heat, because the specific heat of water-vapour is approximately only four-tenths of that of fluid water. Hence with only 40 more kilo-calories you can ' superheat ' the steam to a temperature of 220°C., and increase in this way its specific volume, and therefore its capacity for doing work. It is only a part of the total expense of heat energy, including that for superheating the steam, which is converted into useful work, while most of the 500-odd kilo-calories spent for evaporating the water are accumulated in the steam in the form of its latent heat, and are lost again to the outer atmosphere in the exhaust engines, or to the cooling water in the condensing engines.

Hence the whole expense of heat energy consists of a larger fixed amount needed for boiling and evaporating the water, which is subsequently lost, and a smaller amount, depending on pressure and temperature of the steam, which is converted into mechanical work. With the relatively low pressure and temperature used up to the middle of the nineteenth century the total efficiency could hardly be increased beyond 10 per cent. By continually raising boiler temperatures and pressures approximately 20-per-cent efficiency could be obtained in larger plants by the end of the century, and with modern steam plants using turbines a further increase of nearly 20 per cent can be achieved, making a total now of almost 40-per-cent efficiency.

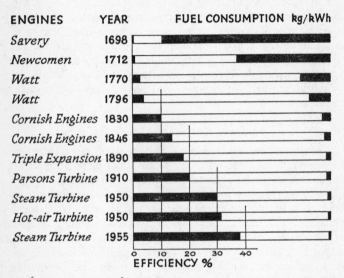

FIG. 16. THE PROGRESS MADE IN THE EFFICIENCY OF STEAM-ENGINES

Fig. 16 gives a graphic representation of this development—from Savery's steam-pump to modern plants. The left-hand side of the diagram gives the efficiencies, the right-hand side the fuel consumption, which is inversely proportional to the efficiency. The figure looks quite encouraging, creating, perhaps, the illusion in optimistic laymen that within the next hundred years the step from 40-per-cent to 100-per-cent efficiency might be taken. Unfortunately, however, this is quite impossible, and we have good reasons to assume that the laws of physics put a definite limit of about 45-per-cent efficiency on all kinds of heat engines. The reason lies in the fact explained in Chapter I. As a consequence of the well-established Second Law of Thermodynamics, the efficiency of any heat engine will be lower than that attained in the Carnot process under optimum conditions. The latter is given by

$$\eta_c = \frac{T_1 - T_2}{T_1} \dots\dots\dots\dots\dots\dots\dots\dots\dots\dots(11)$$

with T_1 and T_2 denoting the highest and lowest temperature, expressed in Kelvin degrees (centigrade + 273°), which are involved in the process. For all engines in practical use the efficiency is given by

$$\eta = \frac{T_1 - T_2}{T_1} \cdot \eta_0 \dots\dots\dots\dots\dots\dots\dots\dots(12)$$

where η_0 is a factor smaller than unity, amounting approximately to between 0·3 and 0·4 for older and to about 0·5 for modern steam plants, if η denotes the overall efficiency considering all losses, including those occurring in the boiler system. It can be seen from the formula that the efficiency will be the better the greater the difference $T_1 - T_2$. Hence, in order to attain high efficiencies, T_1 should be as high as possible and T_2 as low as possible. In the case of steam-engines T_1 is the absolute temperature of the steam entering the high-pressure cylinder, and T_2 the absolute temperature of the steam leaving the low-pressure cylinder. T_2 lies some degrees above the boiling-point of water under atmospheric pressure (100°C. = 373°K.) in the case of non-condensing engines, like those used in locomotives, while lower values of T_2, and accordingly greater efficiencies, can be attained in condenser engines, where T_2 will be some degrees higher than the temperature of the condenser, which theoretically might be almost as low as that of the cooling water.

In the older engines working with so-called saturated steam up to the middle of the nineteenth century the steam was fed from the boiler directly to the cylinder, and therefore T_1 was just a little lower (because of the temperature loss in the connecting pipes) than the boiling temperature of water. This latter, again, depends on the pressure in the boiler. At normal atmospheric pressure of 760 mm of mercury = 1 atmosphere = 1·03 kg/cm² the boiling temperature is exactly 100°C. = 373°K. The variation of saturated vapour pressure with the temperature is shown in Fig. 17, in which, owing to the steep progressive rise of vapour pressure with increasing boiler temperature, three different pressure scales are used in the temperature intervals below 100°C., between 100°C. and 200°C., and above 200°C. The reader will easily discover from the figure that the vapour pressure is as low as one-eighth of an atmosphere at 50°C. = 323°K., and rises to about 16 kg/cm² (= 16 ata) at 200°C. =

473K., reaching finally 225 kg/cm² at the 'critical temperature' of 374°C. = 647°K., above which H₂O cannot exist any longer in the fluid state.

Calculation of Efficiency

While the theoretical calculation of the practical overall efficiency of a steam-engine is a burdensome task with which we do not want to bother the reader, it is quite easy and instructive to calculate from the data given in Fig. 17 the efficiencies of Carnot processes corresponding to different working conditions of steam-engines.

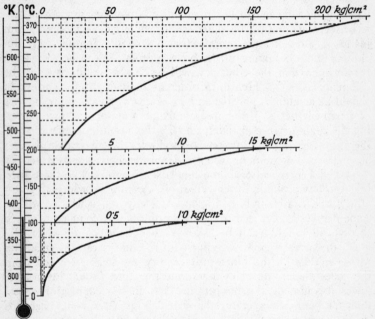

FIG. 17. THE VAPOUR-PRESSURE CURVE OF WATER

Since η_c is roughly twice or three times as great as the η attainable in the real engines, its variation with working conditions gives indications of how to improve efficiencies.

In Table 4, which refers to eighteenth- and nineteenth-century piston engines, T_1 and p_1 denote temperature and pressure of the steam entering, and T_2 and p_2 those of the steam leaving the cylinder and entering the the condenser. Neglecting the temperature-drop in the pipes between boiler and cylinder, as well as between cylinder and condenser, and neglecting some other fairly unimportant causes of losses, a rough estimate of the influence of working conditions on the efficiency can be gained by assuming T_1 to be the boiling temperature corresponding to boiler pressure p_1, as given by Fig. 17, and T_2 to be that corresponding to condenser pressure p_2. (As mentioned above in non-condensing engines which exhaust the steam into the open atmosphere [$p_2 = 1$ kg/cm^2], the exhaust temperature can be considered equal to the boiling temperature at normal atmospheric pressure—that is, $100°$ C. $= 373°$ K.) With these simplifying assumptions, the reader himself can easily calculate the Carnot efficiencies of engines working between a given pressure p_1 in the boiler and p_2 in the condenser by first finding from Fig. 17 the corresponding temperatures T_1 and T_2, and then dividing the difference $T_1 - T_2$ by T_1. With reasonable assumptions of the ratio between the actual overall efficiency of the engine η and the Carnot efficiency η_c, an estimate of the former can be made. A few examples of efficiencies η under various conditions are given in Table 4, and the overall efficiencies, estimated by assuming $\eta = 0.4\eta_c$, are given in the last column.[1]

The advantage of using a condenser—especially in low-pressure engines—emerges clearly from the figures given in the table. Watt and other eighteenth-century engineers were unwilling, for safety reasons, to raise the steam pressure to more than 1·5 ata, or about 7 psi gauge pressure over the atmosphere. As seen from Table 4, the

TABLE 4

CARNOT EFFICIENCY (η_c) AND OVERALL EFFICIENCY ($\eta = 0.4\ \eta_c$)
OF STEAM-ENGINES WORKING UNDER DIFFERENT CONDITIONS

Engine	p_1 (ata)	T_1 (°K.)	p_2 (ata)	T_2 (°K.)	η_c (per cent)	$\eta = 0.4\eta_c$ (per cent)
A	1·5	383	1	373	2·6	1
B	1·5	383	0·2	333	13	5·2
C	6	431	1	373	13·5	5·4
D	8	443	0·125	323	27·1	11
E	8	543	0·125	323	40	16

[1] As will be seen later, the ratio between overall efficiency and Carnot efficiency —assumed here in our rough estimates to be 0·4—is not a constant, but decreases with increasing inlet temperature T_1.

non-condensing engine A with an initial pressure of 1·5 ata has the very poor efficiency of only 1 per cent, while engine B, with the same low initial pressure expanding the steam down to a condenser pressure of 0·2 ata, has about fivefold efficiency. In order to achieve the same η in a non-condensing engine a pressure of nearly 6 ata would be necessary (*cf.* engine C.) Better efficiencies result, of course, from the combination of high initial pressure and low condenser pressure, as exemplified in engine D.

Superheated Steam

Still better results, even without increasing the initial pressure, can be achieved by *superheating* the steam. Engine E in Table 4 is supposed to work with the same boiler and condenser pressures as D, but the steam after leaving the boiler is heated in a separate vessel, the ' superheater,' from 170°C. to 270°C., and only then fed to the cylinder. A rise from 11-per-cent to 16-per-cent efficiency would result unless—as actually happens unfortunately—a part of the gain were recompensated by other factors not taken into account in our simplified theory.

Anyhow, the use of superheated steam removed one of the obstacles against the employment of higher temperatures for increasing the efficiency. As long as saturated steam—that is, steam in thermal equilibrium with the boiling water—was used the fixed bond between temperature and vapour pressure given in Fig. 17 necessitated an ever-increasing rise of pressure. And with the technological development of that time not permitting the application of pressures above ten or twelve atmospheres, the steam temperatures remained rather modest, giving relatively poor efficiencies. To-day, in order to make the relevant temperature difference $T_1 - T_2$ as large as possible, no modern steam plant will give up the use of superheated steam. The limit to T_1 is set to-day not by the pressure, but by the maximum temperature which the moving parts of the engine can withstand in the long run without the risk of damage.

As will be seen later on, the upper limit of safe inlet temperature is estimated at about 730° C., or roughly 1000°K. With an exhaust temperature $T_2 = 320°$K., this would mean a Carnot efficiency of $\eta_c = (1000 - 320) : 1000 = 0.68$, which, multiplied by a reasonably optimistic value of $\eta_0 = 0.5$, gives an overall efficiency of 34 per cent as a probable limit for steam-engines. What had been actually attained in piston steam-engines was, however, scarcely more than one-half of this limit, because neither the high inlet temperature assumed here nor the necessary amount of expansion needed for using low condenser temperatures can be realized in piston engines. Superheated steam, with $T_1 = 573°$K. $= 300°$C. and $p_1 = 20$ ata, increases its volume by a factor of about 90 when it expands down to a pressure of 0·125 ata, corresponding to a condenser temperature $T_2 = 323°$K. $= 50°$C. Assuming a 33-per-cent admission factor (the intake valve being closed at one-third of the stroke), four cylinders, each expanding 3:1, would give only an expansion ratio of 81:1, and the volume of the fourth cylinder would be 27 times greater than that of the first one. Thus even for attaining rather modest efficiencies of 16 per cent or so, unduly large dimensions of the engine would be necessary. Quite apart from the question of efficiency, the dimensions of powerful piston engines, as used in transatlantic liners at the beginning of the twentieth century, grew excessively large, the biggest of them being nearly the size of a three-storey house. As well-known to every motorist, the size of a motor of given power can be reduced by increasing its rotatory speed. While, however, small units as used in motor-cars can easily be designed for 3000 to 4000 r.p.m., large reciprocating engines cannot rotate so fast because of the inertia of the heavy oscillating pistons. Hence only relatively low running speeds could be attained, and, as a further consequence, the dimensions and the weight per horse-power of the engines were relatively great.

The construction of single units of more than 20,000 H.P. utilizing the whole range of expansion from pressures of more than 50 ata down to the low pressure of well-cooled condensers, securing thereby efficiencies of about 30 per cent, was finally made possible by the transition from the piston engine to the steam turbine.

Steam Turbines

The basic principle of the steam turbine, derived from windmills and water-wheels, is much older and simpler than that of the piston engine. And yet it was more than a century after Watt's successful campaign that the first steam turbines came into practical use. Three or four decades later a complete change occurred in power production. While from the beginning of the nineteenth century up to 1860 or 1870 nearly 100 per cent, and even at the end of the century more than 90 per cent, of all the mechanical power in the world was primarily produced by piston steam-engines[1], a sudden drop occurred after about 1900, so that to-day this percentage—as far as new installations are concerned—has fallen to about 10, or even less.

[1] Electro-motors only *re*produce power which has been transmitted to them from the power-stations.

The reason for this radical change is the preponderance of internal-combustion engines in the field of small units up to a few hundred horse-power and the absolutely dominant rôle of turbines in the field of larger units of 10,000 H.P. and more. The use of piston steam-engines is more or less restricted to-day to railway locomotives, smaller steamboats, and to special engines for relatively slow running speeds.

The decisive advantages of the turbine over the reciprocating engine are as follows:

(1)   Fast running speed, and therefore less weight and volume per horse-power.

(2)   Smooth, vibrationless running.

(3)   Higher efficiency.

Why, then, one might ask, did it take a whole century of a rapidly developing age for the well-known ancient principle of the windmill to be adapted to the use of a steam jet for driving a wheel? The reason lies in the difficulty of using steam jets with adequate efficiency. The conversion of heat energy into mechanical work, as performed in a power plant equipped with steam turbines, occurs in three steps:

(*a*) Heat from burning fuel generates in the boiler system super-heated steam under high pressure. The efficiency of this process is called boiler efficiency (η_b).

(*b*) The compressed superheated steam expands in suitably shaped

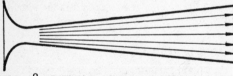

nozzles, producing a high-velocity steam jet. The stored energy of the compressed steam is thus converted into its kinetic energy. The efficiency of this pro-

FIG. 18. SECTIONAL VIEW OF NOZZLES AS USED IN STEAM TURBINES

cess is the thermodynamic efficiency (η_{th}).

(*c*) The kinetic energy of the steam is converted into mechanical work when the jet impinging on the blades of the turbine moves the wheels against the torque exerted by the load. The efficiency of this process is the blade efficiency (η_{bl}).

The overall efficiency will then be given by

$$\eta = \eta_b \cdot \eta_{th} \cdot \eta_{bl} \quad\quad\quad\quad\quad\quad\quad\quad\quad\quad\quad\quad (13)$$

Step (*a*) is identical with that of piston engines. Steps (*b*) and (*c*) differ from those of the piston engines, but (*c*) is somehow related to that of the water-wheels, a quantitative difference between water turbines and steam turbines lying in the far higher speed of the steam jet. This difference is unavoidable, and provides the difficulty in building steam turbines of reasonable efficiency. For in order to secure good thermodynamic efficiency (η_{th}) in any heat engine we must use high inlet temperatures and high pressures. Secondly, expansion of steam from medium pressure even—and still more from high pressure—results in velocities of the order of 1000 metres per second or more. Thirdly, a good blade efficiency (η_{bl}) in converting the kinetic energy of the steam jet into mechanical work will be obtained only when the linear velocity of the blades is not considerably less than half the velocity of the steam jet. These three conditions result in unduly high running speeds of single-wheel turbines, and it was only after many years of experimenting and precise engineering work that steam turbines with suitable revolving speeds could be constructed.

The credit for having constructed the first practically usable steam turbine belongs to the Swedish engineer **De Laval,** whose single-blade ring turbine has been sold commercially since 1883. Its fast running speed of 30,000 r.p.m. was reduced to 3000 r.p.m. by means of a 10:1 reduction gearing, and this simple device held its place among engines of low and medium power for several decades. To-day, however, the Laval turbine has only historical significance, while multi-stage turbines—the first of which, made by **Parsons,** appeared only a year after Laval's—are absolutely dominant in the field of large power engines.

Every steam turbine consists of a system of fixed nozzles in which the steam is expanded and one or more wheels equipped with curved vanes or blades which are set in motion when the steam jet flows over their curved inner surface. As shown in Fig. 19, the steam jet gliding smoothly on and off the blade exerts a pressure as the resultant of the centrifugal forces. Neglecting the small frictional losses, it leaves the blade with the same velocity (relative to the blade) with which it entered. Hence its absolute velocity—that is, the velocity relative to the casing of the engine—will be zero when the velocity of the blade receding in the direction of the central arrow of Fig. 19 is half the absolute velocity of the jet. Zero absolute velocity means that no kinetic energy is left in the steam jet, all its energy being expended to do work on the wheel. Hence, in order to obtain the best values of blade efficiency, the absolute velocity of the exhaust jet should be made as small as possible. As stated above, this is accomplished when the velocity of the moving blade is half that of the steam jet. If, as in the case

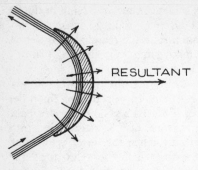

RESULTANT

FIG. 19. FORCE EXERTED BY STEAM
JET GLIDING ALONG THE CURVED
SURFACE OF A TURBINE BLADE

of the Laval turbine, the expansion of the steam and the expense of its energy to the turbine wheel are done in one step each, enormous revolving speeds result.

The work of Sir Charles Parsons and of other inventors in the same field culminated in the successful construction of multi-stage turbines of moderate running speed. Two different methods and combinations of both can be used for solving the problem.

(*a*) *Velocity Compounding.* If the blade velocity is less than half that of the entering steam jet the steam will leave the blade with a considerable remainder of absolute velocity, and therefore kinetic energy. If, as shown in Fig. 20, a ring of fixed blades facing in the opposite direction is mounted to the turbine casing, the steam jet, in backward direction discharged from the blades of the first wheel, is reversed again to its original direction, so that

ENTERING STEAM

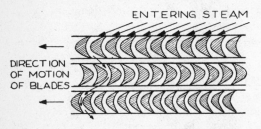

MOVING BLADES

DIRECTION
OF MOTION
OF BLADES

FIXED BLADES

MOVING BLADES

FIG. 20. ALTERNATIVE MOVING AND FIXED BLADES IN A VELOCITY
COMPOUNDING SET OF TURBINE WHEELS

it can drive another bladed wheel mounted on the same shaft as the first. If, after leaving the second set of blades, some further kinetic energy has been left in the emerging steam jet the process of reversing its direction by fixed blades may be repeated. The system of velocity compounding has been introduced by **Curtis** in America. In a purely velocity-compounding turbine the expansion of the steam from inlet pressure to condenser pressure is performed in one step, done in the inlet nozzles, while the conversion of the kinetic energy thus gained into mechanical work done on the wheels is performed in several steps.

(*b*) *Pressure Compounding.* Lower jet velocities, and accordingly lower running speeds, can be attained also by a reduction of the pressure range over which the steam expands. In order to reconcile this scheme with the condition of good thermodynamic efficiency—namely, high pressure-drop between steam inlet and exhaust—the process of expansion must be repeated several times. Imagine, for instance, a turbine in which steam is expanded from, say, 20 ata and 310°C. to 19 ata and 264°C., yielding a jet of reasonable velocity of about 250 m/sec. A considerable amount of thermal energy is left in the exhaust steam, which has to be utilized in order to avoid waste. This could be done by passing the steam to a second turbine in which it is expanded down to a pressure of 18 ata, then to a third with exhaust pressure of 17 ata, and so on till the condenser pressure is reached. In this way moderate running speeds could be reconciled with good thermodynamic efficiency, but at the cost of an enormous expenditure of machinery. It was Parsons' indefatigable endurance which led to the result that, instead of running the steam through a cascade of several turbines, the compounding of pressure can be done in a single turbine containing a multitude of alternate moving and fixed blade-rings, the latter being shaped so that they can act as nozzles in which the steam expands, gaining thereby kinetic energy again. While, therefore, in the velocity-compounding Curtis wheels no pressure-drop occurs in the fixed and moving blades—the former serving only as guide blades for changing the direction of the steam—the pressure is released step by step in the fixed blade-rings of the Parsons turbine. Although the idea of pressure compounding appears to be quite a simple and obvious solution of the problem of reducing the running speed, its implementation was very difficult. For any pressure-drop along the wheels of the engine causes losses by leakage of steam through the slits between moving parts and casing. It required a good deal of precision workmanship to make the blade-wheels exact enough, leaving only tiny slits between their circumference and the casing, and yet avoiding any contact between moving parts, which would soon destroy the gliding surfaces by friction. This task is complicated by the fact that the blades are exposed to the huge centrifugal forces and the powerful driving momentum. Years of development were necessary to reach the degree of exactness which finally secured the success of the multi-stage steam turbine.

(*c*) *Reaction Turbines.* Parsons went a step farther on the way
of securing a gradual pressure-drop along the axial flow of the
steam in the turbine. In the reaction turbine, which was his most
important achievement in this field, the steam expands, and thereby
gains kinetic energy, not only when it flows over the fixed blades,
but also when it passes over the moving blades. When the latter
blades are suitably shaped, as shown in Fig. 21, a pressure-drop and
a corresponding expansion take place in the moving wheel. The
result is that the jet does not, as in the ordinary, or 'impulse,' wheels
(a single blade of which is shown in Fig. 19), leave the blade with
the same relative velocity with which it entered, but with a greater
velocity gained by the expansion. Hence each blade emitting a jet of

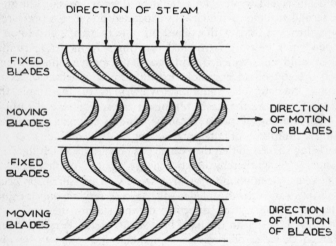

FIG. 21. FIXED AND MOVING BLADES OF A REACTION TURBINE
SUITABLY SHAPED TO ACT AS NOZZLES

higher velocity receives an impulse by reaction, in the same way
as a firing rifle or a rocket recoils. That is why turbines of this type
are called reaction turbines.

Figs. 20 and 21, showing the shape of the blades, are views
of parts of the blade-rings taken in the direction of a wheel
diameter. A sectional view perpendicular to the axis of a simplified
model of the whole turbine is given in Fig. 22. A pair of blade-
rings consisting of a fixed ring of blades and a moving ring is
called a turbine pair. While the number of pairs is only nine in the
model shown in Fig. 22, most turbines in practice are equipped
with many more pairs. In large ship engines the entire installation

consists of separate high-pressure, intermediate-pressure, and low-pressure turbines. All these units are connected through flexible couplings to three separate pinion shafts gearing with a single gear-

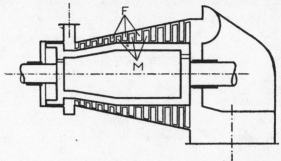

FIG. 22. SECTIONAL VIEW OF A PARSONS TURBINE
(SIMPLIFIED MODEL)
M = moving blades, F = fixed blades.

wheel, which is coupled to the slow-running propeller-shaft. Each single unit has a large number of blade-rings, so that the expansion of the steam from boiler pressure to condenser pressure takes place in many stages. In most modern turbines a combination of velocity compounding and pressure compounding is used.

Parsons' first turbine appeared on the market in 1884, but two decades elapsed before due attention was paid to his work, the importance of which might well be compared with that done by his great compatriot James Watt a century earlier.

The advance of the steam turbine began only a few years before World War I, when it was realized that more powerful, more economic, and more smoothly running engines for large naval vessels could be installed in the form of turbines, instead of the noisy, vibrating, bulky piston engines. The greater economy of the turbine—which is due to the fact that the temperature and pressure interval between inlet and exhaust could be increased in both directions—secured for it the dominant rôle which it plays to-day in large power plants. While up to about 1870 the full range of heat engines of all powers was covered by the piston steam-engine, the situation has totally changed to-day. Fig. 23 gives roughly approximate graphs showing how the global production of mechanical energy is distributed among engines of different types and different powers. The abscissæ in the graphs give the power of the single engines, and the ordinates are drawn proportional to the total power installed in engines of the given class. It can easily

be seen from the lower graphs in Fig. 23 that the low power range
up to about 100 H.P. has become the sole domain of the internal-
combustion engines, while the range from about 10,000 H.P. upward is the field of the turbine. Only a part of the middle class

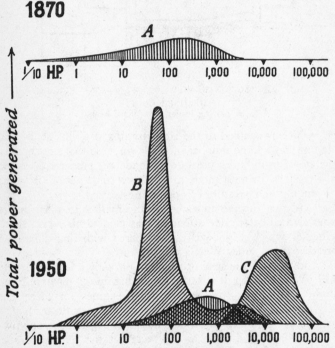

A = Piston steam-engines
B = Internal-combustion engines
C = Steam turbines

FIG. 23. DISTRIBUTION OF POWER PRODUCTION AMONG ENGINES
OF DIFFERENT TYPES AND SIZES

has been left to the piston steam-engine, and its total contribution
to world power generation has fallen from nearly 100 per cent in
1870 to less than about 10 per cent to-day.

Why no Smaller Turbines?

In view of the definitely established superiority of turbines, an
obvious question may be asked. Why is their use restricted to large,
high-power units? When will inventors and manufacturers supply

us with handy small turbines which could be used for silently driving motor-cars and buses?

The objection to building smaller turbines lies in the fact that a certain absolute value of jet velocity, combined with a minimum number of stages of expansion, is prescribed by the conditions of good efficiency. Hence, for a given revolving speed, say, between 3000 and 4000 r.p.m., as used in the motor-car industry, the diameter of the turbine-wheel must have a given size, and the number of pairs cannot be reduced either. While, therefore, a proportionally reduced model of a piston engine with one-quarter the linear dimensions, and consequently one-sixty-fourth the weight, of the original, might work with nearly the same efficiency as the big engine, we cannot arbitrarily reduce the dimensions or the number of stages of a steam turbine without either greatly reducing the efficiency or unduly raising its revolving speed. The Laval turbine, which was a relatively simple engine of small power and small dimensions, could successfully compete at a time in the eighties and nineties of the last century when relatively low steam pressures were used and the standards of efficiency were still low, but it can no longer cope to-day with modern requirements of efficiency.

It may be remarked incidentally that at the start of the motor-car industry about 1900 there was still a kind of competition between the petrol motor and small steam-engines for driving cars, the latter boasting of the advantage of creating less noise and stench. The main reason for the total defeat of the steam-engine lay in the long time interval between lighting the fire in the boiler and obtaining the pressure necessary for driving with full power. The possibility of instantaneous push-button starting was a sufficient reason in itself for preferring the petrol motor—not to speak of its much better efficiency compared with non-condensing steam-engines. Any projects for using turbines for driving road or rail vehicles are concentrated, therefore, on gas turbines, which will be dealt with in Chapter V.

The Limits of the Efficiency of Steam-engines

Let us return to the question raised in connexion with Fig. 16. The progress of efficiency from less than 1 per cent at the beginning of the eighteenth century to records between 30 and 40 per cent to-day seems to justify the expectation that further substantial increases in efficiency, with corresponding savings in fuel, might be achieved in the future. Although negative prophecies in technical matters are generally rather dubious, it can be stated with some degree of certainty that the records reached at present are very near the limit set to the efficiency of heat engines by the laws of physics.

The overall efficiency of a steam turbine, as given by equation (13)—

$$\eta = \eta_b \cdot \eta_{th} \cdot \eta_{bl} \dots\dots\dots\dots\dots\dots\dots\dots\dots\dots\dots\dots\dots\dots (13)$$

—is a product of three factors, each being smaller than unity. While both the boiler efficiency η_b and the blade efficiency η_{bl} can be increased to values over 80 per cent by very careful construction and the exact making of all parts, a much lower limit is set to the thermodynamic efficiency η_{th}. The coefficient η_{th} denotes the efficiency of an ideal perfect engine in which the only energy losses are those of the heat discharged in the condensing steam— an unavoidable loss inherent in the principle of all steam-engines— while all other losses caused by friction, leakage, heat conduction, etc., are supposed to be avoided. It is possible to calculate the thermodynamic efficiencies η_{th} exactly from data obtained by extensive and careful measurements of the thermal behaviour of steam under various conditions of pressure and temperature. The result of these calculations is given in Fig. 24, showing the variation of efficiencies with initial steam temperature. The general law mentioned in a previous section of this chapter states that the efficiency is the greater the larger the interval between inlet (superheater) temperature T_1 and exhaust (condenser) temperature T_2. The lower limit to T_2 is set by the temperature of the surrounding atmosphere, for which an average value of $291°$K. $= 18°$C. $= 64\cdot4°$F. has been taken in drawing Fig. 24. (A variation of plus or minus $10°$C. would not cause any substantial alteration of the efficiency.) Curve A gives the efficiency of the Carnot cycle $\eta_c = (T_1-291):T_1$, rising in the given interval of temperature from about 57 per cent to just over 70 per cent. Curve B gives the thermodynamic efficiency of the ideal engine as a function of inlet temperature T_1, assuming a very high initial pressure $p_1 = 120$ ata and the lowest possible values of condenser pressure and temperature—namely, $p_2 = 0\cdot02$ ata and $T_2 = 291°$K. $= 18°$C. The range of initial temperatures shown in the diagram extends from $400°$C. $= 673°$K., which is near the present record, to $750°$C. $= 1023°$K., which is the limit to which blades made from special expensive alloys may be exposed without the risk of premature wearing out under the heavy strain of the combined action of thermal and mechanical forces.

As may be seen from Fig. 24, it is a clear though regrettable fact —by which the fate of all steam-engines may be ultimately doomed —that the increase of thermal efficiency with rising initial temperature lags far behind that of the Carnot cycle. Even the ideal engine

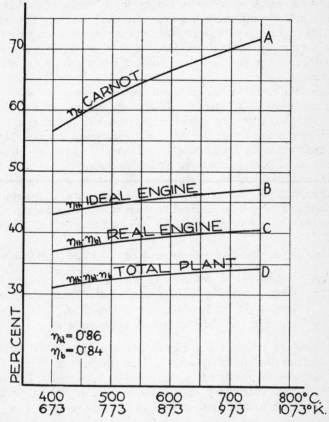

FIG. 24. THE RISE OF EFFICIENCIES OF STEAM-ENGINES WITH INCREASING INITIAL TEMPERATURE OF THE STEAM

working under extremely exacting conditions cannot attain an efficiency of 50 per cent, and what can be realized practically is a good deal less. Curve C shows the efficiency of the real turbine, assuming, rather optimistically, a blade efficiency $\eta_{bl} = 86$ per cent (including all the losses by friction, leakage, etc., occurring in the turbine), and curve D, finally, gives the product of the engine efficiency with an also very optimistic boiler efficiency of 84 per cent. D represents thus the overall efficiency of the whole steam plant from fuel to work done on the generator shafts.

The result is by no means encouraging, because it shows that the triumphant march of steam technique from Savery and Watt to Parsons and his successors has nearly reached its end. For with enormous efforts and technological tricks, and at the expense of using very costly material, we shall be able to save only a small percentage of the present losses, without being able to cut them radically.

Although, therefore, a continuous improvement in the economy of steam-engines, and particularly the replacement of the older plants with low efficiency, is an urgent task which has to be accomplished in the near future, the effect of these measures will be more than made up for by the growing global demand on power. Hence, in spite of all possible improvements, the consumption of fuels for power production will go on rising steadily. If we want, therefore, to reduce as far as possible the accelerated rate of fuel consumption we must find new ways, instead of merely following the path of improving engine efficiencies, which will lead us soon into a blind alley.

The three methods given below may be taken into consideration:

(*a*) Mobilization of new sources of power.
(*b*) Direct conversion of chemical affinities into electricity without the intermediate creation of heat.
(*c*) Utilization of the reject heat of steam-engines.

The first two methods will play an important rôle in the future, and will therefore be dealt with in later chapters of this book. The third method, outlined in the following paragraphs, is being used ever more extensively to-day, and may be considered as the most readily accessible method of making substantial savings in fuel.

Reject Heat must not be Waste Heat

Of all the fuel consumed for power production 80 per cent on the average is uselessly wasted for heating the outer surroundings. The loss is larger in the case of means of transport. It is, on the other hand, reduced to about 65 per cent in some of the better-equipped power stations, and even to 56 per cent in extreme cases reported in the next chapter. In steam plants the heat lost and rejected will scarcely ever amount to less than 60 per cent of the total fuel energy. Seeing that it is not possible to convert more than 40 per cent into useful power, can we not make use of the residual 60 per cent of energy in the form of heat, instead of simply throwing it away? The answer is *yes,* we can do it, and in wisely planned power installations the method of utilizing the reject heat has been applied successfully for years, although it demands some sacrifices as regards efficiency.

Let us imagine as an example a power plant designed for maximum efficiency. The necessary condition, as mentioned repeatedly, is not only high inlet pressure and temperature, but also a good vacuum in the condenser, which, on account of the fixed connexion between vapour pressure and temperature given in Fig. 17, implies low condenser temperature. The condenser must therefore be kept cool by means of a considerable quantity of cooling water, and the reject heat will serve to heat the cooling water from, say, 16°C. to 22°C. In other words, the hundreds of millions of calories of heat rejected daily in the power plant are stored in millions of gallons of scarcely lukewarm water, which cannot be used, except, perhaps, for warming up chilly swimming-pools. If we want to utilize the reject heat we must obtain it in a suitable form. This can be done by means of the ' back-pressure turbine,' in which the steam is not condensed, but leaves the turbine at a comparatively higher temperature and pressure. This looks at first glance like a step backward from the progress made long ago in using condensers for increasing the efficiency. As a matter of fact, the possibility of utilizing the reject heat must be bought at the cost of decreased efficiency. Still, the relative loss of efficiency caused by dispensing with a condenser is much smaller in modern installations than it was in the time of Watt, because of the much higher initial pressures and temperatures we are using to-day. By tapping the steam still under pressure from the turbine we obtain a heat carrier which on account of its pressure is able to flow by itself through well-insulated pipe-lines, and can therefore be transported to places where it is needed. In this way the part of the heat which is left in the steam after doing work in the turbine can be used either for space heating or for industrial purposes. As shown in Fig. 25, taken from an article by F. E. Simon in *Discovery* (April 1952), we have the choice between more power and no residual heat and slightly less power plus practical use of the reject heat. Although certainly mechanical or electrical energy is economically more valuable than energy in the form of heat, the electricity equivalent

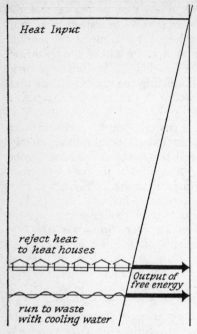

Heat Input

reject heat
to heat houses

Output of
free energy

run to waste
with cooling water

FIG. 25. THE CHOICE BETWEEN
HIGHER EFFICIENCY AND
UTILIZATION OF REJECT HEAT
By courtesy of UNESCO

of usable heat to be gained from tapping steam from the last stages of a back-pressure turbine is so much greater than the loss of power that the net gain of this bargain is considerable. It may be mentioned as an illustration that the reject heat of a moderately large power station of 100,000 kW installed capacity suffices to heat a district of about 10,000 houses.

Both district heating and industrial use of reject heat from power plants are being practised to-day in quite a number of places. We can be assured, therefore, that the turnover from waste heat to usable heat is no longer a problem of science and technology, but merely one of organization. If in designing projects of power plants and in planning new cities or new districts due attention is paid to suitable use of reject heat of the power plants, in the long run fuels worth millions of pounds might be saved.

CHAPTER IV

Internal-combustion Engines

OWING to the widespread use of motor vehicles, a great number of people are well acquainted with the working principle of the petrol engine. In dealing, therefore, with the rôle of internal-combustion engines (I.C.E.'s) in power economy we can restrict ourselves to a survey of the different types of engines, their relative merits, and the expectations as to further developments.

The Different Types of Internal-combustion Engine

The common characteristic of all I.C.E.'s is the combustion of the fuel in the cylinder. Among the variety of engines included under this heading two different main groups may be discerned: the petrol engine, as used in motor-cars and aeroplanes, working on the constant-volume, or Otto, cycle, and the Diesel engine, as used in heavy cars, buses, locomotives, and motor-ships, working on the constant-pressure, or Diesel, cycle. In both groups, again, a subdivision can be made according to whether a four-stroke or a two-stroke cycle is used.

The operating scheme of the very popular four-stroke petrol engine is:

(a) *Intake Stroke* (*or Suction Stroke*). A mixture of air and vaporized petrol is sucked in during the outstroke of the piston. The petrol is vaporized by being drawn in through fine jets in the carburettor.

(b) *Compression Stroke.* The gas mixture is compressed during the back stroke, and is ignited by a spark at the end of the stroke. The heating of the gas by combustion of the mixture occurs very quickly while the piston is close to the top dead centre (TDC)—hence the name constant-volume cycle.

(c) *Power Stroke* (*or Expansion Stroke*). The heated gas expands, doing work on the piston.

(d) *Exhaust Stroke.* The exhaust valve being opened, the burnt gas is driven out during the back stroke.

Compression and expansion stroke are common to all I.C.E.'s. The suction and exhaust strokes are eliminated in the two-cycle engines, and are replaced by a rapid blast of a new charge at the end of the expansion stroke (bottom dead centre; abbreviated BDC), expelling the burnt gas and admitting fresh gas mixture. In order to perform this action the new charge must be compressed in a chamber external to the cylinder. In smaller engines this is done in the crank-case by the piston during its outstroke; in larger ones the compression is done by a separate blower driven off the engine crank-shaft.

In Diesel engines pure air is admitted and then compressed to a very high compression ratio, so that its temperature is raised by the rapid compression over the igniting point of the fuel used. Near the end of the compression stroke a charge of Diesel oil is injected under pressure through a spray-valve in the cylinder-head. In mixing with the hot air the oil starts to burn. The injection of the oil is continued during a small part of the expansion stroke, and is governed so that in spite of the expansion already begun the pressure is kept at a constant high value by the burning mixture. (Hence the name ' constant-pressure cycle ' which is used as an alternative for Diesel cycle.) After the closing of the fuel injection the expanding hot gas continues to exert force on the piston, delivering power to the shaft.

Diesel engines, as well as petrol motors, can be built as four-stroke or two-stroke engines. In the first case the power stroke (or expansion stroke) is followed by the exhaust stroke driving out the burnt gases. In the following intake (or suction) stroke air is sucked in, and during the next stroke compressed as described above.

In the two-stroke Diesel (Fig. 26) the air is supplied by a blower *B* driven by the engine crank-shaft through gears which are not shown in the figure.

The four phases in Fig. 26 are:

(*a*) The piston, after having performed the power stroke, has reached the bottom dead centre (BDC), and has uncovered the transfer ports *T* through which air rushes in from a ring-shaped canal *C*, where it has been delivered under pressure by the blower *B*. The burnt gas is driven out through the exhaust port *E*, the exhaust valve having been opened by a cam-shaft, not shown in the figure.

(*b*) Phase at about 80 degrees crank travel from BDC. The transfer ports have been closed by the piston, and the exhaust valve by the cam-shaft. The air is being compressed.

(*c*) Top dead centre (TDC). The air has been compressed to
the volume of the clearance space, which is about one-sixteenth of
the cylinder volume. If the temperature were kept constant during
the compression stroke the pressure would rise, according to Boyle's
Law, in reverse proportion to the volume—namely, sixteen times

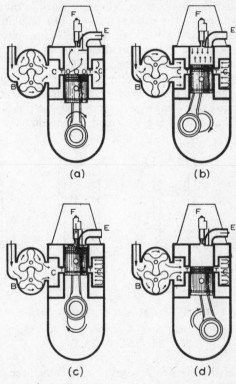

(a)　　　　　　　(b)

(c)　　　　　　　(d)

FIG. 26. FOUR PHASES OF OPERATION IN A TWO-
STROKE DIESEL ENGINE

B, blower; *C*, ring-shaped canal; *T*, transfer ports; *E*, exhaust port;
F, fuel-injector.

the initial pressure. As a matter of fact, however, the rapid compression occurs adiabatically, thus raising the temperature to about
900°F., or approximately 480°C. = 753°K., and the pressure to
about 50 ata. During a very short time interval, while the crank
travels from about 15 degrees before top dead centre to 15 degrees
after TDC, a charge of fuel is injected at high pressure and starts

to burn in the hot air. The timing of the injection is so chosen that on account of the ignition delay of a few milliseconds the combustion starts in time to raise the pressure immediately after TDC.

(*d*) Combustion stroke, or power stroke. The heated gas expands during and after the combustion delivering work to the crank-shaft. In the position shown the exhaust valve is just beginning to open, and very shortly afterwards the upper edge of the piston will pass the transfer ports, and thus admit the expelling blast which makes the engine ready for the next cycle.

Fig. 27 gives the corresponding phases in a two-stroke petrol engine using the Otto cycle. With small units of a few H.P., as in motor-cycles, no extra compressor is provided, the pressure necessary for the expelling blast being delivered by the down stroke of the

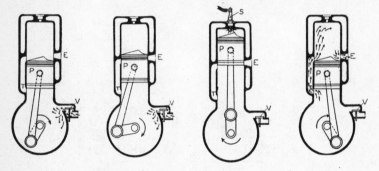

FIG. 27. FOUR PHASES IN A TWO-STROKE PETROL ENGINE

P, piston; *V*, inlet-valve; *T*, transfer port; *E*, exhaust port; *S*, sparking-plug.

piston compressing the air in the crank-case. When the piston moves upward again it compresses the gas in the cylinder, while new mixture is sucked into the crank-case through a valve *V* and an intake-pipe which is connected to the carburettor. The vane attached to the piston-head serves to deflect the gas jet upward in order to expel all burnt gases from the whole cylinder volume. Ignition is done immediately before TDC by a sparking-plug *S* situated in the cylinder-head.

Both Figs. 26 and 27 represent particular engines serving as concrete examples to illustrate the operation of the two-stroke cycle. Many other forms and designs are being used; thus, for instance. the ring-shaped canal *C* shown in Fig. 26, serving to feed the intake ports of the engine, may be replaced in multi-cylinder engines by a big cylindrical container feeding all the cylinders. In other engines the expulsion of burnt gas is done in the reverse direction— namely, from the cylinder-head downward—the position of intake and exhaust ports being mutually exchanged, etc.

FOUR – CYCLE

A diagrammatic representation comparing the action of four-stroke and two-stroke engines during the revolution of the crank is given in Fig. 28. While a complete two-stroke cycle is performed during each revolution of the crank, two revolutions of a four-stroke engine are needed to complete the cycle. The main advantage of the two-stroke cycle is the occurrence of a power stroke in each revolution, causing more uniform running speed and saving, therefore, weight of fly-wheels. This is of particular importance in the case of single-cylinder engines. Four-stroke engines must be

TWO–CYCLE

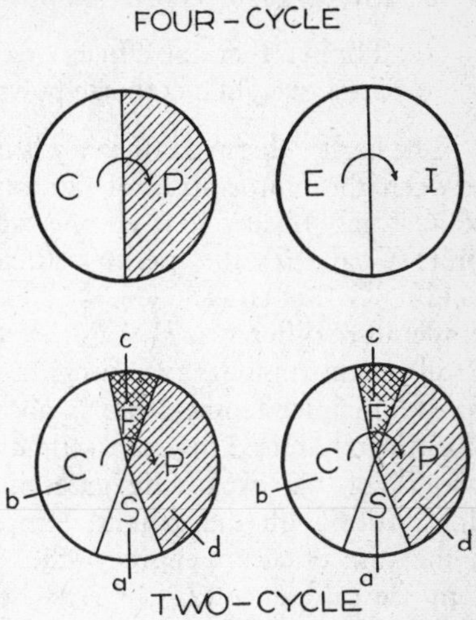

FIG. 28. ACTION OF FOUR-STROKE AND TWO-STROKE CYCLES DURING TWO REVOLUTIONS OF CRANK

C, compression stroke; *P*, power stroke; *E*, exhaust stroke; *I*, intake stroke; *S*, expulsion period; *F*, fuel-injection period. The phases indicated by *a, b, c, d,* correspond to the phases shown in Figs. 26 and 27.

fitted with a comparatively large fly-wheel or with a multiplicity of cylinders. It is instructive to compare the rate of uniformity in different engines as given in the following table:

TYPE OF ENGINE	POWER STROKE PER REVOLUTION
Double-acting steam-engine	2
Two-stroke I.C.E.	1
Four-stroke I.C.E.	$\frac{1}{2}$

Relative Merits of I.C.E.'s and Steam-engines

The striking dominance of internal-combustion engines in the power range up to 100 H.P. is mainly due to the growing popularity of motor vehicles, and this in its turn is the result of a decisive advantage of the I.C.E.'s—namely, the easy and almost foolproof operation, push-button start, and low maintenance labour. Other advantages over the steam-engine are:

(*a*) Higher thermal efficiency.
(*b*) Less weight per horse-power.

The higher thermal efficiency is mainly due to the large difference between the temperature of the burning mixture and that of the exhaust gas. In any heat engine working between the temperature limits T_1 and T_2 the thermal efficiency is a fraction of the corresponding Carnot efficiency $\eta_c = (T_1 - T_2) : T_1$; hence by increasing the temperature difference $T_1 - T_2$ the efficiency can be increased. While in all steam-engines, including modern steam-turbine plants, the maximum inlet temperature is about 800°K. to 900°K., and may be ultimately raised to just a little over 1000°K., the initial temperature T_1 of the expanding gases in I.C.E.'s is well over 2000°K. Hence, although the exhaust temperature T_2 is also higher than in the case of steam-engines—thus compensating part of the gain from the difference $(T_1 - T_2)$—better thermal efficiencies can be achieved in I.C.E.'s than in mobile exhaust steam-engines. This holds particularly in the case of Diesel engines.

Any attentive reader will be tempted to ask: Why was it possible for constructors of I.C.E.'s to reach initial temperatures of more than 2000°K. long ago while to-day, with all the modern technological development, 1000°K. appears to be a kind of unsurmountable temperature barrier for steam-engine designers? The answer is that, most luckily for the I.C.E., the heating to high temperature takes place in the cylinder itself, which is again periodically cooled by the intake of fresh air. In the case of steam, superheating to 2000°K. or more would mean keeping the walls of the superheater permanently glowing white-hot, and in addition exposing the blades of the first turbine-wheel to a temperature which would inevitably destroy them. The reciprocating action of the piston engine, though disadvantageous from the purely mechanical point of view, proves to be an advantage from the thermal standpoint.

Here, again, the attentive reader might ask: Was it not one of Watt's decisive achievements that by providing a separate condenser he could avoid the waste of heat caused by the alternate cooling and heating of the cylinder in the engine of his forerunner Newcomen? Why, then, can we talk of the advantage of the periodical cooling of I.C.E. piston engines when we have been told just before what a nuisance it had been in steam-engines?

The question is well founded and deserves a full answer. It is true that alternate heating and cooling causes a loss. But this loss is considerably smaller in internal-combustion engines than it was in Newcomen's steam-engine. For in the latter case most of the heat transferred from the hot steam to the cylinder-walls was carried away by the condensed water. In I.C.E.'s, on the other hand, the

TABLE 5. WEIGHT PER HORSE-POWER OF HEAT ENGINES

Steam-engines including boiler, stationary plants excluding stack. Dry weight excluding water and fuel. Notations: > more than; ≈ approximately equal to.

Type of Engine	Nominal B.H.P.	lb./B.H.P.	kg B.P.S.
Stationary plants: Steam piston engine plant	10,000	> 150	> 67
Steam turbine plant	48,000	≈ 100	≈ 45
Modern hot-air turbine plant	3000	6·2	2·8
Locomotives: Passenger steam locomotive	2500	≈ 116	≈ 52
Cummins Diesel engine	150	6	2·7
G.E. Diesel engine	1000	24	11
Cars and motor-cycles: Four-cylinder four-stroke engine	≈ 50	≈ 4	≈ 1·8
Single-cylinder two-stroke engine	6	3·7	1·65
Aeroplane engines: Junkers six-cylinder two-stroke Diesel	750	1·7	·76
Wright Cyclone petrol motor	900	1·16	·52

amount of heat transferred to and taken away from the cylinder-walls during a cycle is relatively smaller on account of the much higher frequency of the alternating process, and, besides, in contrast to the steam-engine, a considerable part of the lost heat is recovered by helping to raise the temperature of the compressed gas. The salient point, however, is that with this relatively small loss a greater gain can be bought—namely, the high initial pressure and temperature, which afford better thermal efficiency.

The elimination of the boiler system, together with the high running speed of I.C.E.'s, made it possible to reduce their weight per horse-power considerably below that of steam-engines. Table 5, at p. 79, gives a synopsis in British and Continental units.

These advantages of the I.C.E.'s are partly counteracted by the following advantages of the steam-engines:

(*a*) Possibility of using cheaper fuel.
(*b*) Better uniformity of motion, less noise, and less stench.
(*c*) Greater robustness and longer operating life.
(*d*) Possibility of reversing the motion.

There are piston steam-engines still working which were almost continuously running for half a century, while modern fast-running automotive engines are made to last (in favourable circumstances) from 200 to 400 hours under full load at 4000 r.p.m.

Owing to these advantages a large number of piston steam-engines are still used in smaller stationary plants, as well as in railway and naval services.

The Dieselization of Railways

Although the piston steam-engine is still the most widely used motive-power in railway service, an unmistakable trend away from the steam locomotive is noticeable. Alongside with the progress of railway electrification—which is of the greatest economic importance in countries with ample water-power resources—the most striking development in this direction is the ' Dieselization ' of the U.S. railways which has been going on for a few years, and which is expected to lead to a complete change-over to Diesel-electric locomotives by 1960.

The reason for this radical substitution lies not only in the higher efficiency of the Diesel engine, but also in its prompt readiness for operation. This is of special importance in switching service, where locomotives are used intermittently with long intervals between working periods. As soon as a Diesel engine is stopped not a gramme of fuel is consumed, while the fire under the boiler of a steam locomotive must be kept burning during a stop in order to keep the engine ready for service. Hence, as Ayres and Scarlott point out in their interesting book *Energy Sources*, a given quantity of energy in the form of Diesel oil can do ten times as much useful work in railway switching service, five times as much in freight service, and more than three times as much in passenger service. These facts led to the definite decision to replace steam locomotives by Diesels, a decision which was soon made irrevocable by dis-placed coal locomotives being scrapped, round-houses being dis-mantled, and a considerable number of coal-cars being sent on their way to the scrap-heap. Projects of a gradual Dieselization of British Railways have also been made.[1]

All this is quite sensible and economic as long as the rich flow of crude petroleum continues to run from the oil-wells. The situa-tion will be different when easily accessible oil resources are used up, something which is going to happen in the next or second next generation. It will be a very urgent task by that time to find methods of synthesizing oil from coal which are cheaper and more economic than those which are used to-day.

The Relative Merits of Four-stroke and Two-stroke Engines

The main advantage of the two-stroke cycle, which consists in delivering a power stroke during each revolution, thus furnishing a more even turning momentum on the crank-shaft, at the same time allowing the building of lighter and cheaper engines, has been discussed already, and it has been mentioned that this advantage is of special importance in the case of small single-cylinder engines. As a matter of fact, the original development of the two-stroke engine was made for small motors as used for motor-cycles. Later on the advantage of obtaining more power per dead weight of the engine made itself felt in the high-power class as well, and since the thirties not only small engines, but also larger units, as, for instance, railway Diesels and ship Diesels, are built on the two-stroke principle.

Other advantages of the two-stroke cycle are:

(*a*) The work required to overcome the friction of the exhaust and suction stroke is saved.

---

[1] By these projects eventually 2500 main-line Diesel locomotives will be in service on British Railways.

(*b*) The expulsion of burnt gas, or scavenging, is more complete in low-speed engines, as it does not leave the clearance volume full of burnt gases, as in the case of the four-stroke engine.

The task of replacing the burnt gases by a new charge is, on the other hand, a problem which becomes ever more difficult with increasing running speed of the engine. Assuming the scavenging blast to occur between 30 degrees crank travel before BDC till 30 degrees after BDC, the replacement must be done within $\frac{60}{380} = \frac{1}{6}$ of a revolution, or within $\frac{1}{300}$ of a second in an engine running at 3000 r.p.m. Scavenging at this rate is apparently not done completely enough, for experience shows that the efficiency of two-stroke engines declines at very high speed.

Another disadvantage of the two-stroke cycle is the fact that along with the burnt gases a small part of the new charge escapes through the exhaust port. This matters little in Diesel engines, but in engines working on the Otto cycle part of the fuel is wasted by the escape of unburnt gas mixture.

As a result of these advantages and disadvantages, a balance has been obtained with the two-stroke cycle dominating in the field of small single-cylinder petrol motors and the four-stroke dominating among multi-cylinder engines for motor-cars. In the field of Diesel engines both types are used up to units of the maximum power so far built.

The Relative Merits of the Otto and Diesel Cycle

The outstanding advantage of the Diesel engine is its greater economy. It was a stroke of genius when Dr Rudolf Diesel, a German engineer, in trying to raise the compression ratio without running the risk of premature self-ignition, devised his system of compressing pure air and injecting the fuel near TDC. In doing so he killed two birds with one stone by not only increasing the efficiency, but also making it possible to use cheaper fuel.[1] Diesel oil is a less volatile fraction among the products obtained by the

[1] The world-wide fame of the name of Diesel and its generally accepted association with the system of compression-ignition engines may be attributed to the fact that the co-operation of Diesel himself with the licencee of his patents, the M.A.N. (Maschinenfabrik Augsburg-Nürnberg), led to the development of that engine which first seemed to prove convincingly the advantages of the new principle. It should be borne in mind, however, that British engineers were in the field long before Diesel, and that the earliest practical engine of that type was the Hornsby-Akroyd. Details are given in *Discovery* (August 1953).

distillation of crude petroleum,[1] and needs, therefore, less refinery processing, so that it is considerably cheaper than motor petrol. Hence not only the specific fuel consumption in lb/H.P.h, but still more the price of energy in pence/H.P.h, is quite considerably lower in the case of Diesels than in the petrol engines using the Otto cycle.

Although Diesel's idea had been patented in 1892, and his first engine was built in 1895, it took a couple of decades for extended practical use to be made of his invention, and half a century elapsed between the date of his first patent and the start of the great run of Dieselization which has revolutionized the U.S. railways to-day, and will soon spread to other countries which are not lucky enough to run their railway systems by electricity gained from home water-power. At present Diesel engines are dominant not only among all the I.C.E.'s used for railway service, but also in heavy road service, as, for instance, in motor-lorries and buses.

Another advantage besides that of greater economy is the use of a fuel which is less volatile and less easily inflammable. This would be of special importance in the case of aeroplanes. Within the last thirty years more than a thousand people have been burned to death in the debris of crashed planes when the petrol flowing from the broken tanks had been ignited. Diesel oil does not ignite under similar conditions; hence hundreds of passengers of crashed planes might have survived, after mending some broken bones, if their planes had been equipped with Diesel engines.

As a matter of fact, the Junkers works, Germany's best-known aeroplane manufacturers, built double-piston two-stroke Diesel engines, and these had a very good performance record in their planes. Still, no Dieselization as yet has taken place in aviation, and it is very doubtful whether it will come at all, because the present trend in aircraft engines is away from the piston engine and towards turbo-props and jets.

Another problem is that of motor-cars for private use. Daimler-Benz in Germany in 1937 equipped their Mercédès 200 D with a 2545-cc Diesel engine, and recently Borgward in Germany, Fiat in Italy, and the Standard Company in Britain followed. Delettre in France and Packard in the U.S.A. are expected to follow next, but no wave of Dieselization, as in the case of the U.S. railways, has made itself felt. The reason lies in certain drawbacks inherent in Diesel engines which make them less fit for use in cars for gentle-men drivers. The high compression ratio, advantageous though it may be from the economic point of view, does not ensure the smooth,

[1] See Chapter IX.

noiseless running which the owner of a good car is so proud of.
Another disadvantage of the Diesel engine is its liability to exhaust
smoky and ill-smelling burnt gas. Although a well-kept Diesel car,
serviced by an expert mechanic, can be driven as smokelessly as a
car with a petrol engine, any small negligence in keeping the fuel-
spray valve clean or properly timing the injection mechanism will
result not only in a rise in fuel consumption, but also in a sudden
increase in the smoke and stench of the exhaust gases. People
living near roads with heavy traffic of Diesel cars are always suffering from the excessive soot content of the atmosphere.

The result is that in the hierarchy of motor-cars the Diesel-
driven vehicles have so far played more or less the rôle of economic,
second-class goods, while the smoothly running six- or eight-cylinder
petrol engine represents the superior class. As long as fuel economy
is not the leading requirement in the choice of cars, and as long as
no silent Diesel engine in the class between, say, 30 and 100 B.H.P.
has been constructed, the Dieselization of the vast fleet of privately
owned motor-cars will make slow progress.[1]

The Waste of Energy in Motor-car Traffic

The tendency of well-to-do car-buyers—and consequently also of
car-designers—to put the concern for driving economy behind that
for comfort and the luxurious appearance of cars has induced Ayres
and Scarlott to make some critical remarks which deserve attention,
although they apply mainly to conditions in the U.S.A. We are
quoting from their book *Energy Sources*, pp. 131–132:

> Something about the possession and operation of a motor-car pro-
> vides effective anesthesia for any awareness of economy. We may
> hesitate to spend a few dollars for a book or for a tool (things of
> more or less lasting value), but we think nothing of paying a like
> amount for a tankful of gasoline. In a few hours or days the gasoline
> is gone, and we have left only the memory of a few miles on the road
> and perhaps many exasperating minutes in a line of stalled traffic.
> A few people for technical reasons keep an account of miles per
> gallon of fuel, depreciation per mile, and other costs of motoring,
> but the result is usually so appalling that the accounts are hurriedly
> discontinued and forgotten.
>
> The highly competitive sellers of motor-cars and motor-fuels search
> earnestly for features in their products to advertise. The advertising

---

[1] Quite recently Perkins at Peterborough and the M.A.N. Works at Nuremberg
constructed new silent Diesel engines which might cause a quicker change-over to
Diesel cars in the next decade.

specialists, conscious of the public pulse, do not waste much space and money talking about economy. They talk, instead, of performance, comfort, style, and reliability. Nearly everything said about a new car means lower fuel efficiency—for example, larger body, longer wheel-base, greater weight, softer tyres, more horse-power, more rapid acceleration, higher speed, automatic drive, improved flexibility of control. And now powered steering is almost upon us! No wonder car-miles per gallon of gasoline have shown no improvement over the past thirty years. Passenger-miles per gallon have probably shown a definite decline, because an even larger proportion of motorists like to ride alone in about two tons of assembled steel driven by a silent power plant capable of more than a hundred horse-power. This decline is in spite of engineering changes that have made considerable improvements in ton-miles per gallon.

The motor-car operator is largely to blame. He insists on excessive weight, power, and luxury. He accelerates too rapidly. His speed is excessive. He knows, in a vague way, that he is paying for all this, but he feels that he is getting his money's worth of what may be called the ' amenities ' of motoring.

In spite of the better thermal efficiency of I.C.E.'s, the total all-out fuel efficiency in motor-car traffic is nearly as bad even as that of steam-driven railways. For what we have called overall efficiency of the engine takes into account only the losses in the engine itself, ignoring all the losses occurring before the fuel enters the engine and after work is done on the crank-shaft. The conception of *energy-system efficiency* has been used, therefore, in Ayres and Scarlott's book, a calculation being made of the amount of actually useful work for driving a car on the road that is left over from a given quantity of energy contained in crude petroleum when the corresponding amount of energy for extracting and refining the crude oil and for transporting the petrol to the consumer is subtracted, and when, in addition, there are taken into account all losses which occur within and behind the engine. Ayres and Scarlott quote calculations made by R. J. S. Pigott, past President of the Society of Automotive Engineers, who found from practical experience that the engine mechanical efficiency (considering losses of useful energy by work done on fan, water-pumps, and generator) averaged 71 per cent. Much lower still is the ' rolling efficiency,' representing the ratio of useful work after deduction of the loss between clutch and the road (including losses in the transmission, rear axle, and tyre deformation), which turned out to be only 30 per cent. Hence by multiplying all these factors an extremely poor energy-system efficiency results, as shown in Table 6, taken from *Energy Sources*:

TABLE 6. ENERGY-SYSTEM EFFICIENCY OF THE MOTOR-CAR
IN PER CENT

	Efficiency of Each Step	*Efficiency including all Preceding Steps*
Production of crude petroleum	96	96
Refining of petroleum	87	83·5
Transportation of petrol	97	81
Thermal efficiency of engine	29	23·5
Mechanical efficiency of engine	71	16·7
Rolling efficiency	30	5

The net result is, therefore, that 95 per cent or more of the energy contained in the original amount of crude petroleum is used not to move the car along the road, but to extract and refine the crude oil; to carry the petrol to the filling station; to heat the water in the radiator and the gas in the exhaust; to operate motor-car auxiliaries; and to overcome friction in gears and tyres.

Ayres and Scarlott conclude their discussion of this problem with the following remarks:

The reason such huge amounts of energy are being wasted in the form of liquid fuel (and the reason the idea of wasting still more is being seriously entertained for the future) is that we insist upon operating heavy cars with dangerous and useless potentialities of speed and acceleration. This inclination, which is growing all the time, will do more to advance the end of the fossil-fuel era than any other factor. Higher automotive-engine efficiencies are announced from time to time as resulting from improved engines or superior fuels. But motorists have not realized any increase in mileage, since the potential efficiency increase has been offset by running more powerful engines under lighter partial loads and in hauling more tons of automobile at higher average speed. To take a simple example, a car designed for a maximum speed of 100 miles per hour must have 3·5 times as much installed horse-power per ton as a car designed for a maximum speed of 50 miles per hour. Other things being equal, the fuel consumption is about proportional to installed horse-power.

Histories written a few centuries hence may describe the United States as a nation of such extraordinary technologic virility that we succeeded in finding ways of dissipating our natural wealth far more rapidly than any other nation. At any rate, we are having a wonderful time doing it.

Can We expect Further Increases in Efficiency?

Before answering this question it will be useful to give explanations of the very complex problem of efficiency in general. As we have seen just now, the economy in using fuels does not depend on the engine efficiency alone, but is a product of a number of factors, some of which are quite independent of the properties of the engine or of the car at all. Thus the low rolling efficiency, which is responsible for a good deal of the losses, depends among other things on the condition of the road, and even on the weather. In addition, although we are accustomed to talk bluntly of the efficiency of a certain engine, we ought to bear in mind that the efficiency is by no means a given fixed property of an engine, but is a function of its operating conditions. Fig. 29 shows the variation of specific fuel consumption with the load of an average I.C.E. (curve *A*) and of a 5500-B.H.P. marine Diesel engine, built by Sulzer Brothers in Winterthur, Switzerland, which for a long period held the record of efficiency among all heat engines (curve *B*).

As can be seen from Fig. 29, curve *A*, the specific consumption, which is lowest near full load, rises considerably on both sides of the minimum value in the case of most I.C.E.'s, while the efficiency, which is inversely proportional to the specific consumption, is declining either way. The average efficiency of operation under varying running conditions of an engine is, therefore, in general considerably

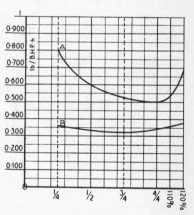

FIG. 29. VARIATION OF SPECIFIC FUEL CONSUMPTION WITH OPERATING POWER

Curve *A*, average I.C.E.; Curve *B*, 5500-B.H.P. eight-cylinder two-stroke Sulzer marine Diesel.

lower than its rated efficiency, which for obvious reasons is taken under optimum conditions. This circumstance makes itself more felt in motor-car driving than in big power plants. Central stations are running their engines most of the time under conditions near optimum efficiency. In motor-car driving on even roads, however, highest thermal efficiency of the engine is obtained when it is working at full power—that is, with high speed of the car (which is in itself an uneconomic method of driving)—whereas at moderate speed, which *per se* is more fuel- and tyre-saving, the engine as such does not work with its optimum efficiency. Hence cars with less H.P. in their engines, running at their rated power with reasonable speed of the car, would ensure the best results in economic use of fuel.

Rising fuel prices and competition between engine-designers may be expected to result in flattening the consumption-load curve (as, for instance, shown in curve *B*, Fig. 29), and in eliminating engines with too low efficiency. In this way the average efficiency gained by using liquid fuel in I.C.E.'s may be improved by just a small percentage. But no radical reduction of the total losses, say, from the present amount of 95 per cent in motor-car driving to below 50 per cent, is likely to occur within a foreseeable future. And also the optimum efficiency achieved so far in single engines seems to have reached a limit beyond which scarcely any considerable improvement can be expected. The reason for this may be explained shortly.

On account of the high temperature of the burning gases in the cylinder and of the greater temperature difference $T_1 - T_2$ through which the working substance passes in internal-combustion engines, the progress to higher thermal efficiencies, inaugurated by the introduction of the Otto motor, started from a higher level than that of steam-engines a century earlier in the time of Watt. By steadily raising the compression ratio, further increases in the efficiency could be achieved, and a kind of sudden jump occurred nearly half a century ago through the development of the Diesel engine. At a time when steam-engine designers were proud of achieving efficiencies of 20 per cent, Diesels were built with efficiencies between 30 and 40 per cent. Yet performances of this kind, achieved during and soon after World War I, were not followed by a similarly steep increase of efficiencies. The record created by the Sulzer 5500-B.H.P. marine Diesel in 1934, with an overall engine efficiency of 41·2 per cent at full load and a comparatively flat consumption curve (Fig. 29, curve *B*), was unbeaten for seventeen years, and it was not until 1951 that still better results were reported from tests made in the very same workshop where Diesel's first engine had been built more than half a century before.

The new engine made by the M.A.N. Works at Augsburg, Germany, achieves an overall efficiency of 44·6 per cent, corresponding to a fuel consumption of clearly below 140 g/B.H.P.h (0·31 lb/B.H.P.h). This progress was achieved by improved and enforced supercharging of the engine. It was known long ago that the power output of an engine of given size—and to a certain degree its efficiency too—could be raised by *supercharging*—that is, by forcing the air under pressure into the intake ports. Provided the same compression ratio, the same fuel-air ratio, and the same intake temperature are maintained as in operating from atmospheric pressure, the effect of supercharging is that the pressure in all phases of the cycle and accordingly also the indicated power output, is raised proportionally to the ratio of supercharging. Since the mechanical losses by friction, etc., are only slightly increased, the result is an improvement of efficiency, provided that the gain is not offset by the power consumed by the supercharger—that is, the device used for compressing the air.

On account of the fact that the power output per unit weight of the engine is also increased by supercharging, racing motor-cars as well as aeroplanes have long been equipped with superchargers. In the latter case supercharging is a necessity for compensating the low atmospheric pressure at high altitudes; in the former case the main purpose was boosting engines of given size and weight. In smaller engines, where fuel efficiency is of minor importance, the supercharger is a blower driven from the engine crank-shaft, like the Roots-type blower indicated by *B* in Fig. 26, which is used for providing the scavenging blast in a two-stroke engine. (As a matter of fact, every two-stroke engine uses a moderate amount of overpressure at the scavenging ports, but we do not speak of supercharging as long as the gauge pressure does not exceed half of an atmosphere or so.) Blowers driven from the main crank-shaft consume a part of the engine-power, thus slightly decreasing the efficiency. In larger units, therefore, and particularly in cases where fuel economy matters, the exhaust gases of the engine are utilized by being expanded from the slight back pressure at the exhaust port to atmospheric pressure in a turbine which drives a rotating compressor.

All this has long been part of progressive engine technique; what engineers of M.A.N. did recently was to extend the limit of supercharging by using higher pressures than ever before. Table 7 gives the data of three engines of high efficiency. The compression ratio is the ratio of the volume of the cylinder to that of the clearance-space. The compression pressure is reached at the end of the compression stroke, and is enhanced immediately afterwards by the combustion of the injected fuel to the combustion pressure, which is between 50 and 70 per cent higher.

In comparing engines of different sizes it must be borne in mind that it is easier to obtain good fuel economy in larger units than in smaller ones. Hence the advance which the big Sulzer marine Diesel had gained could not be made up by moderate super-charging of a considerably smaller engine, and it was only by a bold jump to unusually high pressure that a greater efficiency than the 1934 record could be achieved. Those readers who as motorists are

TABLE 7. COMPARISON OF THREE HIGHLY EFFICIENT DIESEL ENGINES

Type of Engine		*Sulzer* 5500 *B.H.P.* *Marine Diesel*	*M.A.N.* *G6V* 40/60	*M.A.N.* *K6V* 30/45
Cycle		Two-stroke	Four-stroke	Four-stroke
Number of cylinders		8	6	6
Bore	mm	720	40	30
Stroke	mm	1250	60	45
Normal speed	r.p.m.	126	275	375
Compression ratio		12	13	13
Charging pressure	ata	1·16	1·5	2·7
Compression pressure	ata	35	45	82
Combustion pressure	ata	60	67	120
Normal output	B.H.P.	5500	1170	1200
Fuel consumption	g/B.H.P.h	150	170	140
Efficiency	per cent	41·5	36·6	44·6

used to starting their engines by hand when the battery fails will appreciate the meaning of the fact that the force on the piston opposing compression in the last phase of the stroke amounts to 5·6 tons in each of the six cylinders of the highly supercharged M.A.N. Diesel. Incidentally, the turning momentum on the crank-shaft does not rise proportionally on account of the lever action of connecting-rod and crank near dead centre.

With both the temperature and pressure approaching the limits to which the available material can be exposed, nearly all possibilities of increasing the fuel economy seem to be exhausted. That is why we can hardly expect any further radical improvements in the efficiency of internal-combustion engines. Small percentage gains appear to be all that we can reasonably hope for.

One must not forget also that efficiencies of 40 per cent and more could be attained only in large-size engines of more than 1000 H.P., running with a high use-load factor (6000 hours yearly or more), so that the savings achieved by better fuel economy justify expensive construction and higher costs of the engine. It may be mentioned only that the turbo-compressor which is used as supercharger of the M.A.N. engine K6V/30/45 costs alone about as much as a round dozen average car motors. (Further details about the combination of turbines and compressors will be given in Chapter V.)

While, therefore, efficiencies of about 45 per cent might be expected as the best future standard in large stationary plants and in naval service, the average efficiency of car-motors will scarcely be raised to more than 35 per cent. And, even assuming (*a*) the product of thermal and mechanical efficiency of car engines to be raised from 20·6 per cent, as given in Table 6 ($\eta = 0.29 \times 0.71$), to twice its value, and (*b*) the rolling efficiency being also doubled by improved roads allowing the use of harder tyres, the result would be an energy-system efficiency of car-driving amounting to 20 per cent. This means that even with almost Utopian assumptions the total fuel losses in car-driving will not be reduced below 80 per cent.

The way out of this wasteful system of using fuels, as sketched at the end of Chapter III—namely, the utilization of reject heat—does not seem to be practicable in the case of mobile engines. We can use the reject heat of stationary plants, but not that of motorcars or locomotives.

Thus in the long run, with increasing fuel prices, and particularly with liquid fuels becoming scarce and expensive, the present state of affairs with road transport being cheaper than railways will be upset by railway transport becoming more economic than road transport. And within the railway systems electrification of the lines will prove to be economic even in countries where electricity is not produced from home water-power. The Dieselization of railways is at present certainly the best method for the U.S.A. with her relatively cheap oil and not too heavily frequented lines, where large investments would not be profitable enough. The case is different, however, in countries like Britain or Germany, with their shorter and more frequented railway-lines—countries with coal as their natural home-produced fuel. The maximum of fuel economy in the management of British Railways is likely to be achieved by electrification of all main lines[1] and by suitable choice of the sites of the power stations so that their reject heat can be used for household and industrial purposes. By further development of steam and hot-air turbines (described in the next chapter) engine efficiencies of 40 per cent could be achieved in coal-fuelled plants, so that an overall efficiency of electric traction (considering losses in the electrical engines and the transmission lines) of about 28 per cent might be taken into account. Assuming two-thirds of the whole traffic to be handled by electric traction, and assuming further that two-thirds of the reject heat of the railway-owned power plants could be utilized, annual savings of about seven million metric tons of coal might be expected—not to speak of the advantage of the perfect cleanliness and greater comfort of electric railways. For achieving this aim it would not be necessary to solve any more technological problems, but only problems of planning, investing, and organizing. This is a case where more radical progress could be made by a wise administration than by engineering genius.

[1] British Railways propose in the future the virtual elimination of steam locomotives from both main and suburban lines.

CHAPTER V

Gas Turbines

THE transition from the piston steam-engine to the steam turbine secured distinct progress both as regards efficiency and power per unit weight of the engine. Similar progress was made by the transition from steam to internal-combustion engines. There was, therefore, an eagerness to combine the advantages of the turbines with those of the I.C.E.'s, or, in other words, to develop the gas turbine. **Armengaud** in France was the first to construct a gas turbine working on the same principle as modern engines of that kind, without, however, scoring any success. What were the difficulties which delayed the practical use of gas turbines for nearly half a century?

Any heat engine whatever obtains its power from a working substance expanding in gaseous form and doing work by pushing a piston or driving a turbine-wheel. The working substance may be water-vapour in the case of the steam-engine, burnt gas in the case of the I.C.E., or simply pure air in the case of the hot-air engine, which, being the simplest form of heat engine, may serve as an example in some of the following explanations. By compressing air or any other gas we can store energy in it, and can regain the stored energy by letting the gas expand in a cylinder doing work on a piston. This is a similar procedure to winding up a spiral spring and regaining the work when the spring drives the wheels of a watch. Under ideal conditions, avoiding all losses through friction, etc., the work done by the expanding gas would be equal to the work consumed in compressing it, but no power production occurs in this process.

The cycle of compressing and expanding can, however, be modified in such a way that the gas—or air in the particular case of the hot-air engine—delivers more work in expanding than it consumes for its compression. The modification consists of heating the gas after it has been compressed. Since the work done is proportional to the product of pressure and volume of the gas, we can increase the work done in expanding by increasing its volume through heating it. This is done in the I.C.E. in the cylinder itself by burning a mixture of air and vaporized fuel, while in the hot-air engine the air after being compressed passes an air-heater consisting of a bundle of tubes exposed to the flames of the furnace and then expands, doing work in a cylinder or a turbine. For closing the cycle—that is, for restoring the initial state—the air has to be cooled down to its initial temperature in a pre-cooler, whereupon it is admitted to the compressor again. This cyclic process is called the *Joule cycle*, because it was Joule who about a century ago suggested its use for a hot-air engine.

Compression and expansion are done in the hot-air engine between the same limits of pressure, the surplus of work done in expanding being due to the greater volume of the heated air. The net output of power in all heat engines is the difference between the gross work done by the expanding medium and the work consumed for compressing this medium. In steam-engines, too, some work must be done in feeding water from the low condenser pressure to the boiler, with its high pressure. As mentioned, however, in Chapter III, this work is relatively very small, because the specific volume of water is so much less than that of the steam.

The Complication arising from the Relative Amount of Compression Work

Fig. 30 gives a graphic representation of the relative amount of compression work in proportion to the net output of different engines. The illustration compares engines whose net output of power is assumed to be the same if losses through friction, leakage, etc., are neglected. This ' ideal net output ' is given by the length *ab*, identical in all the graphs representing the different engines; *ac* is the gross output of the engine from which the compression work *cd* has to be subtracted, leaving the ideal net output *ab* and the useful output *a′b′*.

While the compression work is almost negligible in steam-engines, it amounts to quite a substantial percentage in internal-combustion engines, especially in Diesel engines, with their great compression ratio. In gas turbines, moreover, the compression work surpasses the net output for two reasons: (i) The working substance has not only to be compressed but also to be delivered at high pressure to the turbine; (ii) since the initial temperature T_1 cannot be raised as high as in reciprocating piston engines, the gross work obtained from a given quantity of the working substance (not from a given quantity of fuel!) is less than in piston engines.

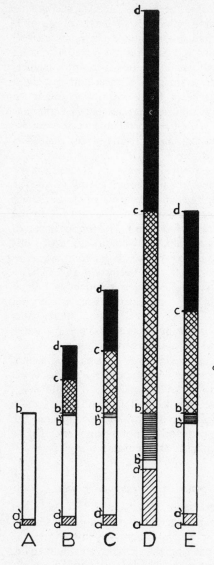

FIG. 30. RELATIVE AMOUNT OF
COMPRESSION WORK

ab, ideal net output; *ac*, gross output of engine; *cd*, compression
work; *bc*, part of engine output
consumed for compression; *bb′*
losses in the compressor, to be
added to *bc*; *aa′*, mechanical
losses in the engine, decreasing
output; *a′ b′*, useful output; *A*, steam-
engine; *B* petrol (Otto) engine;
C, Diesel engine; *D*, first gas
turbine; *E*, modern gas turbine

The high proportion of work consumed for compression, though a drawback by increasing the weight per horse-power of the engine, would not by itself have been a serious obstacle against making gas turbines if the low mechanical efficiency of compressors and turbines at that time had not consumed the relative small difference between turbine output and compression work. Working with small differences between assets and debits of the balance-sheet is always a risky matter. Imagine a merchant buying a certain article at a wholesale price of fivepence and selling it for sixpence. Assuming a large turnover, he may become rich, as long as everything goes right. But if a percentage of the articles he buys is lost or spoilt in transport, and if some of his customers become insolvent, he will be ruined.

That is exactly what happened with the power balance-sheet of the first gas turbines. As shown by Diagram D in Fig. 30, the relatively small margin between output and compression work was used up by the mechanical losses aa' in the turbine and bb' in the compressor. Hence the engines of Stolz and Armengaud, though built on the right principles, were able only to run off load, but could not deliver any amount of power worth mentioning. Gas turbines are in two respects worse off than reciprocating I.C.E.'s: the relative amount of the compression work is considerably greater and the mechanical efficiency of the compressor is smaller. During the compression stroke of an Otto or Diesel engine only a relatively small amount of energy is lost by friction and leakage while the rotating compressor is working with less efficiency.

The way out of these difficulties was clear from the beginning, and was, of course, known to the constructors of the first gas turbines. But several decades of technological and engineering research were necessary to pave that way. Two conditions had to be fulfilled to obtain a sufficient net output: (i) To increase the ratio between gross power output and compression work; (ii) to increase the mechanical efficiency of turbine and compressor.

Technological Research improves Thermal Efficiency

The first task—namely, to obtain more power from a given quantity of working medium, and, therefore, from a given amount of compression work—could be accomplished only by raising the initial temperature of the gas entering the turbine, and this, again, was a matter of the temperature to which the turbine-blades could be safely exposed. The constructors were therefore faced with the very same problem as the makers of steam turbines were in their efforts to raise the efficiency. With thousands of steam turbines working in ships and power plants, and considerable capital being invested in turbine manufacture all over the world, there was enough incentive for research work, and enough funds were available for laboratories, with the aim of developing new alloys for superheater tubes and turbine-blades which could withstand both the thermal and mechanical stress. For although the temperatures used in turbines (ranging from about $300°$C. $= 573°$K. $= 572°$F. in the earlier engines up to about $700°$C. $= 973°$K. $= 1292°$F. in modern plants) are far below the melting-point of steel ($\approx 1530°$C. $\approx 1800°$K. $\approx 2800°$F.), the *creeping* of the metal made itself felt. ' Creeping ' means the slowly progressing deformation of the substance under the combined action of heat and mechanical tension. At moderate temperatures steel under a heavy stress, like high pressure in a tube or the centrifugal forces and the impact of the steam jet on the blades, will suffer a small elastic deformation which will last during the stress and vanish when the tension is released. At higher temperatures, however, the elastic deformation is accompanied by the slowly progressing creeping deformation, which in the long run may completely alter the shape of the stressed piece of the engine, or even destroy it. Hence, in order to ensure sufficient operating life of the turbines, it is necessary to keep the stresses of superheater tubes, blades, etc., below the creep limits of the material employed. The creep limit of a metal is defined as the maximum stress which might be permanently applied without the risk of a continuously progressive deformation; it is a function of the working temperature, decreasing with increasing temperature. Fig. 31 shows the progress which has been made in raising the creep limit of steel alloys between 1925 and 1940.

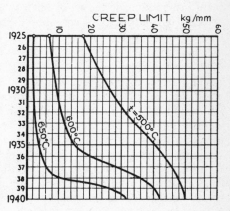

FIG. 31. PROGRESSIVE RAISING OF THE CREEP LIMIT OF AUSTENITE STEELS BETWEEN THE YEARS 1925 AND 1940

As can be seen from the diagram, at the beginning of the period, in 1925, a stress of 20kg/mm² could not be maintained at a temperature of 500°C., while in 1937 it was admissible at 600°C., and in 1940 at 650°C.

The consequence of this purely technological progress in the field of steel metallurgy was the possibility of raising the initial temperature, improving in this way the thermal efficiency of steam turbines and at the same time reducing the ratio between compression work and output of gas turbines, thus removing one of the grave obstacles against the practical use of such engines.

Aero-dynamical Research improves Mechanical Efficiency

Quite independent progress was made in reducing the mechanical losses of rotating compressors and in improving the design of turbines. Considering the fact that one of the advantages of turbines over piston engines consists in eliminating all oscillatory movements from the engine, it is natural (though not absolutely necessary, as will be shown later) to use rotating compressors which do not need reciprocating pistons, cranks, or valves. The so-called axial compressor is simply a kind of inversed turbine consisting of a series of bladed wheels mounted on a common axle between which fixed blades, mounted in the casing, are inserted. Just as pressurized gas impinging on the blades of turbine-wheels puts them into motion, so the set of blade-wheels of a compressor, when driven from outside, will compress the air. The ordinary rotating fan or blower used as a desk ventilator is an example of the most simple device of this kind, although furnishing only a very slight amount of pressure.

The problem, therefore, was not to make rotating compressors at all, but to make them operate efficiently enough. By simply adjusting blades on the rotating shaft and inserting deflecting blades anyhow on the casing we should obtain a device which, when set in rotating motion, would whirl the air around a great deal, converting thereby part of its energy into heat, failing, however, to do its compression work with due efficiency. The development from relatively primitive blowers to efficient enough compressors, which finally led to practicably usable gas turbines, was made possible by long-term aero-dynamic research on laminar and turbulent motion of air flowing with great speed over aeroplane wings, as well as over blades of propellors, turbines, and compressors. The theoretical and experimental study of the influence of the shape and position of the blades on the air-stream helped to reduce the mechanical losses both of turbines and compressors, so that to-day figures of 88 per cent for the mechanical efficiency of the turbine and 85 per cent for that of the compressor have been realized. The stage reached in this way is represented in graph *E,* Fig. 30.

Turbo-props and Jet Propulsion

The recent development of gas turbines was fostered and accelerated by the rising demand for high-speed aircraft. During World War II a gradual change-over from piston-engine to turbine-driven fighters and bombers took place. Aircraft designers had long

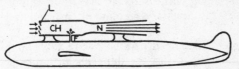

FIG. 32. PROPELLING MECHANISM OF THE VI
CH, combustion chamber; *L*, loosely suspended lids; *F*, fuel injection pipe; *N*, nozzle.

been regretting the fact that the task of propelling a plane by pushing a blast of air from fore to aft was accomplished by machinery making a wide detour, obviously dissipating energy by additional friction, and requiring an extra amount of engine weight. The blast of the exploding air-petrol mixture is first used to push a piston whose oscillating movement is transformed by means of a connecting rod and a crank into rotation of the main shaft, which, again, by means of a propeller, produces finally the air-blast giving the necessary thrust to push forward the plane. Quite a number of links, gliding parts, and bearings are required for each of the many cylinders of the two or four or six engines of a plane, each item causing frictional losses. Is there no short cut for transforming the blast of the burnt gases into the final blast pushing on the plane?

As a matter of fact, a very direct method of using the blast of the burnt gases for producing the thrust is practicable, but it turns out not to be the way leading to greater fuel economy, though certainly affording some advantages. Fig. 32 shows the very simple principle of propulsion used for the German V1 weapon of 1944, which is not a rocket but an automatically controlled unmanned aeroplane. The propelling mechanism consists mainly of a combustion chamber *CH* mounted on top of the plane, terminating aft in an open nozzle N, and being alternately opened and closed in front by a shutter consisting of a series of loosely suspended lids L. The plane is launched by a catapult, and once in flight the lids are opened by the air-stream delivering fresh air into the chamber. Fuel oil is then injected and ignited in the usual way by means of a sparking-plug. The explosion closes the lids with a bang, and the burnt gases are emitted under pressure through the nozzle, producing a rocket-like thrust. When the exhaust is complete the pressure in the chamber is low enough to allow the lids to open again, so that the whole process is repeated again and again several times a second. (The difference between the V1 and a rocket as used in the V2 consists in the fact that the latter, carrying all the exhaust matter along with itself, can also fly in empty space, while the V1 inhales the necessary amount of air, and is also airborne by wings.)

The V1 motor described here is, owing to its simplicity, cheap price, and light weight, just the thing for an unmanned war-plane used for a single short flight, after which it is lost. But its efficiency is low, because the pressure, achieved before combustion by the *Staudruck* (" back-thrust ") of the air-stream, is much lower than the compression pressure in the usual I.C.E.'s.

Hence the transition to jet propulsion, as actually used in modern fighters, was made by providing means for compressing the intake air. The work required for compression must be done by the energy released in combustion, and this leads logically to the combination of turbine and compressor, as known already from the exhaust-gas-driven supercharger.

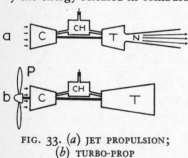

FIG. 33. (*a*) JET PROPULSION;
(*b*) TURBO-PROP

C, compressor; CH, combustion chamber; T, turbine; N, nozzle; P, propeller.

A schematic representation of the total arrangement of jet propulsion is given in Fig. 33*a*. A compressor C driven from the turbine T delivers air to the combustion chamber CH, where fuel is burnt, increasing thereby temperature and volume of the gas. The energy released by the combustion is used partly for driving the turbine T, in which it is, however, not fully expanded. The exhaust gas leaving the turbine with a considerable amount of back-pressure expands in the nozzle N as a high-speed jet producing a reaction in the form of the thrust which propels the aircraft directly without making the detour over air-screws.

In spite of its attractive simplicity, jet propulsion has not become so far the universally applicable means for propelling all sorts of aeroplanes. Its main advantage of dispensing with propellers is of unique importance in the case of extremely fast fighter aircraft, and breaking through the sound barrier—that is flying with speeds greater than the velocity of sound—has been accomplished only by jet-propelled planes. Still, the efficiency is low, especially at low travelling speeds, so that the stock of fuel carried on board is used up sooner than in the case of propeller-driven planes. For this reason the turbo-prop shown in Fig. 33*b* is still more widely used for long-distance air-liners. The difference from the jet propulsion is simply that the burnt gas streaming from the combustion chamber is fully expanded to atmospheric pressure in a larger turbine T, whose power suffices not only to drive the compressor (as in case *a*), but also to drive a propeller P. The residual thrust exerted by the exhaust gas is insignificant.

Both jet propulsion and the turbo-prop have certain advantages which make them particularly suited for use in aircraft. Discarding pistons, turbines have a smaller cross-section than reciprocating engines of equal power. They have no valves, no cam-shaft or other quickly oscillating parts the wear of which may give rise to breakdowns. For this reason the percentage of turbine-driven planes is steadily rising both in military and civil aviation. It would not be wise, however, to use engines like those represented in Fig. 33*b* for stationary power plants, because their efficiency is still too low.

Gas Turbines for Stationary Plants

One of the reasons for the low efficiency of the simple aircraft gas turbines is the loss of energy in the form of the heat carried away by the hot exhaust gases. Although the burnt gas is cooled by the expansion in the turbine, its temperature on leaving the turbine at atmospheric pressure is still rather high. Another disadvantage of the simple design (Fig. 33*b*) lies in the fact that the nearly adiabatic compression done by the compressor C causes a considerable rise in the temperature, and this, again, increases the compression work, thus decreasing the proportion of net power output.

Two important modifications were therefore necessary to improve the efficiency of gas turbines enough to compete with the steam-engine. One is the *recuperation* or *regeneration* of heat by transferring a part of the heat of the exhaust gases to the compressed air entering the combustion chamber. This is done in a heat-exchanger, which, as shown in Fig. 34, is inserted between compressor and combustion chamber. (The arrangement of the single parts of the engine in Fig. 34 is different from that in Fig. 33.) What is symbolically represented in Fig. 34 under *HE* as a single tube surrounded by a wider tube is in the actual engines a bundle of tube nests containing several hundred small tubes through which the compressed air flows to the combustion chamber while a large cylindrical container surrounding the system of tube nests carries the hot exhaust gases. In this way a part of the heat of the exhaust gases is recuperated and used for raising the temperature of the air *after* compression, but before combustion. (It would be quite wrong to heat the air before compressing it, because this would involve more power consumed for compression!) Although recuperation is a very suitable means of increasing the turbine efficiency, it cannot be used in aircraft engines on account of the size and weight of the heat-exchanger. It may be mentioned as an example that the heat-exchanger of a particular 2000-kW test plant, to be described later, whose power corresponds approximately to that of the combined engines of a reasonably big plane, has a length of 56 ft and a weight of several tons.

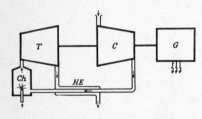

FIG. 34. RECUPERATING GAS TURBINE

T, *C*, and *Ch*, as in Fig. 33; *HE*, heat-exchanger; *G*, generator driven by turbine.

The other modification for improving the efficiency is the use of intercoolers in the compressor. As stated before, the compression work (which is on the debit side of the energy balance-sheet) is increased by the temperature rise caused by the compression. In order to reduce this detrimental heating the compression is done in two or more stages—that is, by compressors arranged in series. Between each two of these compressors an intercooler is inserted, which consists of tube bundles similar to those of the heat-exchanger, but surrounded by a wider tube, through which cold water flows.

The modifications distinguishing stationary gas-turbine plants from those used in aviation may be shortly summed up as cooling during compression and heating by recuperation between compression and combustion. By making use of these improvements, along with those incorporated already in aircraft turbines, a stage was reached where competition with steam turbines could be seriously taken into consideration. Two further advances aiming at better performance of gas turbines are noteworthy, and will be dealt with, therefore, in the next sections of this chapter.

The Free Piston Engine

Logical deductions from two well-known facts led to a development which may result finally in engines particularly adapted for use in railway and air service. The facts are:

1. Dispensing with oscillating pistons is mechanically an advantage, thermodynamically, however, a disadvantage, because, as explained earlier in this chapter, the combustion temperature cannot be raised as high in turbines (with their blades uninterruptedly exposed to the blast) as in piston engines, where piston and cylinder-walls are periodically cooled.

2. The gross power output of any I.C.E. is partly expended for compression work, and only partly spent usefully as net output.

Combining these two facts leads very naturally to the following idea: Let us do the compression work by a piston engine working efficiently from a high temperature level of the burnt gases and leaving enough heat and pressure energy in the exhaust gas to drive a turbine which performs the net output work. Once having grasped this idea a further step for simplifying the device is close at hand. By choosing the compressor also of the piston type, and using the piston alternately as working and compressing means, as in the conventional I.C.E.'s, we need no longer make the detour via turning a crank-shaft.

The result of this line of thought is simply a kind of Diesel engine whose pistons are not coupled to a crank-shaft and do no useful work, so that enough energy is left in the exhaust gas to drive the turbine. Fig. 35 shows a free piston engine in which twin pairs of pistons are linked together in such a manner that they are moving in opposite directions, thus yielding the perfect mass balance which is necessary for vibrationless motion of an engine. The turbine T drives, in addition to the shaft coupled to the load (not shown in the figure), a supercharger S which feeds air to the two compressor cylinders C, C. Owing to the large area of the two pistons (P), the small pressure yielded from the supercharger

suffices to drive back the small pistons (*p*) which produce the high
pressure in the combustion chamber *CH*. Near the inner dead centre
of the pistons fuel is injected and burnt as in any conventional
Diesel engine, pushing back the pistons so that a new charge is

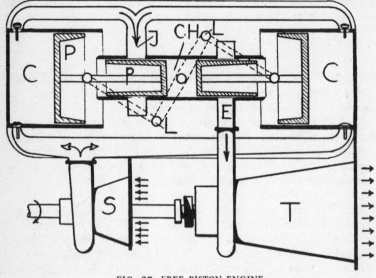

FIG. 35. FREE PISTON ENGINE

T, turbine; *S*, supercharger; *C*, compression cylinder; *J*, intake-pipe; *CH*, combustion
chamber; *E*, exhaust-pipe; *P*, large piston; *p*, small piston.

delivered from *C* through the intake pipe *J* to *CH*, while the burnt
gas exhausts under high pressure through the pipe *E* to the
turbine *T*.

It is quite possible for the smooth running of a turbine and
the high efficiency of a Diesel engine to be combined in the free
piston engine, which, though still in the early stages of its development, has already achieved remarkable performances.

The Escher-Wyss-AK Closed-cycle Gas Turbine

Tentative experiments made with internal-combustion engines
fuelled with pulverized coal were not really successful; hence practically all I.C.E.'s have been dependent so far on liquid fuels. On
account of the fact that the calorie and the kilowatt-hour cost
considerably more in the form of Diesel oil—and still more in that

of petrol—than in the form of coal, better fuel efficiency of I.C.E.'s does not inevitably make them more economic than steam-engines, or, in other words, less lb of fuel per kilowatt-hour do not necessarily imply less pence per kWh if engines using different sorts of fuel are compared.

Considering further that shortage and rising prices are to be expected sooner with liquid than with solid fuels, coal-fired engines with equal efficiency would be preferable to those fuelled with oil. Gas turbines of the design shown in Figs. 33 and 34 might on principle be fuelled with pulverized coal instead of oil, but their operating life would be rather short, because particles of ash and incompletely burnt coal would cause erosion and premature wear of the blades. Even when using liquid fuels the wear of the blades is greater in internal combustion turbines than is the case with steam turbines.

A promising way out of these difficulties, which offers, moreover, some additional advantages, has been paved by the Swiss scientists Ackeret and Keller in their construction which has been developed by the Escher-Wyss Engineering Works at Zürich. The so-called Escher-Wyss-AK closed-cycle gas turbine is the successful and modernized revival of the hot-air engine which a century earlier had been tried as an alternative to the steam-engine with the aim in view of avoiding the losses caused by the latent heat of steam. As explained in Chapter III, a great deal of the heat energy consumed for producing steam is not converted into mechanical work, but discarded as the latent heat of the steam, both in exhausting and in condensing engines. This kind of loss can be avoided by using a working substance which remains throughout the cycle in the gaseous form, as, for instance, air which is alternately compressed, heated, expanded, and cooled, as explained at the beginning of this chapter (p. 94). The price we have to pay for this advantage is the amount of compression work, which is incomparably greater than that of the water-feeding pumps of steam-engines. As a consequence, large and heavy engines are required. More than a century ago, in 1853, a motor-ship, the *Ericsson*, driven by a hot-air engine crossed the Atlantic. With its four cylinders of 427 cm = 14 ft diameter each, it seems to have made a record for size unbeaten yet, but still its output of useful power was only 220 H.P.; four modern aircraft turbines with 1000 H.P. each could easily be placed within a single cylinder of the *Ericsson* engine. This is a striking example of the disproportion between gross power and net power output shown in Fig. 30 *D*.

It was a bold enterprise, therefore, when the Escher-Wyss Engineering Works, trusting in theoretical calculations and the aerodynamical reseach of Ackeret and Keller, started developing a modern version of the hot-air engine which, as a piston motor, had been discarded long ago. Fig. 36 gives a schematic representation of a model of this engine, which differs from the recuperating gas turbine shown in Fig. 34 only in the following two points. (i) the working substance (air or any other permanent gas) describes *a closed cycle*. While in the open-cycle turbines, as used in aviation, fresh air is inhaled by the compressor and the burnt gas exhausted, the AK turbine operates like a condensing steam engine, using the same working medium over and over again. (ii) Instead of the combustion chamber *CH* of Fig. 34, an air-heater *AH* is inserted between compressor and turbine. Hence the closed-cycle gas turbine, though not belonging to the category of steam-engines, is not an internal-combustion engine either. The burnt gas, after giving off its heat in the air-heater and in a heat-exchanger used for pre-heating the combustion air (not shown in Fig. 36), escapes through the stack without being used as the working substance. This involves certain advantages as well as disadvantages.

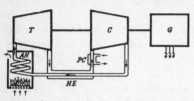

FIG. 36. SIMPLIFIED MODEL OF THE ESCHER-WYSS-AK CLOSED-CYCLE GAS TURBINE

AH, air-heater; *T*, turbine; *HE*, heat-exchanger; *PC*, pre-cooler; C, compressor; G, generator.

The operation of the Escher-Wyss-AK closed-cycle air turbine will be understood best by using the so-called Sankey diagram of the engine given in Fig. 37, in which the shaded areas represent the energy flow. The width of these areas is drawn proportional to the amount of energy (heat as well as mechanical energy) transmitted per unit time through each part of the machinery. The flow of the working medium (air or any other permanent gas) is indicated by broken lines. In explaining the operation of the device we shall use the abbreviation l.p. for low pressure and h.p. for high pressure. We start from the air-heater *AH* in the centre of Fig. 37 to which compressed air is fed after being pre-heated in the heat-exchanger *HE* by the air exhausted from the l.p. turbine. The air-heater consists mainly of a burner surrounded by bundles of tubes through which the air flows. As indicated in Fig. 37, the air, after receiving heat from the burnt fuel, expands in the h.p. turbine T_1, where a part of its energy is converted into mechanical work used for driving the compressors. This is represented by the energy stream flowing from T_1 to C_2 and C_1 on the left. The rest of the energy flows with the exhaust air of T_1 to the l.p. turbine T_2, where the useful work is done which drives the generator issuing the useful electrical energy output. At the exhaust of T_2 the temperature

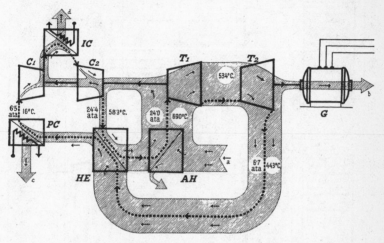

FIG. 37. SANKEY DIAGRAM OF ESCHER-WYSS-AK CLOSED-CYCLE TURBINE

Shaded areas represent energy-flow; broken line shows the flow of air, horizontal unbroken lines the engine-shafts. *AH*, air-heater; T_1, h.p. turbine, T_2, l.p. turbine; *G*, generator; *HE*, heat-exchanger; *PC*, pre-cooler; C_1, l.p. compressor; *IC*, inter-cooler; C_2, h.p. compressor; *a*, heat input from fuel; *b*, power output; *c*, heat discarded in pre-cooler; *d*, heat discarded in inter-cooler.

of the air is still 443°C.; hence, as seen from the width of the shaded strip leading downward from T_2, it carries a considerable amount of energy along, which is recuperated by being transmitted in the heat-exchanger *HE* to the air entering the air-heater *AH*.

(The diagonals in the rectangles of Fig. 37, denoting heat-exchangers, coolers, or heaters, are symbols representing tube-walls which are permeable to the heat flow, but impermeable to the gases.) After leaving the heat-exchanger the air is still rather hot, and the rest of the heat left in it has to be withdrawn in the pre-cooler *PC* in order to reduce as far as possible the amount of the compression work. The reject heat discarded with the cooling water is indicated by the arrow directed downward from *PC*. After being duly cooled the air is fed to a two stage compressor C_1 and C_2, with inter-cooling *IC* between the low- and high-pressure stage. The reason for inter-cooling is the same as that for pre-cooling—reduction

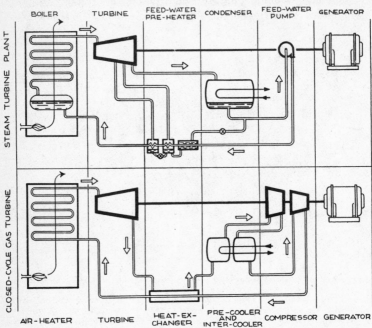

FIG. 38. COMPARISON BETWEEN A STEAM TURBINE AND A CLOSED-CYCLE GAS TURBINE

of the compression work. As seen from Fig. 37 the h.p. compressor C_2 is traversed by two energy-flows in opposite directions. One, indicated by the shaded strip along the shaft, is the mechanical work transmitted through the shaft of C_2 supplied from T_1 to drive C_1. The other shaded strip, going from left to right through C_2, is the small residue of heat energy left in the air after inter-cooling.

After compression the air traverses the receiving side of the heat-exchanger *HE*, where it is heated to about 430°C., and is then fed to the air-heater, thus closing the cycle. The total temperature drop of the working medium which determines the Carnot efficiency of the cycle amounts to about 670°C. and is kept almost constant, independently of the load.

In some respects, and particularly as regards its field of application, the closed-cycle gas-turbine installation resembles more a steam turbine, as shown in Fig. 38, in which each part of a steam plant is drawn opposite the corresponding part of a closed-cycle turbine plant.

Differing from steam-engines, the ranges of temperature and pressure can be chosen independently of each other, and high efficiencies can be achieved in closed-cycle gas turbines without expanding down to very low pressures. Thus in a 2000-kW Escher-Wyss test plant the ratio between inlet and outlet pressure of the turbine was chosen at the fixed value of about 3·6, irrespective of the power output, while the absolute values of both pressures are varied according to the load. This is illustrated by Table 8, giving the data of measurements taken by H. Quiby, of the Federal Institute of Technology, Zürich, at the test plant with different loads.

At full load the pressure of the cycle varies between 6·75 ata and 24·12 ata, resulting in a relatively high density and good heat transfer of the working medium, and permitting, therefore, the use of fewer stages of the turbines and smaller dimensions of the heat-exchanger and the inter-coolers. This fact is of particular importance in large plants, where the drawback of using an air-heater is more than fully compensated by the reduction of the size of the other parts.

The co-ordination of load and circuit pressure, which, as shown in Table 8, is approximately proportional to the power output, is effected by the usual centrifugal speed governor alternately opening and closing connexions to a high-pressure and a low-pressure accumulator. With increasing load the whole pressure-level is raised, while the pressure ratio, the temperatures, and also the jet velocities remain almost unchanged. In this way the angles of the flow of the jet over the turbine-blades remain at their optimum value at all loads, and no valves or throttles at the highly heated high-pressure turbine inlet are necessary. This is a distinct advantage over the steam-engine.

Further progress has been made recently in building smaller units up to 6000 H.P., in which compressor and turbine are united on the same shaft and fitted in a single casing. Fig. 39 shows a sectional view of a 3000 H.P. set with the compressor on the left and the turbine on the right. The whole equipment has a diameter of only about 1·1 metres (43 inches) and a total length of 6 feet. Corresponding to the small dimensions, the weight of the engine is also relatively small. The figure given in Chapter IV, Table 5, for the weight per horse-power of the hot-air turbine refers to this

TABLE 8. TRIALS WITH A CLOSED-CYCLE GAS TURBINE
(h.p.=high pressure; l.p.=low pressure)

Test No. Chronological Order		3	2	1	4
Useful output at coupling	kW	446	1005	1644	2111
Quantity of air in circuit	kg/sec	4·75	9·65	15·38	19·40
Pressures:					
Inlet h.p. turbine	ata	5·89	11·84	18·83	24·12
Outlet l.p. turbine	ata	1·65	3·35	5·31	6·75
Pressure ratio		3·57	3·53	3·55	3·57
Temperatures:					
Inlet h.p. turbine	°C.	698	693	691	687
Outlet h.p. turbine	°C.	543	540	538	534
Outlet l.p. turbine	°C.	452	449	447	443
Inlet l.p. compressor	°C.	13·3	14·2	15·5	16·4
Outlet h.p. compressor	°C.	56·5	56·3	57·5	58·3
Specific fuel consumption at the generator coupling	g/H.P.h	247	206	199	194
Total efficiency at coupling	per cent	25·5	30·5	31·6	32·6
Idem, subtracting power consumed for auxiliaries	per cent	24·4	29·5	30·5	31·5

engine, which with only 6·2 lb/H.P. compares favourably with other stationary plants.

In comparing the relative merits of open-cycle and closed-cycle gas turbines we may start with the disadvantages of the latter. Although in the schematic representation given in Figs. 34 and 36 the air-heater is only twice as large as the combustion chamber, their real sizes differ enormously, the volume of the air-heater actually used being more than a thousand times greater than that of the combustion chamber of an open-cycle turbine of equal power. Hence the advantage gained by the I.C.E. over the steam-engine in dispensing with a boiler, thus saving costs and weight of the engine

and avoiding thermal losses, is partly lost again by the introduction
of the air-heater. The additional weight and volume caused by this
part of the installation make the closed-cycle gas turbine less fit
for railway and air service, while the advantages to be discussed now

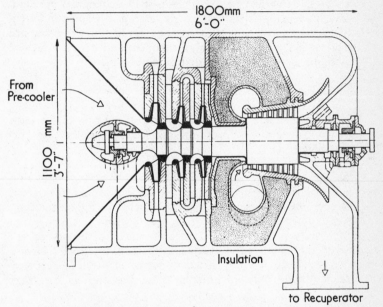

FIG. 39. COMPACT SINGLE-SHAFT ARRANGEMENT OF A 3000-H.P. ESCHER-
WYSS-AK CLOSED-CYCLE GAS TURBINE

may open a wide field of application in power stations and in plants
for marine propulsion.

The advantages of the closed-cycle gas turbine are:

(1) The separation of combustion air and working substance
permits the use of any gaseous, liquid, or solid fuel,
including coal, lignite, and peat. For the same reason
the AK turbine will be particularly suitable for use in
atomic plants.

(2) The working medium, which may be air or any other
permanent gas, can be kept perfectly clean, and therefore
the wear of the turbine-blades is less than in open-cycle
gas turbines or steam plants.

(3) The absence of valves and throttles in the hot parts of the
plant is also favourable for long operating life.

(4) There is the possibility of raising the total efficiency by further development to more than 40 per cent.

(5) Good part-load efficiency.

(It may be remarked here that the best Diesel engines are—and will probably remain—clearly superior as regards efficiency and evenness of the load-consumption curve. But considering the difference of the prices of Diesel oil and coal, the economic efficiency, expressed in pence/H.P.h, will be better in turbine plants.)

(6) Smaller size of installations in the power class of 50,000 kW and more.

(7) A considerable part of the reject heat emerges in the form of water at a temperature high enough for heating purposes.

This last advantage is of particular importance for combining power plants with district heating. For all endeavours to improve the efficiency of conversion of heat into power—including the combination of internal combustion with turbines which at first glance appeared quite promising—fell short of reaching the 50-per-cent mark, and the likelihood of ever reaching it is very small. Hence, as stated repeatedly, a really effective method of avoiding waste of fuel can be found only in utilizing as fully as possible the reject heat.

CHAPTER VI

Heat Pumps

WHILE the conversion of heat into mechanical work or electric energy inevitably involves losses—the optimum efficiency attained with heat engines is 45 per cent—the inverse process of converting power into heat is done with 100-per-cent efficiency. For all practical purposes, however, we have to make a distinction between the physical notion of heat *generation* and the economic notion of heat *supply*. As pointed out in Chapter I, all the electrical energy consumed in a boiler is turned into heat, but only a part of it is used for its proper purpose of heating the water, while a certain percentage is lost by conduction through the walls.

More than 100 per cent Heat Equivalent from Mechanical Work

With regard to heat supply effected by consuming other forms of energy, it turns out that not only losses occur, as in the example just mentioned, but gains as well. By consuming an amount of one kWh of electrical energy we can, under certain conditions, not only supply its full equivalent of 860 kcal for heating a room or a quantity of water, but much more—three times or even ten times as much.

As mentioned in Chapter I, this can be done by means of heat pumps, which are mechanically driven devices for drawing heat from a colder body and transmitting it to a hotter one. Heat pumps and refrigerators using compressors work on the same principle, devised by Lord Kelvin a century ago. The difference lies only in the purpose of the device. Both are consuming power; both draw heat from a colder body and deliver it to a warmer one. In the case of the refrigerator we are interested in cooling a body, and do not care what happens to the heat which is discarded. In the case of the heat pump we are interested in heating a body, and do not care where the heat is taken from.

The Working Principle of the Heat Pump

Both devices—the compressor-driven refrigerator and the heat pump—can be considered as inverted heat engines. This can best be made clear by illustrating the following explanations with the Sankey diagram of a steam-engine and its inversion, shown in Fig. 40, in which, as explained in Chapter V, the width of the shaded areas is drawn proportional to the amount of energy transmitted through each part of the engine. In a steam turbine the steam, generated in the boiler, expands in the turbine, and, in turning the turbine-wheels, loses a part of its heat. This is first converted into mechanical work, and then, in the generator, into electrical energy. The latent heat of the steam, however, which is more than half of the total heat input supplied by burning the fuel, is given off to the cooling water of the condenser. The net result of the process is that a quantity of heat taken from a higher temperature level (the furnace) is partly converted into mechanical work and partly discharged to a lower temperature level in the condenser. It is, of course, only the conversion into power which needs machinery: the transmission of heat to a lower temperature level is something which occurs by itself through heat conduction, radiation, or convection wherever temperature differences exist. The inverse process of heat transmission from cooler to hotter bodies, on the other hand, does not occur by itself, but needs an aid like the heat pump, the operation of which is shown in the lower half of Fig. 40. Here, too, heat is drawn from the surroundings of a boiler by an evaporating fluid, and subsequently discharged in the condenser as the latent heat of the fluid. But, differing from a steam-engine, the heat pump must be able to draw the heat from cool surroundings—say, from the atmospheric air or from the water of a lake or a river. Since heat flows by itself only from hotter to cooler bodies, the evaporator, being in contact with any medium, will draw heat from it only when the temperature of the evaporating fluid is lower than that of the surrounding medium. As a result of this condition, it follows that, instead of water, other substances with a much lower boiling-point must be chosen as working substances of the heat pump. Ammonia (NH_3), with its boiling temperature of $-33 \cdot 4\,°C$. at atmospheric pressure, is one of a series of suitable substances, the thermal data of which will be given later. Thus, if we pour liquid ammonia into a vessel and insert it into a lake, the ammonia will start to evaporate, drawing the necessary amount of heat from the surrounding water, which is cooled thereby. This, however, is only one side of the task of the heat pump. What we want it to do is to transmit this heat to a higher temperature level—for instance, to that of the circulating water of a central-heating plant. This cannot be done simply by bringing the ammonia vapour into close contact with the heating system, because its temperature is lower than that of the lake water,

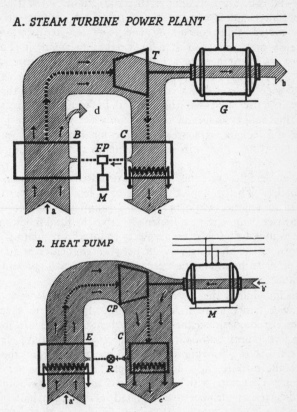

FIG. 40. SANKEY DIAGRAM OF STEAM PLANT AND HEAT PUMP

Shaded area, energy-flow. *B*, boiler; *T*, turbine; *G*, generator; *C*, condenser; *FP*, feed-pump; *M*, motor; *a*, heat supplied from fuel; *b*, electrical energy output; *c*, heat discarded with cooling water; *d*, heat lost in stack; *E*, evaporator; *Cp*, compressor; *R*, regulating throttle-valve; *a'*, heat drawn from river; *b'*, energy input for driving compressor; *c'*, useful heat output.

and accordingly much lower still than that of our heating system. Since it cannot discharge heat to any surroundings unless its temperature is higher than that of the surroundings, it follows that its temperature must be raised before it condenses, giving off its latent heat. This can be done by compressing the vapour instead of allowing it to expand. Unlike the steam-engine, therefore, where the steam on its way from boiler to condenser expands—doing work thereby and decreasing its temperature and pressure—the heat pump is no producer, but a consumer, of energy. A compressor driven from a motor—thus consuming energy—raises temperature and pressure of the vapour, so that, in condensing again, it is able to discharge its latent heat to a higher level of temperature, as, for instance, the water circulating in our heaters. The decisive point of the device—which makes it so much more economic than, for instance, electric stoves—is the fact, to be seen from the shaded area of the Sankey diagram, that not only the work done in compressing the vapour is recovered in the form of useful heat, but also the latent heat drawn from the low temperature level in evaporating the fluid.

To complete the cycle the fluid must be re-fed from the condenser to the boiler. This is done by means of a feed pump in the case of the steam-engine, while in the heat pump the overpressure in the condenser is more than sufficient to drive the fluid medium back into the evaporator. It is even necessary to insert a regulating device in the form of a throttling-valve R between condenser and evaporator in order to maintain the necessary differences of pressure and temperature.

In steam-engines the water re-fed from the condenser is at a much lower temperature than the boiler. In order to save fuel, therefore, in modern steam plants a heat-exchanger is inserted between feed pumps and boiler. The process of ' bleeding ' is used, which consists of drawing a part of the steam from the later stages of the turbine and using it for pre-heating the water fed to the boilers. This step, which is analogous to the recuperation used in closed-cycle gas turbines, is illustrated in the upper half of Fig. 38.

In the heat pump the fluid re-fed to the evaporator enters at a higher temperature, and is, therefore, more readily evaporated again. It might be surprising at first glance that the interior of the evaporator is kept permanently at a temperature below that of the surroundings, although the continuous input of fresh matter consists of nothing else but a warm fluid. The explanation lies in the fact that the evaporation, accelerated by the sucking action of the compressor, consumes so much heat that a balance is reached at a temperature low enough for a plentiful heat supply by conduction from the surrounding water. In other words, the action which keeps the temperature of the evaporator below the level of its surroundings is the same as that which cools your soup when you enhance evaporation by blowing over its surface.

Instead of inverting steam-engines, one can construct heat pumps and refrigerators also by inverting the hot-air engine working on the Joule cycle. As a matter of fact, the earliest types of refrigerators, first used towards the end of the nineteenth-century in ships carrying frozen meat, were working on the Bell-Coleman cycle, named after the designers of the engine, which uses a reversed Joule cycle.

Fig. 41 contrasts the Sankey diagrams of the Joule cycle and its inversion. Both operations can be executed by using a turbine and a compressor and two heat-exchanging devices between them. In the closed-cycle gas turbine these devices are the air-heater and the pre-cooler, in the heat pump the heat-exchanger HE_1 cooling the working medium by giving out heat to the surroundings, and the heat-exchanger HE_2 heating the working medium by withdrawing heat from the surroundings. Corresponding to the reversed action of the Bell-Coleman cycle, the heater HE_2 is at the lower, and the cooling device HE_1 at the higher, temperature. Accordingly, the compressor of the heat pump also consumes more energy than the turbine produces, the deficit being covered by the work of the motor which replaces the generator

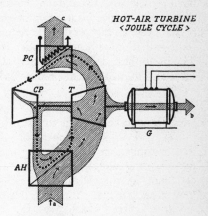

HOT-AIR TURBINE
< JOULE CYCLE >

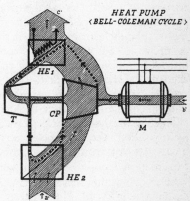

HEAT PUMP
< BELL-COLEMAN CYCLE >

FIG. 41. SANKEY DIAGRAMS OF JOULE CYCLE AND BELL-COLEMAN CYCLE

AH, air-heater; *CP*, Compressor; *PC*, precooler; *T*, turbine; *G*, generator; *a*, heat input from fuel; *b*, power output; *c*, heat discarded with cooling water; HE_1 and HE_2, heat exchangers; *M*, motor; *a′*, heat input from river; *b′*, power input from mains; *c′*, useful heat output.

of the turbine plant. Referring to Fig. 41, we can follow the single steps of heat-pump operation working on the Bell-Coleman cycle. The working medium (for instance, air or any other permanent gas) is heated by compression to a temperature sufficiently high to give off useful heat by conduction in the heat-exchanger HE_1. After having done so it expands in the turbine, doing work which partly covers the power consumption of the compressor. During this stage of the cycle its temperature is reduced below that of the surroundings, so that in the heat-exchanger HE_2 it receives heat from outside. The useful heat delivered in HE_1 is the sum of the heat received in HE_2 and the equivalent of the work done by the motor.

The main advantage of the gas turbine over all steam-engines, which consists of avoiding the losses caused by delivery of the latent heat in the condenser, is no longer an advantage in the heat pump, with its inverse operation. For it is just the heat delivered to the condenser which is the useful output, and therefore heat pumps using working substances which alternately evaporate and condense are theoretically more efficient than those using permanent gases like air. There is, however, one advantage of the Bell-Coleman cycle, which was the reason for its early use as a refrigerator. With air as the working medium, the cycle can be performed as an open cycle instead of a closed one—that is to say, the circulating air can be taken from, and exhausted again to, the room which has to be either cooled or heated. If the device is intended, for instance, to be a refrigerator we can omit the heat-exchanger HE_2 and suck the air with the intake-pipe of the compressor directly from the storage-room which we want to be cooled. After having run through the cycle (heating by compression, discharge of the heat in HE_1, cooling below the initial temperature by expansion) the well-cooled air is returned to the storage-room. That was the method which first allowed the safe transport of frozen meat from overseas to Europe.

For using the Bell-Coleman cycle as a heat pump the heat-exchanger HE_1 in Fig. 41 can be dispensed with and the air drawn from and exhausted again to the room to be heated. In this case the pressure range of turbine and compressor must lie between atmospheric pressure and a lower pressure corresponding to a moderate vacuum, the establishment of which involves no difficulty. The possibility of dispensing with a heat-exchanger is a certain advantage because it saves not only weight and costs of the machinery, but also the temperature-drop in the heat-exchanger which decreases the performance-energy ratio.

Returning to heat pumps working on the vapour-compression cycle, it may be added that some modern installations are so designed that the water-pipes of the central heating can be alternately connected to the condenser and the evaporator, so that the same device can act as a heat pump in winter and as an air-conditioner in summer.

A purely quantitative difference between heat pumps and refrigerators lies in the temperature range. Most heat pumps draw the heat from water at some degrees above freezing-point, and deliver it to the heating system at temperatures between, say, 120°F. = 49°C. and 150°F. = 65.5°C. The whole range is shifted downward for about ten degrees centigrade in the case of household refrigerators, and still more in that of industrial refrigerating plants.

The Working Substances

The condition which gases have to fulfil to be usable as working substances in heat pumps or refrigerators is that their boiling-point at normal pressure shall lie in the interval between, say, —10°C. and —40°C., or, more properly expressed, that their vapour pressure at the limiting temperatures T_1 and T_2 can be easily handled by compressors of normal, not too expensive, design. Fig. 42 shows the vapour-pressure curves of four substances, ammonia, sulphur dioxide, methyl chloride, and Freon 12, having suitably low boiling-points. The horizontal unbroken lines correspond to the temperatures T_1 of the condenser and T_2 of the evaporator in heat pumps and refrigerators. The broken lines denote the temperature and pressure ranges.

In the initial state of a refrigerator, before the compressor is set working, the pressure and temperature will be the same in the evaporator as in the condenser, the latter simply being the room temperature T_0. The pressure will be the saturation vapour pressure of the working medium p_0 which corresponds to T_0. With ammonia as the working medium and $T_0 = 15°C.$, the saturation pressure will be $p_0 = 7.4$ ata = 105 psia, as indicated in Fig. 42 by an asterisk at the intersection of the vapour-pressure curve of ammonia with $T_0 = 15°C.$ When the compressor begins to work its sucking action at the intake-pipe will reduce the pressure in the evaporator, while that in the condenser will rise. By regulating the speed of the compressor and the throttling action of the regulator R (*cf.* Fig. 40) the pressures can be set at suitable values—for instance, at p_1 in the condenser and p_2 in the evaporator, with $p_1 > p_0 > p_2$. While originally at T_0 and p_0, with saturated vapour in contact with the fluid medium, full equilibrium between fluid and gaseous phase was reached with no more evaporation taking place, the equilibrium will be upset by lowering the pressure from p_0 to p_2, and the working substance will evaporate, cooling down

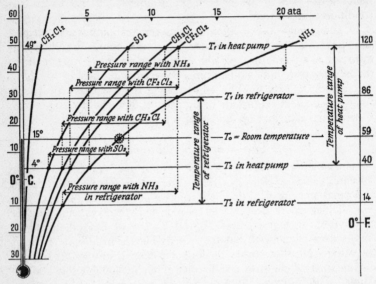

FIG. 42. SATURATED VAPOUR-PRESSURE CURVES OF AMMONIA (NH₃), SULPHUR DIOXIDE (SO₂), METHYL CHLORIDE (CH₃CL), AND DIFLUOR-DICHLORMETHANE (FREON 12) (CF₂CL₂)

T_0, room temperature; T_1, temperature of condenser; T_2, temperature of evaporator. Broken lines indicate temperature and pressure ranges as used in refrigerators and heat pumps.

at the same time till its temperature is T_2, to which p_2 is the corresponding saturation pressure. In the same way, the temperature in the condenser will rise to T_1, the temperature to which p_1 is the corresponding saturation pressure. The saturation pressures which belong to temperatures as used in heat pumps and refrigerators can be gathered from Fig. 42. If, for instance, a refrigerator is designed to operate with $T_2 = -10°$C. and $T_1 = +30°$C. and is using ammonia (NH₃) as the working medium, its compressor has to compress the evaporated substance from 3 ata to 11·9 ata. Although ammonia is one of the most commonly used working substances in refrigerators, it is not the most suitable medium in the case of heat pumps on account of its inconveniently high saturation pressure at the condenser temperatures needed in heat pumps. As can be seen from Fig. 42, with $T_2 = 4°C. = 40°F.$ and $T_1 = 49°C. = 120°F.$, the saturation pressures of ammonia are $p_2 = 4\cdot7$ ata $= 67$ psia and $p_1 = 20\cdot3$ ata $= 294$ psia, thus requiring a great deal of strength from the compressor.

Freon 12 (CF_2Cl_2) seems to be so far the best-suited working medium for heat pumps, its saturation pressures at the same temperatures $T_2 = 40°F.$ and $T_1 = 120°F.$ being $p_2 = 3\cdot3$ ata $= 47$ psia and $p_1 = 12\cdot1$ ata $= 172$ psia.

The Performance-energy Ratio

Corresponding to the inverse operation and the inverse purpose of heat engines and heat pumps, the conceptions of efficiency of both types of devices are reciprocal to each other. While a heat engine should produce as much mechanical energy as possible from a given quantity of heat, the heat pump should deliver as much heat as possible when driven with a given quantity of mechanical work, and the refrigerator should remove as much heat as possible from a body at a lower temperature level. Accordingly, the ratio expressing the efficiency of a heat pump is reciprocal to that of the heat engine. The word efficiency is, however, reserved for the latter engine, while other expressions, which will be defined at once, are used to characterize the economy of heat pumps and refrigerators. The relevant ratios of the three types of devices are:

Heat Engine

$$\text{Efficiency } (\eta) = \frac{\text{work done}}{\text{heat supplied from fuel}}$$

Refrigerator

$$\text{Coefficient of performance (C.O.P.)} = \frac{\text{heat removed from cooler body}}{\text{work done}}$$

Heat Pump

$$\text{Performance-energy ratio (P.E.R.)} = \frac{\text{heat delivered to hotter body}}{\text{work done}}$$

The ideal process for performing all three tasks would be the Carnot cycle working with an ideal gas which acts as a heat engine by expanding the gas at the higher and compressing it at the lower temperature, but as a heat pump or a refrigerator by operating in the opposite direction. Denoting the absolute temperature of the higher temperature level by T_1, and that of the lower one by T_2, the efficiency of the Carnot cycle as a steam-engine is given according to equation (10) by

$$\eta_c = \frac{T_1 - T_2}{T_1} \dots\dots\dots\dots\dots\dots\dots\dots\dots(10)$$

Inversely operated, the Carnot cycle would act as a refrigerator with a coefficient of performance given by

$$C.O.P. = \frac{T_2}{T_1 - T_2} \dots\dots\dots\dots\dots\dots\dots(14)$$

or as a heat pump with a performance-energy ratio of

$$P.E.R. = \frac{T_1}{T_1 - T_2} \dots\dots\dots\dots\dots\dots\dots(15)$$

Real engines are less efficient than an ideal engine performing the Carnot cycle would be. Still, the qualitative deduction which may be drawn from the three equations given above hold for the real devices also: the difference between highest and lowest temperature of the cycle, $T_1 - T_2$, is the numerator of the expression (10) giving the efficiency of heat engines, and the denominator of those giving the C.O.P. of refrigerators and the P.E.R. of heat pumps. To obtain the greatest economy, therefore, the old rule, stated repeatedly in the foregoing chapters, that in heat engines the temperature interval should be chosen as large as possible, holds inversely in the case of heat pumps and refrigerators: the smaller the temperature range through which the heat has to be pumped, the better the performance-energy ratio. The consequence of this fact is the further rule that central-heating installations designed for heat-pump operation should use large-area radiators or embedded panel heaters which can maintain the desired room temperature with relatively moderate temperature of the circulating water. It has been found that in the British climate $T_1 = 49°C.$ $= 120°F. = 322°K.$ can be assumed as sufficient during the greater part of the heating period, while approximately $T_2 = 4°C. = 40°F.$ $= 277°K.$ can be taken as the winter temperature of river-water. Inserting these values into equation (15), we obtain for the performance-energy ratio of an ideal heat pump working on the Carnot cycle:

$$P.E.R._{Carnot} = \frac{322}{322 - 277} = 7.15$$

The performance-energy ratio actually obtained in the real heat pumps is, of course, lower, just as the efficiencies of heat engines are always lower than those of the Carnot cycle operating between the same temperature limits. Still, it has been found that in existing heat-pump installations using electric motors for driving the compressor between three and four times as much heat can be supplied to the rooms as by electric stoves consuming the same amount of power. In large central-heating plants, therefore, as used in great buildings or factories, quite considerable savings can be achieved, especially if the heat-storage capacity of large, well-insulated water-tanks is used. For in that case the heat pumps may be driven and the heat may be stored during the off-peak periods, when the cost of the kilowatt-hour is lower.

Savings achieved by the Use of Heat Pumps

Table 9, taken from *Heat Pumps and Thermal Compressors,* by S. J. Davies, gives a comparative synopsis of heating costs based on data presented by J. A. Sumner, the City Electrical Engineer of Norwich, who designed and built the first large-scale heat-pump installation in Great Britain.

The figures given in Table 9 are based on the assumption that the compressors are driven from electric motors connected to the mains, which in their turn are fed from a coal-fired central station. The drawback of this system lies in the fact that even a $4:1$ performance-energy ratio—which is a rather optimistic estimate—is more than compensated by the $1:4$ or $1:5$ overall efficiency of the power plant. The difference between the conventional central heating with coal-fired boilers and the heat-pump installation is an increase of fuel efficiency from 55 per cent to something near 80 per cent at the cost of quite a considerable amount of rather expensive machinery. Still, even under these less favourable conditions there are other important advantages of the heat pumps which make their employment recommendable in all planning of new city districts or industrial establishments. Centralization of coal consumption, under skilled direction, results, as Davies points out in his book, not only in fuel-saving, but also in reduction in pollution of the atmosphere and saving in transport of fuel and ashes. And, last but not least, there is a reduction in low-grade labour.

A greater amount of fuel economy could be achieved by other methods of running heat pumps. One which might be suitable for large installations is to drive the compressors by heat engines the reject heat of which is directly added to the heat supplied from

TABLE 9. COMPARATIVE ANNUAL COSTS OF HEATING A LARGE
BUILDING

Seasonal total heat supplied: 20,000 therms (5·04 × 10⁸ kcal).
Coal of heating value 12,000 B.Th.U. (3030 kcal.) per lb at 65s. per ton.
Average combustion efficiency, 55 per cent. Cost of electricity: (*a*) loads
on peak, £4 per kVA per annum plus 0·6*d*. per kWh; (*b*) loads off peak,
0·6*d*. per kWh. Average performance-energy ratio, 4 : 1.

	Coal-fired Boilers	Heat Pump	
		Alone	With Thermal Storage
Capital cost (£)	1500	4000	4500
Annual capital charges (£)	225	280	315
	(15 per cent)	(7 per cent)	(7 per cent)
Cost of coal or electricity (£)	440	601	367
Attendance (£)	230 (including coal- and ash-handling)	—	—
Repairs and maintenance (£)	150	50	50
Replenishing working substance (£)	—	25	25
Total annual cost (£)	1045	956	765
Cost per therm (per 25,200 kcal)	12·5*d*.	11·5*d*.	9·1*d*.

the heat pump. In this way the detour over electric engines, causing
losses and additional expenditure on machinery, could be avoided,
and, in addition, the high efficiency of Diesel engines, further
enhanced by utilizing the reject heat, could be exploited. Let us
tentatively assume that the efficiency of the heat engine be 40 per
cent, and that five-sixths of the residual 60 per cent—that is, 50
per cent—of the fuel consumption could be recovered as useful
heat in the water circulating in the heaters. With a performance-
energy ratio of 4 the useful output would be

$$0·4 \times 4 + 0·5 = 2·1 = 210 \text{ per cent}$$

Compared with the consumption efficiency of 55 per cent in conventional central heating, this would mean a gain of nearly 300 per cent over the present heat output. Installations of this kind, if economically developed, would be particularly suitable in countries with cheap Diesel oil.

Still greater savings—even a total emancipation from fuel—can be achieved by heat pumps driven electrically from water-power plants. Even now widespread use is made of electric heating in countries like Norway with ample supply of hydro-electricity. But, considering the fact that conversion of heat into mechanical and subsequently electric power is done with an average efficiency of only 20 per cent, we have reason to consider electric power as a ' high-grade ' sort of energy, while heat is only ' low-grade ' energy. An expression of this fact is the assessment of only one-fifth of the heat value of fuel as its electricity equivalent. And within the field of heat, again, we can distinguish between high-grade and low-grade heat, depending on the temperature. For with steam or gas at very high temperatures we can obtain greater efficiencies in producing power than with the same media at lower temperatures. And, inversely, by using heat pumps the cheaper low-grade heat of water at 120°F. or so circulating in our radiators can be obtained with a much better performance-energy ratio than high-grade heat. As a consequence thereof we must consider it an economic sin to convert valuable electrical energy into low-grade heat by the direct process of electric heating with a conversion factor of 1 : 1 or even less. (While writing these lines I notice that I am guilty of this very sin myself by warming my feet on a small electric radiator standing behind my desk. I may offer as an excuse, first that my radiator is switched to mains fed from water-power, and, secondly, that small-scale use of electric heating as a transitional measure on days when it is not worth while to operate the central heating might be permissible.)

All kinds of direct conversion of electricity into heat, as, for instance, by the Joule heat produced in wires, or in high-frequency induction stoves, or by the electric arc, should therefore be reserved for the production of high-grade heat, as necessary for melting or welding, etc., while large-scale electric supply of low-grade heat, as used in households and also in certain branches of industry, should be done as far as possible by heat pumps. A glance at Fig. 5, Chapter II, shows that on the world average more than two-thirds of the vast global energy consumption is used for domestic and industrial heat. Hence countries having abundant water-power and lacking resources of mineral fuel might in the long run secure very considerable savings in their national economy by fostering the installation of ever more heat-pump plants in order to cover as fully as possible their needs of domestic and industrial low-grade heat from home-made electricity.

There are some countries which are favoured by their natural wealth of water-power to such a degree that, after full development of all economically justifiable power sites, and after the erection of a sufficient number of heat-pump installations, they are not only self-sufficient as regards mechanical power and low-grade heat, but also able to cover their residual needs of fuel for road traffic by the export of electricity. Certainly, however, two conditions must be fulfilled to realize a far-reaching emancipation from fuel. One is the existence of hydro-electric plants of sufficient capacity for the supply of cheap power, and the other is the availability of a source of heat to maintain the evaporation of the working medium at the lower value of the temperature range. Flowing water is the best heat source provided that its temperature remains safely far from freezing-point. Unfortunately the water of several streams like the Danube, which during the greatest part of the year is well suited for feeding heat pumps, cools down to temperatures very near freezing-point in severe winters. It cannot be used, therefore, at times when heat pumps would be most urgently needed. For as soon as a considerable amount of heat is withdrawn from water near its freezing-point it would start to freeze, and would thus, by forming an ice-crust round the evaporator, prevent further circulation and supply of heat.

Instead of water, air can also be used as a heat source which is almost free from the difficulties inherent in water. But on account of its much lower heat capacity very large volumes of air would be required to supply the necessary amount of heat, and this would involve the use of large motor-driven fans, which in their turn consume power, and decrease, therefore, the performance-energy ratio.

The best conditions for economical use of heat pumps exist, therefore, in countries with ample water-power, and with cities and industrial plants on the borders of big rivers, lakes, or on sea-shores with moderately warm water. In such areas extensive capital investment in power plants and heat-pump installations will prove to be of equal importance for the future prosperity of these districts as roads, railways, and the like.

Recent Development of the Heat Pump in Great Britain

The disproportion between British coal production and the ever-growing fuel demands has drawn the attention of the competent authorities to heat pumps for domestic heating in the post-war period. Although the heat pump is a British invention, with two exceptions it had never been used practically in Britain before 1945. It was only quite recently that in accordance with the general trend for better utilization of fuels systematic research was carried on as to the best methods and available heat sources for heat pumps under the climatic and other conditions of the country. Observations made on some of the representative rivers of England have shown that their temperatures were sufficiently high above freezing-point to serve as a source of heat during the greater part of the year. To be quite safe, however, river-water-fed heat-pump installations should be fitted with fuel-fired additional heating, which can be used under extreme conditions. When using ground-water as the heat source this precaution is not necessary, so that it is advisable to start digging sufficiently deep wells when constructing new buildings planned for installation with heat pumps. However, ground-water is suitable only for small installations. Atmospheric air can in principle serve as a heat source in all circumstances, but the performance-energy ratio drops sharply when outside temperatures are low. Besides, proper equipment for the prevention of ice formation on an air-heated evaporator would have to be provided. Best yields may be achieved, of course, where waste heat of some kind can be used as the low-grade heat source.

The two largest British heat-pump installations use Merlin aeroplane engines adapted to run on town gas and driving centrifugal compressors. By utilizing the waste heat of the engine a total heat output of 2700 kW can be obtained with an engine consuming 525 kW. Both installations can be used for heating in winter and for cooling and air-conditioning in summer; one is installed in the Royal Festival Hall in London, while the other is on board a big liner.

An installation built by the Electrical Research Association for a laboratory in Shinfield, Berkshire, operates only on the soil as a source of low-grade heat. The heat is extracted by means of a pipe system of one-inch pipes with an over-all length of 500 feet, laid horizontally three feet deep in the ground. A calcium chloride mixture circulating through the pipe system transfers the heat to the evaporator. The compressor is driven by a 7·5–kW electric motor; the heat output is quoted as 28 kW. Owing to the low input temperature, the outlet temperature of the warm water is only 120°F., but it suffices for heat distribution with large-area radiant panels.

Since 1952 special attention has been paid to the development of smaller dual-purpose units This does not only mean heat pumps which heat in winter and cool in summer, but also units employed all the year round for larder-cooling and simultaneous domestic water-heating. By extracting heat from an uninsulated larder, quite a small heat pump driven by a fractional horse-power motor can supply 300 litres of water daily at a temperature of 60°C (= 140°F).

A similar dual-purpose heat pump might well be installed in dairies to provide refrigeration for milk and hot water for washing bottles. Smaller heat pumps with 0·75 h.p. motors and a P.E.R. of 4·0 are intended for heating single rooms. A house in Crawley New Town is to be heated by one hundred of these small units. The British Motor Corporation has been carrying on experimental work for the last two years for the development of somewhat larger heat pumps driven by 2- and 5-h.p. motors for the heating of medium-sized residential houses. Production on a commercial scale will not be taken up, however, before thorough tests have proved the machines reliable and suited to their purpose.

Thermal Compressors

A device which is not identical with, but somehow related to, the heat pump is the thermal compressor used for the recovery of the latent heat which in conventional plants is lost. There are quite a number of industrial operations, like distilling, drying, concentration by evaporation, etc., in which very large quantities of heat are lost to the atmosphere in the form of the latent heat in the water vapour which is driven off in these processes. One of these operations, used nearly all over the world, is the production of rock-salt, which is withdrawn from the mines with water. The salt dissolves in the water, and the resulting brine is led through pipe-lines to the salt-pans, where the water is driven off again by evaporation, leaving crystallized salt in the pan. The origin of this method dates from antiquity, and in former times, with cheap rural labour and plenty of wood for fuel, nobody cared about the calories dissipated into the atmosphere with the water vapour. Considering, however, the everlasting global demand for rock-salt, the total waste of fuel caused by this method all over the world is quite considerable. The concentration of the salt in the brine is about 27·8 per cent, so that 2·6 kg of water have to be evaporated for each kilogramme of salt produced. We need about 85 kcal/kg to heat water from the ambient temperature to boiling-point, and another 539 kcal/kg to evaporate it, giving a total heat consumption of 624 kcal for each kilogramme of water, and about 1620 kcal net per kilogramme of salt. Since, however, the efficiency of old-fashioned salt-pans can be roughly estimated to be about 55 per cent, the gross heat consumption in this process is approximately 3000 kcal per kg of salt. The coal equivalent of this amount of heat is 0·417 kg of coal. With an annual average *per capita* consumption of 7 kg salt for food and 2 kg for feed and industry the world consumption is approximately $2 \cdot 10^{10}$kg $=20$ million metric tons. Assuming that about half of this quantity is produced from brine by artificial evaporation (the other half being produced in open salt-beds dried by the sun), we obtain a fuel consumption of about 4 million tons of coal equivalent. This represents an expenditure of roughly £14 million yearly, which is made to perform a process in which a fraction of 539:624, or about 86 per cent, of the heat transmitted to the brine is lost—not inevitably lost, but in a form in which it might easily be recovered by provision of suitable means.

The described process of salt production is, however, only a single example of a large number of analogous operations used in the foodstuff industry, for distilling, for the manufacture of sugar and jam, in the chemical industry, for manufacture of pharmaceutical products, for concentrating dyes, and for evaporating solutions and mixtures of two or more substances. Taking all these processes into account, the figure given above for calories wasted by dissipating latent heat to the atmosphere may be increased very considerably.

Unlike the case of steam locomotives, in which, except for heating the trains in winter, utilization of the latent heat of the steam is practically not feasible, the recuperation of the latent heat can and should be performed in stationary plants like distilleries and similarly operating installations. This economic necessity was recognized long ago, and towards the end of the nineteenth century improvements had already been made by using the vapour to preheat the brine. If the vapour, instead of being allowed to escape, is drawn through tubes surrounding the brine pipes it condenses, giving off a part of its latent heat to the brine flowing into the pan. Full recovery of the heat is, however, possible only by employing a compressor, which establishes the temperature difference necessary for heat conduction from the condenser to the boiler.

The operation of the thermal compressor will be understood best by comparing Figs. 43 and 44, which show the Sankey diagrams of a primitive conventional distillation plant and of a plant equipped with a thermal compressor. In the old-fashioned plant the large heat input (*a*) supplied from the fuel is partly lost through the chimney (*b*) and partly used for heating and evaporating the fluid. The latent heat of the vapour (*c*) is discarded with the cooling water of the condenser, and a small residue (*d*) is (uselessly) left in the distillate.

In the thermal compressor distillation plant the still and the condenser are united in the same container (*SC*), the liquid to be distilled serving as the cooler for condensing the vapour, whose latent heat in its turn serves for evaporating the fluid. In order to permit the spontaneous transition of heat from the vapour to the liquid by heat conduction, the temperature of the former must be higher than that of the latter; hence a compressor *Cp* driven by the motor *M* is inserted into the flow of the vapour from the still to the condenser tubings. To start the operation of the plant a relatively small heat input supplied by an auxiliary electric heater *AH* suffices, which heats the fluid to a temperature near boiling-point at normal pressure. When the compressor is set working its sucking action at the intake pipe reduces the pressure in the still, so that the fluid begins to boil. The vapour, after being heated by the compression, is admitted to the windings of the condenser, where in condensing again it discharges its latent heat to the boiling fluid, and maintains in this way a continuous evaporation. The condensate emerges at a temperature just a little above that of the liquid, which owing to the low pressure in the still boils at a temperature below its normal boiling-point. In flowing through the windings of the heat-exchanger *HE* the distillate discharges most of its residual heat to the supply fluid, which after being pre-heated is admitted to the still through a throttling-valve *R*. The latter is necessary to maintain the pressure-drop between the atmosphere and the still.

The point of the thermal compressor is the full recuperation of the latent heat, which, in addition to the avoidance of chimney losses, has the effect of reducing the energy consumption of the plant to about *one-tenth* of that of the primitive still. As can be seen from the width of the shaded areas of the diagrams, which are drawn proportional to the energy-flow, the old plant loses large amounts of heat (*b*) and (*c*) to the surroundings, so that a great heat input (*a*) is required. In the thermal compressor plant the large heat-flow circulates within the plant. The expenditure of

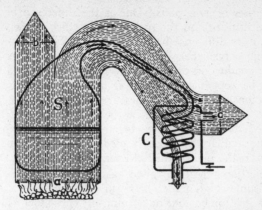

FIG. 43. SANKEY DIAGRAM OF PRIMITIVE DISTILLATION PLANT

S, still; *C*, condenser; *a*, heat input supplied from burning fuel; *b*, heat escaping through chimney; *c*, latent heat discharged with cooling water of condenser; *d*, heat of distilled water.

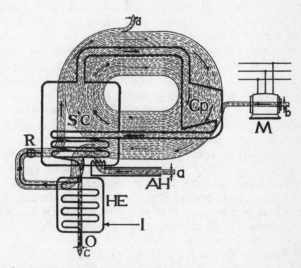

FIG. 44. SANKEY DIAGRAM OF DISTILLATION PLANT USING THERMAL COMPRESSOR

SC, combined still and condenser; *Cp*, compressor; *M*, motor; *AH*, auxiliary electric heating; *HE*, heat-exchanger; *I*, inlet pipe; *O*, outlet pipe; *R*, throttling-valve; *b*, energy input driving compressor, supplied from electric mains; *a*, heat supplied from auxiliary heater; *c*, heat of distilled fluid; *d*, heat losses by conduction through the walls.

energy consists of two parts. One is the difference between the
heat of the distilled and the supply fluid, which is proportional to
the temperature difference between the two. This can be kept low
by providing a good heat-exchanger. The other part is the losses
(*d*) by heat conduction through the walls of the still and the pipes,
which can be kept low by providing adequate thermal insulation.
In installations which fulfill these requirements in a sufficient
measure the whole expenditure can be kept so low that the power
input (*b*) of the motor suffices to cover all energy expenses. In
that case the auxiliary heating *AH* used for starting can be switched
off as soon as the plant is in full operation.

The objection might be raised against the thermal compressor
that it consumes high-grade energy in the form of electric power
for a task which can be done by burning any cheap fuel. The answer
is that even considering the 1:5 ratio in average production of
electricity from heat, considerable savings can be achieved on account
of the fact that in well-built thermal compressors only about one-
tenth of the energy or one-half of the electricity equivalent of the
consumption of the conventional plants is needed. This argument
is of particular weight in countries with ample water-power, where
the use of imported fuel should be reduced as far as possible. A
thermal compressor plant was installed for the United Swiss Rhine
Salt Works Company, at Ryburg, Switzerland, by the Escher-Wyss
Company in 1941–42. The output of the installation is 40,000 tons
of salt per annum, and the yearly saving amounted to 14,000 tons
of coal, so that a second installation was ordered very soon.

Considering the fact that the salt production, which in its present
state of development consumes approximately 4 million tons of coal
equivalent annually (wasting 90 per cent thereof) is only one of a
great number of industries operating with large-scale evaporating
plants one can easily surmise the scope of fuel-saving which could
be made by the general introduction of thermal compressors. Compared with heat engines and heat pumps, the thermal compressor
excels by achieving the greatest percentage of saving, which amounts
to 90 per cent and even more.

Apart from the fuel-saving and the cleanliness of operation, a
further advantage of the thermal compressor lies in the possibility
of shifting the boiling pressure, and consequently also the boiling
temperature, to any value which is best suited for the purpose of the
plant. In the example explained above it was assumed that the
pressure in the still lies a little below atmospheric pressure. By
regulating the power of the compressor and the throttling action

of the regulating-valve *R* it might also be set considerably below atmospheric pressure, or, on the other hand, by replacing the throttling-valve with a feed-pump, raised to higher pressures.

Concentration of solutions and colloidal mixtures by evaporation plays a rôle in the foodstuff industry, and in preparing condensed milk or fruit juices the method of evaporating at lower pressures and temperatures leads to higher-quality products by better preservation of their vitamin content. In this very important branch the thermal compressor is particularly suited to meet the demands corresponding to the trend of progressive industry: more machinery, less fuel, less attendance and maintenance costs, better quality of products.

CHAPTER VII

Electricity

How Faraday's Discovery changed our Life

O N October 17, 1831, **Michael Faraday**, continuing with
admirable perseverance his researches on the interaction
between electrical and magnetic phenomena, made a striking discovery with quite simple means. He wound several feet of
thin insulated wire around a hollow paper cylinder, and connected
the ends of the coil to a galvanometer. When he thrust a bar magnet
quickly into the cylinder the galvanometer needle was deflected to
one side, and to the other side when he withdrew the magnet.
No deflection was observed, however, when the magnet remained
in a fixed position inside or outside the coil. Faraday concluded
rightly that it was the change of the magnetic field passing the
interior of the coil which induced a current in the coil, the direction
of the current depending on whether the change was an increase
or decrease of the field.

Faraday's Law of Electromagnetic Induction, emerging from this
experiment, was the key for the invention of the dynamo machine
and the transformer. Hence all devices for generating and transforming electric current are based on electromagnetic induction; and
what these devices mean for our civilized life can easily be illustrated
by fancying that by a whim of the Creator it was decreed one day
that all the devices on our globe which operate on the induction
law should cease to work. If this happened in the evening while
we were in the cinema the first result would be that we should be
sitting in the dark with the loudspeakers failing to work. Having
lit a match, we should walk into the street, and find there total
blackout, no Tube, no taxis. For along with the electric motors
all petrol motors would be stopped, because the ignitor of the
sparking-plugs would fail to work. No light, either, at home, no
telephone, no radio or television; nobody could tell you what had
happened. When the morning dawned we should be disappointed
at getting no milk and no newspapers, and what started apparently
as a series of mere inconveniences would grow during the following
days into serious troubles, ending with a real world catastrophe.

The food-supply to all big cities is based entirely on rail and motor traction, and railway traffic in its turn, even if not electrified, is dependent on electricity, because the whole system of signals and intercommunication uses devices based on electromagnetic induction. Imagine the difficulties of organizing an emergency service with Diesel buses and horse-driven cars for the needs of a ten-million population when all means of organizing anything—that is, all means of communication, such as Press, radio, telephone, telegraph —were paralysed, and when at the same time nearly all possibilities of travelling were blocked. Apart from this, the industries of all civilised countries would stop working, so that, with millions of unemployed and with a total cut in the production of goods, unprecedented and incurable misery would occur, killing perhaps three-quarters of the population, and leaving the rest in a deplorable state.

What can be learnt from this sombre vision? The consequences of the discovery of this one man, Michael Faraday, son of a blacksmith, former apprentice of a bookbinder, have changed our civilization and our pattern of life so radically that cancelling them would mean the disruption of our economy and the destruction of the material basis of our life. Can anything like that be said of any one of the so-called Great Men of History? Nothing at all would happen if the good God decreed *ex post facto* all the battles of Cæsar or Napoleon undone. Seen from a greater perspective, the deeds of these vain and conceited conquerors have not left any deep vestiges, apart from a few ugly scratches, in the development of humanity. The men who really ' make history,' by leaving enduring marks in the fabric of time, are not those to whom innumerable pages of our history-books are devoted, but those who, far from the hustle and bustle of their days, are creating things helpful to the welfare of mankind.

Thoughts like this, which sound commonplace in the Atomic Age, were realized only by few in the nineteenth century. A story is told of a visit of Mr Gladstone to Faraday's laboratory at a time when the famous statesman was not yet Prime Minister, but Chancellor of the Exchequer. After having listened to Faraday's explanation of one of his pieces of apparatus Gladstone asked him, " But, after all, what good is it? " Faraday, doing his best to bring things home, replied, " Why, sir, one day you will tax it."

Survey of Notations and Conceptions used: Direct Current and Alternating Current

The symbols D.C. and A.C. denote direct current and alternating current respectively. Both in the case of D.C. and A.C. circuits the power *P*—that is the energy produced, transmitted, or consumed per second—is proportional to the product of current *I* and voltage *E*. While, however, the power of a D.C. circuit is directly equal to this product, the current being expressed in amps, the voltage in volts, and the power in watts,

$$P \; = \; IE \dotfill (16)$$

the corresponding relation is a little more complicated in the case of A.C. circuits, and needs some explanation.

Single-phase A.C. Circuits

Fig. 45 represents voltage and current of an A.C. circuit as a function of time during one complete cycle.

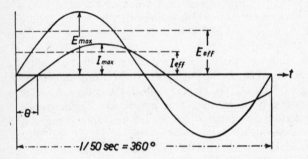

FIG. 45. VARIATION OF VOLTAGE *E* AND CURRENT *I* WITH TIME IN A.C. CIRCUIT

The frequency of A.C. supplies is standardized at 50 cycles per second in Europe and 60 C.P.S. in America. Voltage and current change their direction twice during each cycle, oscillating in the form of a sine wave between positive and negative peaks. By dividing each cycle into 360 degrees any fraction of the cycle can be expressed in degrees. Fig. 45 represents a case in which the current is lagging behind the voltage. This happens when the load inserted in the circuit is a so-called *inductive load*—that is, one, such as an electric motor, for instance, containing coils in which the current produces a magnetic field. The fraction of the cycle by which the current lags behind the voltage is called the *phase angle θ*; it amounts to 30° in the particular case represented in Fig. 45. A phase angle of 90° would mean that the voltage reaches its peak value at the very moment when the current passes through zero.

The figure given as the voltage of the mains of an A.C. circuit —as, for instance, your house supply, which may be 230 volts— does not mean the peak value E_{max} of the periodically varying voltage, but the so-called *effective voltage*. This term means the equivalent voltage of a D.C. circuit which, being applied to any ohmic resistance, or, more particularly, to the filament of an incandescent lamp, develops the same Joule's heat as your A.C. circuit. It can be proved by calculations that both for voltage and current the relations between peak value and effective value are given by the equations

$$E_{max} = \sqrt{2}\, E_{eff}$$
$$I_{max} = \sqrt{2}\, I_{eff} \quad\dots\dots\dots\dots\dots\dots\dots\dots\dots\dots(17)$$

A graphic representation of this relation is given in Fig. 45 by the horizontal dotted lines, whose distances from the zero line are the effective values of voltage and current.

Power of A.C. Circuits

The power P consumed in an A.C. circuit is given by the expression

$$P = E_{eff} \cdot I_{eff} \cdot \cos\theta \dots\dots\dots\dots\dots\dots\dots\dots(18)$$

which is equivalent to

$$P = \tfrac{1}{2}\, E_{max} \cdot I_{max} \cdot \cos\theta \dots\dots\dots\dots\dots\dots\dots(19)$$

The characteristic difference from the corresponding expression of the power of a D.C. circuit (17) is the factor $\cos\theta$, which varies from one to zero when the angle θ increases from $0°$ to $90°$. If we insert a purely ohmic, non-inductive resistance in the circuit— that is, one causing no (or no appreciable) magnetic field—the phase angle θ is zero, $\cos\theta = 1$, and the power consumption is the same as in a D.C. circuit of equal voltage. If, on the contrary, we insert a purely inductive, non-ohmic resistance in the A.C. circuit, as, for instance, a coil of thick copper wire wound on a ring-shaped iron core, the phase angle will be nearly $90°$, and $\cos\theta$ nearly zero, so that practically no power is consumed, although quite an appreciable amount of current is flowing through the coil. We are speaking of a ' wattless current ' in this case, which is realized, for instance, in a well-built transformer when its primary winding is

connected to an A.C. voltage source while its secondary circuit is
kept open. Of course, no work is done in this case either, and as
soon as we connect the secondary to a load, such as a lamp or a
motor, the phase angle will sink below 90°, and we shall have to
pay for the consumed kilowatt-hours.

Three-phase A.C. Circuits

What has been explained so far about A.C. circuits refers to
the so-called *single-phase alternating currents*, represented by Fig.
45, which are generally used for the house supply. The wholesale

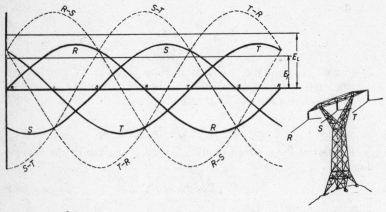

FIG. 46. PHASE VOLTAGE AND LINE VOLTAGE IN A THREE-PHASE
A.C. CIRCUIT

production and distribution of electricity, however, as well as most
of its use in industry, is done in the form of *three-phase A.C. circuits*.
When you look at a high-tension transmission line, like that of the
national grid, you will generally see three or six wires, sometimes
with an additional thinner wire connecting the tops of the masts,
which is not kept under tension, but serves merely to keep the
masts at equal potential. If six wires are used they represent three
pairs of parallel connected twin leads such that the line consists
essentially of three leads—which we shall distinguish by the letters
R, S, T—carrying alternating currents of equal strength and equal
frequency, but with a mutual phase-shift of 120°. The curves representing the variation of the voltages of the three leads are, therefore, sine waves of equal frequency, displaced by one-third of a cycle.
The voltage against earth of each of the three leads is called the

phase voltage; it is represented by the unbroken lines *R*, *S*, and *T* in Fig. 46. The respective voltage differences between two of the three leads are called *line voltage*; they are represented by the broken lines *S–T*, *T–R, and R–S* in Fig. 46. The variation with time of both the phase voltage and the line voltage being given by a sine curve, the relation between peak value and effective value is the same as in a single-phase A.C. circuit—namely, $E_{max} = \sqrt{2}E_{eff}$. The heights of the two horizontal thin lines over the zero line represent the effective values of the phase voltage E_f and the line voltage E_L. The relation between them is given by

$$E_L = \sqrt{3}E_f \dots\dots\dots\dots\dots\dots\dots\dots\dots(20)$$

The figure which is given as the tension of a three-phase transmission line means always the effective value of the *line voltage*.

Star Connexion and Delta Connexion

The peculiar property of the three-phase system is the fact that, while the phase voltages *R*, *S*, and *T* are oscillating, their algebraic sum is constant and equal to zero. Hence no return lead is necessary if care is taken that the loads are equally distributed between the three phases. Fig. 47 (*a*) shows how three windings, or groups of windings, of a three-phase motor or a three-core transformer can be connected to the line. The outer terminals of the windings are connected to the leads *R, S, T*, while the inner terminals are united in

a common zero-point which can be, but is not necessarily, earthed. This kind of connexion is known as *star-connexion*; the voltage supplied to the windings is the phase voltage. The other kind of connexion, shown in Fig. 47 (*b*), which supplies the line voltage to the load is called *delta connexion*.

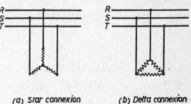

(a) Star connexion   *(b) Delta connexion*

FIG. 47. STAR CONNEXION AND DELTA CONNEXION

Use of Three-phase and Single-phase Circuits

The main advantage of the three-phase system lies in the simple and robust construction of three-phase motors. More mechanical power is supplied all over the world by these motors than by any other type of engines.

Certainly the three-phase system is not used in electric railways or in road vehicles, because of the inconvenience which the tapping of currents from three separate leads would cause. Railways use either single-phase alternating currents of particularly low frequency ($16\frac{2}{3}$ cycles per second) or D.C. circuits, as, for instance, in the case of the Italian State Railways and underground railways in cities. Only one wire is needed to supply the current, while the rails are used as the return lead.

Power of Three-phase Circuits

In three-phase as well as in single-phase A.C. circuits a phase-shift θ exists between voltage and current when a motor or any other inductive load is inserted. The curves representing the three currents in R, S, T, lagging behind the respective voltages, have not been drawn in Fig. 46, because the diagram would then be rather confusing. It suffices to say that the conditions are quite analogous to the case of a single phase, represented in Fig. 45, but repeated three times with a phase-shift of 120°. We have to distinguish between the fixed phase-shift of 120° between the three phases R, S, T and the phase angle θ between voltage and current in each phase, which depends on the kind of the load.

The power of a three-phase circuit is given by the equation

$$P = 3E_t I \cos\theta \quad \dots\dots\dots\dots(21)$$

Considering equation (20) and putting $\sqrt{3} = 1.73$, the power can be expressed also by the line voltage:

$$P = 1.73 E_L I \cos\theta \quad \dots\dots\dots\dots(22)$$

Joule's Heat and Transmission Losses

When a current flows through a conductor some amount of heat is developed, which is called Joule's heat. If the current is I amps and the resistance of the conductor is R ohms the amount of heat produced per second is

$$H = I^2R \quad \dots\dots\dots\dots(23)$$

H being expressed in joules per second = watts. If H is expressed in gramme-calories per second the relation is

$$H = 0.239\, I^2R \quad \dots\dots\dots\dots(24)$$

The production of heat is in many cases the proper object of using electricity; examples are electric stoves, ranges, irons, welding machines, and the like. In other cases heat is not an end in itself, but an intermediate aim, as, for instance, in incandescent lamps, where the Joule's heat is a means of attaining the temperature of bright glow.

Unfortunately, however, heat is always produced according to equations (23) and (24) when a current flows through a conductor, except in the case of superconductivity, which is reached in pure metals when they are cooled down to a temperature near the absolute zero (– 273 °C.). The resistance of superconductors is so infinitesimally small that it was never possible to measure it or to observe a development of Joule's heat. In all other cases, however, Joule's heat is developed whether it is wanted or not, and frequently it is a very inconvenient accessory to the proper process. First of all it is a loss: the amount of energy turned into heat is lost for conversion into mechanical power, chemical energy, and the like. We speak in this connexion of the *dissipation of energy* into heat. Secondly, too much heat can be harmful to the motor, transformer, or any other electrical apparatus. In many cases the Joule's heat contributes more to the wear of electrical engines than the mechanical stress.

From the standpoint of power economy the problem of the thermal losses is of importance. In most of the dynamos and motors the ohmic resistance R is kept low enough by proper choice of the cross-sections of the conductors to reduce the Joule's losses to a few per cent, or even less, of the power. Since, however, the resistance of a transmission line is proportional to its length, the thermal losses can amount to quite a considerable percentage, and set, therefore, a limit to the distances over which electric power can be economically transmitted. In view of the importance of the problem of power transmission, which will be discussed in this chapter as well as in Chapter XII, the basic formulæ will be given here. Denoting by p the fraction of the power P which is dissipated into heat H, we get $p = H/P$. In calculating the relative loss in a three-phase transmission line we have to consider that in each of the three leads Joule's heat is developed such that the total dissipation per second is given by $3I^2R$, where R denotes the resistance of a single lead. Inserting for P the right-hand side of equation (22) and considering that $1 \cdot 73 = \sqrt{3}$, we obtain

$$p = \frac{H}{P} = \frac{3I^2R}{1 \cdot 73\,E_L\,I\cos\theta} = \frac{\sqrt{3}\,IR}{E_L\cos\theta} = \frac{\sqrt{3}IE_r\cos\theta \cdot R}{E_L{}^2\cos^2\theta}$$

or, using equation (22) again:

$$p = \frac{P.R}{E_{\mathrm{L}}^{2} \cos^{2}\theta} \quad \dots\dots\dots\dots\dots\dots\dots\dots(25)$$

The lesson we can draw from equation (25) is that in order to reduce the relative losses we have to make the resistance as small as possible and the voltage as high as possible. It is opportune, further, to make the phase angle θ very small, because $\cos\theta$ reaches its maximum value $\cos\theta = 1$ with $\theta = 0$. Although the load at the far end of the transmission line consists mainly of electric motors, and, being, therefore, an inductive load, would cause quite an appreciable lag of the current, it is possible by means of a trick to make the phase angle small enough. For it can be shown from the theory of alternating currents—upon which we shall not dwell here—that a capacitive load has the inverse effect of an inductive load by causing the current to advance, instead of lagging behind the voltage. Hence by inserting batteries of large condensers parallel to the line a balance between inductive and capacitive load can be established, keeping the phase angle so low that $\cos\theta$ lies between 0·9 and 1·0.

With $\cos\theta$ near enough to its maximum value 1·0, all that could be gained by further reduction of the phase angle is insignificant compared with the saving of losses which can be achieved by raising the transmission voltage E_L as far as the technical development of transformers and insulators will allow. This point will be discussed later on.

Installed Capacity, Annual Production, and Load Factor

In giving data on electrical generator plants we have to distinguish between their power, expressed in watts or multiples of watts, and the energy produced by them, expressed in watt-seconds, in kilowatt-hours, or multiples. In order to avoid the use of very large numbers, the units chosen here for the description of big power plants and their production are

$$\mathrm{MW} = \text{megawatt} = \text{one million watts}$$

for the power and either

$$\mathrm{MWh} = \text{megawatt-hours} = \text{one thousand kWh}$$
$$\text{or } \mathrm{GWh} = \text{gigawatt-hours} = \text{one million kWh}$$

for the energy produced (or consumed) in a given period. The conventional unit of the tension in transmission lines is kV = kilovolts = one thousand volts.

The term 'installed capacity' (not to be confused with the synonymous word 'capacity' of a condenser as used in electrostatics!) of a power plant, or a group of power plants, means the sum-total of the powers of all the generators of the plant or the group. It is the maximum power which can be delivered when all the generators are running with full load. This, however, may happen only during the hours of peak consumption, while at other times of the day or the night they are running at part load only.

Fig. 48, reproducing load curves recorded by a Municipal Electricity Supply Undertaking, gives an example of the variation of load with time. The diagram (*a*) gives the load curve of a single day, and (*b*) the course of the daily average load during a year. The ordinate of each point of the curve represents the power— that is, the energy supplied per unit time—and the shaded area below the curve gives the energy supplied within a day in diagram (*a*) and within a year in (*b*). If the plant operated all the time with the full power of its installed-capacity of P MW the energy delivered in N hours would be NP MWh. The actual energy production is, however, only nP MWh, with $n < N$. The ratio

$$f = n/N \quad\quad\quad\quad\quad\quad\quad\quad\quad\quad\quad\quad (26)$$

is called the *load factor*. It may be expressed also with reference to Fig. 48 by the fraction

$$f = \frac{\text{shaded area}}{\text{area of rectangle } ABCD}$$

Applying this relation to Fig. 48 (*b*) we get the yearly average of the load factor. Making the rectangle *AEFD* equal to the shaded area below the load curve—that is, equal to the annual energy production —the distance *AE* represents the number n, which is called the number of *use hours of the installed capacity*, while

$AB = N =$ number of hours in the year = 8760.

From equation (26) it follows

$$n = 8760 \, f_y \quad\quad\quad\quad\quad\quad\quad\quad\quad\quad\quad (27)$$

with f_y representing the average load factor of the year. The instantaneous value of the load factor is simply the ratio between the actual power delivered at the given moment and the full power of the installed capacity. As can be seen from Fig. 48, the instantaneous load factor is subject to hourly changes, and its average over periods of days, weeks, or months is subject to seasonal varia-

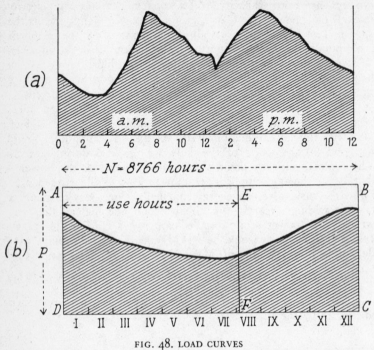

FIG. 48. LOAD CURVES

(*a*) diurnal variation of load; (*b*) seasonal variation of daily load averages.

tions. The monthly average f_m of the load factor for a specific supply net can be found by adding the readings for the month of all electricity meters in the net, and dividing the sum—which represents the monthly energy consumption—by the product of the installed capacity of the supplying plant and the number of hours in the month. Assuming, for instance, that the power plant feeding the net has an installed capacity of 50 MW = 50,000 kW, and the consumption during a thirty-day month to be 16·2 GWh = 16,200 MWh, we obtain for the load factor

$$f_m = \frac{16,200}{50 \times 720} = 0·45 = 45 \text{ per cent}$$

and the number of use hours during this month will be $n = 324$.

The installed capacities of all plants and the total electricity production of thirteen countries in 1950 are given in Table 10, based on data contained in the *Statistical Year Book No. 6* of the World Power Conference.

TABLE 10. INSTALLED CAPACITIES AND ELECTRICAL ENERGY
PRODUCTION IN 1950

Country	Installed Capacity (MW)	Energy Production (GWh)	Load Factor (f_y)	Use Hours
Ireland	265	926	0·399	3,500
Turkey	335	790	0·269	2,360
Austria	1,914	6,365	0·383	3,360
India	2,301	6,574	0·339	2,970
Australia	2,330	8,402	0·413	3,620
Switzerland	2,976	10,479	0·507	4,440
Norway	3,296	17,675	0·612	5,360
Italy	7,488	24,681	0·376	3,300
France	9,230	33,319	0·414	3,630
Germany	10,505	42,241	0·459	4,020
Japan	10,535	44,890	0·436	4,260
Great Britain[1]	15,083	51,911	0·443	3,880
U.S.A.	82,415	387,924	0·538	4,720

It is desirable to make the load factor as large as possible, because a great part of the costs of the electric current is due to the interest and the amortization rates of the capital invested in the installation. Higher load factor permits better economy of energy production, and in many places, therefore, cheaper tariffs are granted for the night current in order to avoid depressions in the load curve. Among the countries listed in Table 10 Norway has at the same time the highest average load factor and the highest percentage of hydro-electric production. This will be more fully discussed in Chapter XII.

The Rôle of Electricity as a Distributor of Power

Most of the world's sum-total of mechanical work, apart from road traffic, is delivered by electric motors at the present day. It may

[1] Only electricity-supply undertakings, without industrial establishments and electric railways.

appear surprising that in spite of this fact scarcely anything will
be said in this book about electric engines. There are two reasons
for this apparent neglect of an important factor. One is the fact
that electric motors are not *prime movers*; instead of *producing*
power they are *reproducing* it by retailing mechanical energy which
once before has been produced wholesale from fuel or water-power.
The other reason is similar to the reason for more being talked and
written about naughty boys and difficult children at school than
about the good ones. In contrast to all heat engines, with their
poor efficiency, there are no radical improvements to be expected
in electric motors or dynamos. The efficiency of well-built engines
of that kind is well over 90 per cent, so that it is hardly worth
while troubling about the few per cent that might still be gained.
They are running with less noise and rocking than the heat engines,
so that all that could be gained in the field of mutual conversion
of electrical and mechanical energy is of much less importance
than the possible gains in the production of electricity and of mechanical work through conversion into these from heat or chemical
energy.

The main rôle of electricity in power economy is its function as a
distributor of power. Since the development of the technique of the
three-phase alternating current, with its relatively simply built and
efficient engines, the methods of power-supply began to change
completely. In the earlier primary stage of the machine age the
power-supply of industry was fully decentralized, each factory, and
in some instances even each workshop, producing the power for
its lathes, spindles, etc., in relatively small steam units of low
efficiency. This system has been abandoned in the course of time,
and has been replaced by centralization of the production of electricity in larger and more efficient power-plants, and by decentralization of the reconversion of electricity into mechanical work.
The rotating drive-shafts along the walls and ceilings of the work-
shops to which the single lathes or other machines were coupled
by transmission belts have more or less disappeared, and have been
replaced in modern installations by individual electric motors, each
driving its single unit.

Electric Power Transmission

Electricity is therefore quite a unique means of distributing centrally produced power over an arbitrarily large number of smaller
and greater motors or other energy-consuming devices. Apart from

the retailing of energy within the relatively small area of a factory or even a town, electricity is used as well for transmitting power to greater distances. Electric power transmission made its debut during the first Electric Exhibition at Frankfurt-am-Main, in Germany, in 1891, where motors were driven and electric lamps were fed from a current generated by a water-power plant at Lauffen-am-Neckar, about 110 miles away. The possibility of transmitting electric power by means of high-voltage alternating currents (A.C.) opened the door for full-scale exploitation of water-power. The use of water-wheels as a source of mechanical energy was already known in antiquity, but the amount of work obtained from this source was very modest, because it could be consumed only within the immediate neighbourhood of the river or waterfall. Thus mills and iron forges were built on the borders of rivers, using quite a tiny part of the energy of the flowing water, but nobody saw a method at that time of employing water-power for requirements of a big factory situated at a distance from a river.

A change came after 1891, when the Lauffen-Frankfurt experiment showed that by making the détour via electricity the power generated by water-wheels could be distributed over a fairly large area. The expectation of making cheap water-power available on a much larger scale than before was a strong stimulus for technical improvements both on the mechanical and the electrical side of the problem.

One was the replacement of the old, clumsy water-wheels by modern turbines, which will be described in Chapter XII. Another is the use of alternating currents, which can be transformed to higher or lower voltages- as required. Thus the current produced in the A.C. generators of the power-station at a voltage of about 10,000V, or 10 kV, is stepped up to 110 kV, or even 220 kV, in a transformer, is then transmitted through the high-voltage line, and afterwards stepped down again to a safe value by means of another transformer at the point where the power is to be used.

The reason for employing high-voltage transmission lines has been explained at p. 142 in connexion with equation (25), which shows that the relative transmission losses can be reduced to one-quarter by using twice the voltage. Improvements of the insulators made within the last half-century permitted step by step the use of ever higher tensions in the transmission lines. From 16 kV, used with the Lauffen-Frankfurt line in 1891, the tension was gradually stepped up to 50 kV at the beginning of our century, to 110 kV in the twenties, then to 220 kV in the thirties, when a stage was reached which appears to be quite a reasonable standard for distances up to about three hundred miles.

A further voltage rise up to 400 kV is more or less in an experimental stage; there is little doubt, however, that about 400 kV will be the standard voltage for future long-distance lines. Even with the existing 220-kV lines it proved to be reasonably economical

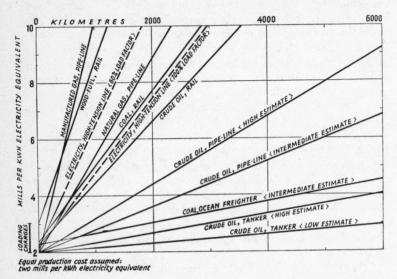

FIG. 49. ENERGY TRANSPORTATION COSTS GIVEN IN MILLS PER KWH

A mill is one-thousandth of a U.S. dollar, and according to the December 1954 exchange rates one penny is worth 11½ mills.

From " Energy Resources of the World," by courtesy of the U.S. Department of State.

to feed electric power which was generated in Vorarlberg, in the Austrian Alps, into the Rhineland grid in Western Germany, over a distance of 300-odd miles.

Power transmissions over distances of 400 miles and more could be made economic either by using higher voltages or by using wires with a greater cross-section, in order to reduce their electric resistance. Owing to its low operating and maintenance costs electric-power transmission is certainly the ideal system for transporting energy within an area of, say, 50 miles diameter. How costly it is, however, at long distances can be seen from Fig. 49, taken from the Guyol Report, comparing the energy transportation costs of electric power with that of fuels by different transportation means. While the saving of loading costs makes electric-power transmission over small distances distinctly cheaper than fuel transportation, the curve representing electric-power transmission costs rises much more steeply with increasing distance than, for instance, that of overseas transport of coal or oil. The distance at which the price of the energy unit (either directly supplied by electric power or by its equivalent amount of fuel) is twice that at the source can easily be derived from Fig. 49, and is given in Table 11 below in round figures:

TABLE 11. DISTANCES DOUBLING THE PRICE OF ENERGY

	km
Wood-fuel, rail	165
High-voltage line (60 per cent load factor)	420
High-voltage line (100 per cent load factor)	700
Natural gas, pipe-line	580
Coal, rail	581
Coal, ocean-going freighter	4,500
Crude oil, pipe-line (high estimate)	1,700
Crude oil, pipe-line (intermediate estimate)	2,540
Crude oil, tanker (high estimate)	5,800
Crude oil, tanker (low estimate)...	12,400

It may appear surprising that pumping oil through pipe-lines is a cheaper method of long-distance energy transportation than sending electricity through wires. One might be tempted to argue quite reasonably that electricity is flowing by itself, while the flow of the oil must be forced by pumps consuming energy and causing installation investments. Besides, the expenditure of material seems to be much less in the case of the electric transmission lines, with their thin wires, as compared with a pipe-line like that between Kirkuk, in Iraq, and Banias, in Syria, with pipes of 30 inches diameter, 555 miles length, and a weight of 165,000 tons. The apparent paradox can easily be cleared up by closer numerical considerations. Transportation of fuel is equivalent to power transmission when we replace the fuel delivered per unit time by its electricity equivalent, which, according to Table 1, is 2470 kWh per ton of mineral oil. (This holds under the assumption of an over-all efficiency of 20 per cent of the conversion of heat into power.) The flow of the oil in the pipe-line is about 2000 tons per hour, which corresponds to an energy flow of 2000 × 2470 kWh/h, equivalent to a power transmission of approximately 5 million kW.

No electric transmission lines handling an amount of power such as this have ever been built, although in the course of future development such lines might be installed.

It is very instructive to compare the ' power ' of the Iraq pipe-line, covering a distance of 555 miles = 893 km, with that of existing electric transmission lines. The supply-line from the Austrian water-power plants to Vienna transmits a power of 100 MW = 0·1 million kW at a voltage of 220 kV, using the thickest of the standardized steel-cored aluminium cables, with 340 square millimetre cross-section. If this line, instead of transmitting only 100 MW, were to be adapted for carrying 5000 MW—that is, carrying fifty times the present power—to the greater distance of 893 km, each of the three phases would require a lead bundle consisting of fifty such cables, and even then the losses would amount to nearly 20 per cent of the power transmitted. The electric transmission lines are, therefore, comparable to leaking pipe-lines with losses which are far greater than those caused by the energy consumed for pumping oil through the pipe-lines. Certainly a part of the oil sent from Kirkuk to Banias has to be used to drive the engines in the seven pump-stations, but this fraction consumed for pumping is hardly more than 1 per cent of the oil transported, while in an equivalent electric-power transmission line over the same distance, consisting of three times fifty wire ropes of the biggest conventional size, the losses due to Joule's heat would amount to nearly 20 per cent.

It would, of course, be possible to design on paper, regardless of costs, high-voltage transmission lines for 5000 MW over nearly 900 km distance with a loss as low as 1 per cent of the power. But such a project could scarcely ever be taken seriously into consideration on account of the enormous investments involved. Considering the fact that copper leads are much more expensive than aluminium ones, the following estimates of a 5000-MW, 220-kV transmission line over a distance of 900 km are made under the assumption that aluminium is the conductor material. Using formula (25), and inserting

$$R = sl/q \qquad \ldots\ldots\ldots \ldots\ldots\ldots\ldots\ldots\ldots\ldots (28)$$

with $s = 28\cdot5$ ohms.mm^2/km as the specific resistance of aluminium and l and q denoting respectively the length (in kilometres) and the cross-section (in square millimetres) of the lead, or bundle of leads, of a single phase, we find that in order to reduce the losses to 1 per cent of the power—that is, to make $p = 0\cdot01$—we should have to use a total conductor cross-section of about three and a half square feet for each phase. This would result in a weight of approximately 2·4 million tons of aluminium for the 900-km line.

The full implications of such enormous requirements of conductor material can be realized best by comparing the energy needed for manufacturing this amount of aluminium with the annual energy losses in the transmission line. The price we have to pay for aluminium is largely dictated by the energy consumed in its production. The conversion of alumina to aluminium metal by the commonly used electrolytic process requires about 20,000 kWh per ton of the finished product, and, adding another 5000 kWh for the preparations of alumina and other ingredients used in the process, we obtain a total of 25,000 kWh, representing the energy equivalent of one ton of aluminium. Denoting by m the weight, expressed in tons, of the conductor material of a transmission line, and by K its total energy equivalent, we get

$$K = 25{,}000m \text{ kWh} \dots\dots\dots\dots\dots\dots(29)$$

The total energy equivalent of our 900-km project is obtained by inserting in this equation $m = 2\cdot4$ millions, which gives $K = 60{,}000$ GWh.

The datum which interests us in this connexion is the ratio between K and the annual loss. The total energy transmitted yearly is given by nP, with n the number of use hours and P the full power for which the line is built. With p representing the relative loss, the total annual loss is given by

$$L = npP\dots\dots\dots\dots\dots\dots\dots\dots(30)$$

We denote the ratio which we are interested in by α, putting

$$\alpha = K/L = \frac{25{,}000m}{npP}\dots\dots\dots\dots\dots\dots(31)$$

Inserting $K = 60{,}000$ GWh, $n = 5000$ hours, $p = 0\cdot01$, and $P = 5$ GW, we find

$$\alpha = 240$$

which means that the energy equivalent of the conductor material would be equal to the energy lost in 240 years. Considering, moreover, that the total installation of the line costs twice or three times as much as the energy equivalent of the conductor material alone, we find that the capital investment for such a project would be much too high. It may be mentioned incidentally that with copper or any other suitable metal as conductor the ratio of investment to losses would be higher still.

The consequence is that electric-power transmission lines cannot compete with oil pipe-lines as regards distance, power transmitted, and smallness of transmission losses. While, therefore, the conversion of electrical power into mechanical energy and *vice versa* can be done with far greater efficiency than the conversion of heat into mechanical work (about 95 per cent efficiency, or more in the case of electricity, and between 20 and 40 per cent in the case of heat), the efficiency of long-distance fuel transportation is decidedly greater than that of electric-power transmission.

In transporting power by electricity we have, therefore, to put up with larger relative losses, or, expressed in terms of the ratio between investment and loss, with smaller values of α. As a matter of fact, in existing transmission lines the value of α, instead of lying above 200, as in the example given just now, is actually less than unity. In the 220-kV transmission line of the Austrian grid, for instance, with 340 mm^2 aluminium cross-section in each phase, the relative loss over a distance of 160 km with $\cos\theta = 0.9$ is 3.4 per cent, so that assuming $P = 100$ MW and $n = 5000$ h, the annual loss is 17 GWh, while the energy equivalent of the aluminium of the line is $K = 11$ GWh. Therefore

$$\alpha = \tfrac{11}{17} = 0.647$$

A low energy ratio like this represents the opposite extreme to exaggerated high values of α. For doubling the line—as planned already—by installing a second equal lead for each phase would reduce the resistance to one-half of its present value, and would therefore reduce the losses from 3.4 to 1.7 per cent. The saving achieved in this way would suffice to compensate within sixteen months the energy equivalent of the additional expenditure of aluminium. And, even taking into account the fact that the additional installation may cost a multiple of the energy equivalent of the second cables, the doubling of the line will be paid for by the savings within a few years. Generally speaking, most of the existing long-distance transmission lines are still ' under-developed,' but some time in the future they will be reinforced to meet the requirements of better power economy.

From the standpoint of far-sighted economic planning the optimum value of the energy ratio might be somewhere near $\alpha = 2$, and by combining amply dimensioned conductors with further increase of the voltage the distance over which currents can be transmitted efficiently might be extended beyond the present limits. If we express the lead resistance R in terms of the ratio α, and substitute the result in equation (25), we obtain

$$p = \sqrt{\frac{3\,k\rho s}{n}} \cdot \frac{l}{\sqrt{\alpha}} \cdot \frac{1}{E\,\cos\theta} \quad\dots\dots(32)$$

with k denoting the specific energy equivalent of aluminium ($k = 25{,}000$ Wh/kg), ρ and s denoting its specific weight and specific resistance, l the length of the line in km, α the energy ratio, and E the line voltage in volts. Assuming $n = 5000$ hours and $\cos\theta = 0.9$, we get

$$p = \frac{37.8}{\sqrt{\alpha}} \frac{l}{E} \quad\dots\dots\dots\dots\dots\dots\dots\dots(33)$$

The variation of the relative loss p with the distance l is plotted in Fig. 50 for large, but still reasonable, values of α and E.

The diagram shows clearly that 110-kV lines with relatively poor conductor material, as, for instance, $\alpha = 0.5$ or less, are not suited for transmission over distances beyond 100 or 200 km on account of the large losses involved. With 400 kV, however, and $\alpha = 2$ or more, distances up to 1000 km could be bridged with just a few per cent losses. Hence the combined losses of converting water-power into electricity, transmission over distances of 1000 km or so, and reconverting into mechanical energy will still be low compared with the losses in converting heat into power. This means that as soon as the technique of 400-kV transmission lines is sufficiently mastered to be in general use, and as soon as economic conditions would permit making of adequate investments for reducing losses, the range of supply from big water-power plants might be extended to distances up to 800, or even 1000, kilometres. It is unlikely, however, that such large distances will be covered without transition to high-tension D.C. lines, as described in the following section. Besides, the rapid progress pushed on by the international " Atoms for Peace " movement may result in making most projects of super-long-distance power transmission illusory. For, although nuclear power will scarcely ever be competitive with power from well-situated hydro plants, it may still become more economic than hydro-power transmitted over a line of several hundred miles.

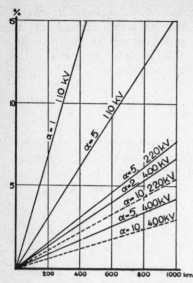

FIG. 50. RELATIVE LOSSES IN LONG-
DISTANCE HIGH-VOLTAGE LINES
$n = 5000\text{h}$; $\cos\theta = 0\cdot90$.

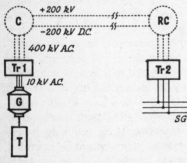

FIG. 51. HIGH-VOLTAGE D.C. TRANS-
MISSION LINE

High-voltage D.C. Lines

If, however, from any reasons whatsoever, the need for building very long transmission lines should arise, the problem might be solved ultimately by D.C. lines. By means of mercury-vapour rectifiers it is possible to convert A.C. currents into D.C., and *vice versa*. Thus after transforming alternating currents to high tension they can be converted into high-tension direct currents, transmitted over the long-distance line, reconverted into A.C., and then stepped down again to the required working voltage. The layman might rightly ask, what is the use of such complications? The advantage of using high voltages, and therefore the necessity of transforming, is clearly shown in Fig. 50. But why more costly devices just for having the transmission lines fed with D.C. instead of A.C.? In Fig. 51 the unbroken lines show the conventional sort of high - voltage transmission, while the dotted lines represent the supplementary inventory for D.C. transmission. Certainly the installation of a converter and reconverter is costly, and involves in addition some loss of power. But three important advantages are gained. One is the avoidance of the phase angle θ, with its cosθ increasing, according to equation (25), the relative losses. The second advantage lies in the fact that the requirements of breakdown strength of the insulators are much less in the case of D.C. lines than with A.C. lines of equal voltage. The insulators must be dimensioned to withstand the peak voltage against earth. In a three-phase 400-kV line the nominal value of the voltage, 400 kV, means the effective value of the line voltage E_L. The voltage against earth is the phase voltage E_f, given according to equation (20) by

$$E_f = \frac{1}{\sqrt{3}} E_L$$

But its peak value is

$$E_{f\ max.} = \frac{\sqrt{2}}{\sqrt{3}} E_L = 0.817\ E_L$$

Hence with $E_L = 400$ kV the insulators should safely withstand a voltage of 0.817×400 kV $= 326.8$ kV, while in the case of a D.C. line of the same voltage each of the conductors will have a voltage of only plus or minus 200 kV against earth, so that the requirements as to insulating strength of the armature are considerably more modest. The third advantage is a reduction of conductor material, because instead of three leads only two are necessary. Here are the expressions for the relative losses in both kinds of transmission lines:

Three-phase A.C.   *D.C.*

$$p = \frac{PR}{E^2\cos^2\theta} \qquad p = \frac{PR}{E^2} \quad \dots\dots\dots\dots\dots\dots(25)$$

$$p = \sqrt{\frac{3k\rho s}{n}} \cdot \frac{l}{\sqrt{a}} \cdot \frac{1}{E\cos\theta} \qquad p = \sqrt{\frac{2k\rho s}{n}} \cdot \frac{l}{\sqrt{a}} \cdot \frac{1}{E} \quad \dots\dots(32)$$

and particularly with $n = 5000$:

$$p = \frac{37.8}{\sqrt{a}} \frac{l}{E} \qquad\qquad p = \frac{27.75}{\sqrt{a}} \frac{l}{E} \quad \dots\dots\dots\dots(33)$$

(with $\cos\theta = 0.9$)

According to these equations, the loss in a 400-kV line over a distance of 1000 km with a conductor expenditure corresponding to $a = 2$ would be reduced from 6.7 per cent to 4.9 per cent by using D.C. transmission, and at the same time insulators for only 200 kV, instead of about 330 kV, would be needed. We must not forget, however, the other side of the balance-sheet, on which appear the losses caused in the twofold process of converting and reconverting A.C. and D.C. The question whether the difference between gains and losses through transition to D.C. will be large enough to justify the additional complications involved is not yet settled. It may be possible that the full advantages of high-voltage D.C. transmission will make themselves felt in future lines over distances which have not so far been spanned.

During World War II, when Norway was occupied by the Germans, projects were made for supplying electric power to the Rhineland and the Ruhr area from the big Scandinavian water-power plants about 1000 km away. These projects were based on the use of 400- or 440-kV D.C. lines. Although they have been dropped in the meantime, the problem of wider distribution of electric power from rich sources will arise again with ever-growing importance, and it is quite possible that then the high-voltage D.C. technique will play an important rôle unless decentralized production of electricity in atomic power plants proves to be more economic.

Specific Uses of Electricity

We have dealt so far with the rôle electricity plays in distributing power which has been centrally produced. As a matter of fact, of all the uses of electricity, mechanical work delivered by electric motors and heat of electric stoves, ranges, furnaces, etc., represent the most significant items in the balance-sheet of electrical power. Still, most of the heavy mechanical work could be done in principle without the détour via electricity as well; the use of electric motors and heating devices is just a more convenient way of performing useful work which might also be done in other ways.

There are, however, many other services to modern civilization which could never have been performed without electricity. The most striking examples are our modern methods of keeping our homes and streets bright at night, and, in addition, the different methods of telecommunication and mass communication—telegraph, telephone, radio, television, and sound film. Among other specific uses, which are less popular, but economically important, electrolysis ought to be mentioned in the first place, because the technique of light metals on which modern aviation is based has its origin in the electrolytic process of making aluminium. Electricity plays further a basic rôle in modern health services; we may mention only X-ray diagnosis, X-ray therapy, galvanization and faradization, diathermy, short-wave therapy, and ultra-violet radiation among the direct applications of electrical apparatus in medicine. Apart from that, modern chemotherapy, with its efficient medicaments like the sulfonamides or penicillin, streptomycin, and others, could never have reached its level of development without all the electric measuring and testing instruments used in chemistry.

As pointed out at the beginning of this chapter, electricity is, therefore, quite apart from the enormous amount of mechanical work performed by electric motors, a vital factor in modern civilization. It is a characteristic feature, however, of the manifold variety of applications of electricity that among all the above-mentioned specific

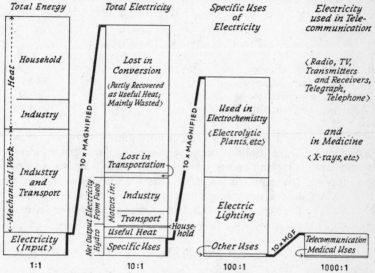

FIG. 52. REPARTITION OF ELECTRIC POWER AMONG ITS DIFFERENT USES

uses of electricity—important though they may be from the sociological and cultural standpoint—not a single one represents a significant item in the total energy balance-sheet of mankind. Or, to put it in other words, in terms of fuel or power economy, all these benefits we get from electricity are quite cheap, and add only a small additional expenditure to the vast global consumption of energy for heat and mechanical work. In Fig. 52 the global repartition of electric power among its different uses is shown graphically. The specific uses which consume only a small percentage of the total production are redrawn on the right side of the figure in a tenfold magnification, and it can be seen from this detail that among the specific uses lighting and electrolysis are the greatest consumers of power, while the consumption for telecommunication or medical uses is comparatively quite small.

The knowledge about the tiny amount of power consumed in radio transmitters may serve to destroy some widespread misconceptions about the influence of radio waves on the climate. In the following table the total energy transmitted within a year is compared with the total annual electricity production, the total human energy production, and the total annual energy income which the earth receives from solar radiation. The data are given in approximate round numbers, expressed in TWh. We note again that 1 TWh = 10^{12} Wh = one thousand million kilowatt-hours.

TABLE 12. ANNUAL ENERGY

	TWh
Radiation from all radio transmitters	0·3
Total electricity production	1300
Total energy production of mankind	24,000
Global income from the sun	900,000,000

The solar radiation penetrating our atmosphere is not only more than two thousand million times stronger than the power transmitted from all radio stations on the earth, but consists also of specifically much more efficient radiations, like light rays and ultraviolet. Compared to the sun and its mighty influence on the climatic conditions of the earth, all our apparently strong transmitter stations are infinitesimally tiny dwarfs.

Fuel Cells

More than 99·99 per cent of all the world's electric power is generated in dynamos driven either by heat engines or water-wheels. Only a tiny fraction of electricity—namely, that used in electric torches, portable radio-sets, etc.—is contributed from primary voltaic cells. In our conventional dry cells chemical energy, in particular that used in synthesizing the compound $ZnCl_2$ from its constituent elements zinc and chlorine, is directly converted into electric energy. With zinc a rather expensive fuel, and a considerable expenditure on costly additional material, like the manganese dioxide used as a depolarizer in the cell, the costs of electric power derived from dry cells are of the magnitude of about £4 per kWh—that is, several thousand times the rate charged by our electricity undertakings.

If, on the other hand, a direct conversion into electricity of the chemical energy of conventional cheap fuels like coal or a hydrocarbon could be realized in a ' fuel cell,' we might expect to obtain higher efficiencies than in conventional steam plants. The detour via heat involves the well-known losses of 65 to 80 per cent caused by thermodynamic laws (see Chapters I, III, and IV). In order to avoid the détour we must try to perform the oxidation of coal or of hydrogen in the form of a cold combustion, in which the energy released in each single elementary process is not used for imparting kinetic energy to the molecules, but is turned into electric energy. If we succeed in doing so with the well-known reactions

$$C + O_2 = CO_2$$

or
$$O_2 + 2H_2 = 2H_2O$$

then the heat losses might be avoided, or at least considerably reduced.

This problem has been a challenge to physico-chemists for many years; it was more than a century ago that the first tentative experiments were made in England. **Davy,** the teacher of Faraday, tried to use the oxidation of coal as the source of a galvanic cell, and **Grove** devised in 1839 a battery in which the cold combustion of hydrogen, with platinum as a catalyst, is used as the energy source. However, with the knowledge available at that time the economy achieved with these cells was so slight that they could not even compete with the conventional galvanic batteries using zinc cathodes (and still less, of course, with the dynamo generator).

Although some minor improvements over Grove's first cell were reported during the nineteenth century, no real success was scored, and none of the numerous patent specifications on fuel cells could be used commercially.

Recent development was started in 1933 by Baur and Tobler in Switzerland, and after the Second World War systematic work of specially trained teams was begun in several countries, with the ultimate aim in view of devising fuel cells for large-scale power production.

Unfortunately, it is not possible to make fuel cells of such a simple construction as the conventional dry cell, in which the cathode (negative electrode) is in itself the reacting substance supplying the electric energy. In contrast to zinc, carbon does not form ions (electrically charged atoms), and therefore solid carbon as the fuel of a cell can be used only in combination with a separate electrolyte in contact with each electrode. The so-called Redox cell consists of two different electrolytes separated by a semi-permeable diaphragm, one of them circulating between the negative electrode and fine-grained coal, which is oxidized during the process. The other electrolyte circulates between the positive electrode of the cell and a vessel through which the air is bubbling. Certain ion-forming reactions take place in both electrolytes, with the net result that the oxidation of the carbon and hydrogen contained in the coal maintains a voltage between the electrodes, while a current flows through the load in the external circuit.

In spite of the considerable efficiency of the process, it is doubtful whether the Redox process can be made economical enough to compete with the steam plants, because the output of the cell decreases rapidly before half of the coal has been oxidized. Pretreatment of the fuel by an additional process would probably more than compensate for the theoretical advantage of higher efficiency.

Greater economy might be gained by the use of gaseous fuel, like CO or H_2, as is done in various other types of fuel cells. The fuel gas under pressure is absorbed in the porous negative electrode, and air or oxygen in the positive electrode. The oxidation process occurs indirectly by the means of intermediary reactions taking place in an electrolyte between the electrodes. Suitable catalysts are required to keep the reaction going. The output of the cell can be increased by operating at high temperatures and pressures. In some cases cells using fused electrolytes are operated at temperatures between 500° and 900°C. and pressures up to about 50 ata.

In the following list the performances of some newly constructed cells are described by giving the voltage obtained at different current densities.

Redox Cell

Current density (mA/cm²)	0	10	20	50
Voltage (volts)	0·91	0·62	0·51	0·20

In 1947 O. K. Davtyan in Russia constructed a cell operating on hydrogen or carbon monoxide at 700°C. which gave 1 V on open circuit and 0·75 V at 30 mA/cm². The fuel efficiency is claimed to be 80 per cent.

E. Justi in Germany developed a hydrogen-oxygen cell operating at 90°C. and a gauge pressure of 1·5 atmospheres which is reported to yield 700 mA/cm² when short-circuited—that is, at zero voltage.

A high-temperature cell using carbon monoxide as a fuel was developed at Birmingham. At a temperature of 820°C the output was:

Volts	1·02	0·84
mA/cm²	0	33

A report on " Recent Research in Great Britain on Fuel Cells " has been given by F. T. Bacon and J. S. Forrest to the Fifth World Power Conference (1956). Following an initial effort made by the Electrical Research Association at Cambridge in 1946, the Central Electricity Authority decided to conduct experimental research in this field, with two aims in view. The report says:

> Broadly speaking, research has followed two lines, depending on the ultimate practical application. In the first place the aim is to develop a cell suitable for bulk electricity generation, using coal, coke, or industrial gases as the fuel. Secondly, research has been directed toward the development of a cell suitable for energy storage and special applications such as traction. Pure gases could be used in this case and the functions of generation and storage combined. Plants of this type used in sufficient capacity would lead to an improvement in the load factor of the supply system.

Experimental work was carried out with solid-fuel cells of the Redox type, and with both low-temperature and high-temperature cells using carbon monoxide or hydrogen as a fuel. As far as can be judged from the present state of investigations, the possibility of large-scale power production from coal, using fuel cells instead of thermal plants, seems to be remote as yet. Efficiency of conversion of chemical into electrical energy, though of first-rate importance, is only one of a number of factors determining the economy of a power plant. The cost of the plant per kilowatt installed capacity, maintenance costs, and reliability of performance must be taken into consideration as well. Experiments made with a low-pressure cell of the Davtyan type showed that two cells, suitably adapted for longer use, could be operated intermittently over a period of nineteen months, but that at the end of this period their voltage had fallen from 0·75 to 0·2 volts.

The most promising success has been scored at Cambridge, in experiments jointly financed by the Electrical Research Association and the Ministry of Fuel and Power. With a high-pressure hydrogen-oxygen cell the following performance was obtained at 200°C. and 42 atmospheres pressure:

Current density (mA/cm^2)	0	10	100	250	500
Voltage (V)	1·10	1·02	0·905	0·805	0·677

The progress made can be seen by comparison with the figures of the Redox cell, and by considering that the current density of the conventional lead accumulator is only 10·50 mA/cm^2 at 2 volts.

The power output per unit of internal volume of the high-pressure cell is estimated at 300 kW per cubic metre, and a power battery including fuel and fuel containers (steel cylinders) would weigh 7 kg per kilowatt-hour of stored energy.

This means that the high-pressure hydrogen-oxygen cell is less bulky and heavy than a lead accumulator of equal capacity. Besides, reloading of the fuel cell can be done much quicker by merely re-filling the storage cylinders. Compared with the internal-combustion engine, however, the cell is inferior with respect to power per unit weight, but superior in its fuel efficiency.

A promising field of application may be found in the combination of the hydrogen fuel cell with devices for using solar energy. If the experiments for obtaining hydrogen and oxygen by photo-chemical decomposition of water under exposure to sunlight should be successful, a combined plant producing hydrogen and using it for generating electric power in fuel-cell batteries might overcome the difficulties of energy-storage encountered in other solar power plants. This will be discussed in Chapter XIII.

CHAPTER VIII

Coal

The Share of Coal in Power Economy

COAL has been the main source of power since the advent of the machine age, and it is still maintaining its dominant position in world power economy in spite of the rapid increase of oil consumption for motor-fuel and of water-power for generating electricity.

Even long before the invention of the steam-engine in the eighteenth century coal was the key which opened the door to the development of our material civilization. It was the use of coal which made glass cheap enough to be in general use for windows in every home, when in the Middle Ages only the rich could afford the luxury of glass windows. And it was coal as well which enabled men to turn iron ore into iron and steel on a large scale, thus producing steel in sufficient quantities to build big machines and ships.

Besides being the chief source of power, coal is a very important raw material. Since the American War of Independence, when supplies of pitch from America for British shipbuilders were cut off, people took an interest in processing crude tar, which was then a more or less unwelcome by-product of coke-making. Later, from the second half of the nineteenth century, this tar-chemistry made rapid progress: most dyes used to-day, innumerable drugs and germ-killers, and, in addition, margarine, plastics, fibres, liquid glues, cosmetics, fertilizers, and a hundred other useful things are all derivatives of coal and coal products.

Thus a very considerable part of the progress of our civilization is based on the use of coal as fuel and as a raw material, so that a sudden stop in global coal production would cause a catastrophe. How long will it be till all our coal resources are used up, and how will civilized nations maintain their way of life after that date? Are there any other sources of energy on which we can fall back after having exhausted all coal reserves?

An illustration of the relative contribution of the main sources of energy—namely, coal, oil, and water-power—to the global energy supply can be given by comparing the energy available in the form

of the combustion heat of a year's global output of coal and oil
with the annual energy production of hydro-electric plants. It may
be mentioned in this connexion that in spite of the widespread use
of coal as a raw material it is only a very small percentage of all
the coal mined that is turned into the derivatives mentioned above,
the greatest part being used as a fuel.

As can be seen from Fig. 6, Chapter II, in some countries, like
Great Britain and Germany, about 90 per cent of the entire energy
consumption is supplied from coal, while the global average is about
60 per cent. (It must be noted that the quotas of the different energy
sources represented in Fig. 6 are given in electricity equivalents. If
we consider the heat value, instead of the electricity equivalent, the
share of water-power—represented by the blank sectors—is reduced
to one-fifth of its relative amount. The global quota of coal is,
therefore, somewhat greater than the black sector of Fig. 6 *d*.)

Half a century ago the preponderance of coal among the
sources of energy was still more marked than it is to-day. Fig 53
shows the annual contribution of the three most important energy
sources—coal, oil, and water-power—in the years 1913, 1926, 1938, and
1950. The data given there are only approximate, for two reasons:
no official production figures have been published lately from some
of the Eastern countries, so that we have to work with estimates
extrapolated from data recorded in the past. Secondly, even if the
total quantities of coal and oil production in tons or barrels were
given exactly, the conversion of
these data into energy supply
could be done only approximately, because the heat value
of different sorts of coal varies
between wide limits. Numerical data have been given in
Table 2, Chapter I. In computing the annual energy supply
contributed by coal and oil, as
represented by the vertical
columns of Fig. 53, an average
value of 6900 kcal/kg has been
assumed for coal and 9500
kcal/kg for oil.

In an analogous figure
representing the electricity
equivalent, instead of the heat

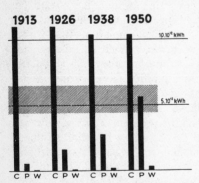

FIG. 53. GRAPHIC REPRESENTATION
OF THE RELATIVE CONTRIBUTION OF
THE THREE MAIN SOURCES OF WORLD
POWER

C = coal, P = petroleum (oil), W =
water-power.

value, the contributions of the two fuels would be reduced to one-fifth of the size given in Fig. 53.

Two striking facts can be gathered from Fig. 53;

1. Although the yearly energy income in the form of coal is still the greatest, it has slightly declined since 1913, while that of the other two has risen tenfold in the same interval.

2. The contribution of water-power, though rising rapidly, is no more than just 3 per cent of that of coal.

In answering the obvious question whether in a distant future coal might be replaced by oil and water-power two more facts must be considered:

(*a*) We could, of course, increase the annual output of oil so that its energy equivalent would surpass that of coal, and a steady increase will doubtless take place during the next years. But the result of this development would be the total exhaustion of the world reserves of oil at a much earlier date than those of coal. With the present trend of coal remaining stationary and oil production doubling every twelve years, the oil resources of the earth will be exhausted in less than a century, and the coal resources in less than a millennium. Hence, instead of lengthening our coal reserves by using oil, within a few decades we shall have to synthesize the petrol needed for cars and planes by hydrogenization of coal.

(*b*) In contrast to the fossil fuels—coal and oil—water-power is not exhaustible as long as the earth is inhabitable at all. So these resources will be all right till the next glacial age. Unfortunately they do not suffice for the increasing demand of world power. As will be explained in Chapter XII, the energy gained from hydro-electric power plants can scarcely be increased beyond a limit amounting to about ten or twenty times the present output, according to the opinion of experts. This means that after having fully developed all economically justifiable hydro-electric projects the energy gained would be between 30 and 60 per cent of the present global energy consumption, and, of course, a much smaller percentage of the demand to be expected in a few centuries. This explains the interest in atomic energy.

The Different Uses of Coal

The methods of coal production—*i.e.*, the art of coal-mining—are fully and vividly described by Dorothy Rowlands in her book. What we are mainly concerned with are the ways of using and consuming coal. The percentage of consumption for household, industry, transport, shipping, etc., varies widely from country to country, and it is hardly possible to give average figures for the whole world. The following table is based on data given by Ruhemann on coal consumption in Great Britain in 1938:

TABLE 13. DISTRIBUTION OF COAL CONSUMPTION IN GREAT BRITAIN IN 1938

Industrial heating + general manufactures	34 per cent
Household heating and cooking	21·9 ,, ,,
Blast-furnaces and iron-works	10·6 ,, ,,
Gasworks	10·4 ,, ,,
Power stations	8·5 ,, ,,
Locomotives	7·1 ,, ,,
Colliery engines	6·8 ,, ,,
Coastal shipping	0·7 ,, ,,

As the author remarks, the coal used for long-distance shipping is not included in the list, as it is counted as exported coal, and not as a part of home consumption.

In most of the uses of coal, heat and/or power is gained by direct combustion. But there are very important branches of coal consumption in which coal is converted first by certain processes into more suitable forms of fuel. Two of these processes—namely, carbonization and gasification—have been used for more than a century, while a third—hydrogenation—is still in its initial stage to-day, and will attain its real importance only as soon as the world oil reserves are exhausted.

Even the direct combustion of coal (without any previous chemical transformation) is preceded by the mechanical process of milling in some cases, and of briquetting in others. The automatic firing of big boilers can be accomplished better by using liquid or pulverized fuels. This fact, together with the possibility of the quicker lading of liquid fuel, explains the preference given to oil-firing for naval craft which, along with nuclear power, is going to replace coal on modern ships. In stationary plants, however, coal-firing is kept up for economic reasons, and the use of pulverized coal is more and more widespread to-day in larger plants.

While pulverized coal is used directly in suitable furnaces and boilers, part of the coal-dust accompanying the output of the mines is pressed, with pitch or tar as a binding substance, into blocks called briquettes, which are a handy and relatively clean material for household use.

Carbonization

Coal-seams are the fossil relics of primeval forests which have been buried under heavy pressure for millions of years. Owing to the composition of the organic substances which were finally turned into coal, the main constituents of coal are carbon and carbon-hydrogen compounds. A small percentage of a series of other elements, including sulphur, nitrogen, oxygen, silicon, and other impurities, is mixed with the main constituents.

Carbonization is a kind of dry distillation at temperatures sufficiently high for boiling off the carbon-hydrogen compounds, while the solid residue, called coke, is enriched in pure carbon. The process is somewhat analogous to that of making charcoal by smouldering wood in kilns under conditions in which air is excluded. As a matter of fact, the technique of carbonization, or coking, was developed originally in the search for a substitute for charcoal, which up to the beginning of the eighteenth century had been the standard fuel for iron production.

Iron ore is mainly a compound of iron and oxygen, and the process of producing iron by smelting is, chemically speaking, a combination of melting and reducing the ore. Reduction means the splitting of compounds; in the case of iron ore it means the removal of the oxygen. Carbon is the element which can act simultaneously as the fuel supplying the necessary heat for melting the ore and as a reducing substance binding the oxygen from the ore by forming carbon dioxide, which escapes into the air. Neither wood nor raw coal—though consisting mainly of carbon—is fit for this purpose, because the fuel + reductor must be smokeless, or the iron is spoilt by fumes and soot. Charcoal, which is wood enriched in pure carbon by evaporating and burning off the hydrogen compounds in the charring process, is an excellent reductor for iron, and it is still used to-day for producing high-quality Swedish iron. Its disadvantage is the high price: to make enough charcoal for smelting a ton of iron involves cutting down three acres of trees. That is the reason why, in spite of a sufficient supply of ores, the extensive use of iron in quantities of many million tons was not possible until the revolutionary idea of subjecting pit-coal to a process analogous to charring wood was put into practice. Experiments made in this direction as early as the sixteenth century had proved to be a failure, and even when they succeeded a century later the results in smelting iron ore were rather poor, because coke as a fuel needs a higher oxygen-supply than charcoal, and the blast-furnaces used up to that time had an insufficient amount of draught. It was only in 1735 that a new tall blast-furnace with a fierce draught, invented by **Abraham Darby**, was able to do the job. Since then the price of iron has come down with a rush—the beginning of the real Iron Age.

While the production of coke was the immediate aim of the first cokeries, it was discovered later on that coking supplied valuable by-products, especially gas, which from the middle of the nineteenth century has been used at a steadily increasing rate for heating and lighting in household and industry. Several different sorts of gas can be made from coal, but when we are speaking of ' gas ' simply we mean gas supplied by the gas-mains. This gas consists of elementary hydrogen (H_2) (about one-half of the quantity of the gas), methane (CH_4) (about one-third), and smaller percentages of ethylene (C_2H_4) and carbon monoxide (CO), which is the poisonous constituent of the substance. Gas soon proved to be a very convenient kind of fuel, because it is clean, smokeless, and easy to handle when supplied by the mains. In addition, it was very commonly used for illumination in houses, as well as in street lamps, before it was replaced by electricity.

The growing demand for gas caused the development of a separate gas industry, and to-day there are in Great Britain alone more than a thousand greater and smaller gas undertakings in which the carbonization of coal is done with the chief aim of producing gas, with coke as the by-product. Hence there are two different industries performing the process of carbonization, with different economic aims and working methods—the gas industry, with its main object of producing high-quality gas and selling their by-product, coke, to any consumers who may need it as a high-quality fuel for stoves, boilers, central heating, etc., and, on the other hand, the coking industry, specializing in producing that sort of coke which is best for the demands of the iron-works, generating at the same time gas as a by-product, without particular regard to its quality, and consuming a part of the gas in its own ovens.

The quantities of coal carbonized in Great Britain can be seen in Table 14, which is based on data given by Ruhemann:

TABLE 14. PRODUCTION AND CONSUMPTION FIGURES FOR GREAT BRITAIN
IN 1938 IN MILLIONS OF METRIC TONS

Total coal production	228
Home consumption	178
Export	50
Gas industry	19
Coking industry	18

Gasification and Hydrogenation

In places where the demand for gas outweighs that for coke it is economic to increase the ratio of gas to coke output. This can be done by the process of *gasification,* which turns a part, or even the whole, of the coke into combustible gases. One method consists of passing hot steam over the glowing coke. When the temperature is properly chosen a thermal dissociation of the water-vapour occurs whereby the molecule H_2O is split into free hydrogen (H_2) and oxygen, which by an incomplete oxidation process combines with the carbon to form carbon monoxide (CO), a poisonous but also combustible gas. The formúla of this process is

$$H_2O + C \rightarrow H_2 + CO$$

The resulting mixture of hydrogen and carbon monoxide is called water-gas.

If, therefore, more gas and less coke is desired the gas output of the carbonization process can be increased by gasification of the residual coke. If necessary, all the coke can be gasified, leaving only a small residue of ash. Another, still more simple method of gasification consists of carrying out the dry distillation of coal or wood not under exclusion of air, but admitting a restricted amount, just sufficient to burn the coal to carbon monoxide, which escapes along with the distillation products. The resulting gas mixture, which, like ordinary coal-gas, but with different ratios of the components, consists of hydrogen, hydrocarbons, CO, and also nitrogen (N_2), is called *producer* gas. Its heating value is only about one-fourth of that of high-quality gas; still, the total energy content of gaseous fuel generated in this way is greater than that obtained by carbonization alone, because much more producer gas than ordinary coal-gas can be obtained from a given quantity of coal.

Producer gas can also be made from wood in special ovens (generators), in which incomplete oxidation plus dry distillation of wood takes place. A mixture of air and producer gas is, of course, explosive, and can therefore be used in ordinary car-motors, instead of the usual mixture of air and petrol-vapour. During the War, when petrol was short, buses—and in the countries of the Axis Powers many private cars also—were fitted with generators making producer gas from wood. But this method provided only a rather poor substitute for petrol, and as soon as the oil shortage was over after the War the bulky generators disappeared from the motor-vehicles.

Table 15 gives the heating values of different sorts of gas. The low-temperature gas, which is the first in the list, is produced by carbonization of coal at lower than the usual temperatures, whereby a kind of cream of all the coal-gases of the most volatile constituents with specifically high heating value is liberated.

TABLE 15. HEATING VALUES OF DIFFERENT SORTS OF GAS

Type of Gas	Th.U./cu. ft.	kcal/m³
Low-temperature gas	780	6930
High-quality gas	570	5070
Coke-oven gas	540	4800
Water-gas	300	2670
Producer gas	130	1560

By the carbonization and gasification processes *gaseous* fuel is obtained. What is, however, more urgently needed in our age of motor-vehicles and aeroplanes is *liquid* fuel for driving internal-combustion engines. Apart from the fact that the natural reserves of petroleum will be exhausted within the next few generations there are highly industrialized countries like Britain and Germany with ample coal resources, but practically no oil-fields. A procedure, therefore, which could efficiently turn coal into petrol, instead of gas, would save millions of dollars for oil imports, and secure at the same time fuel for cars and planes for a certain period after the exhaustion of our oil resources.

The problem, therefore, popularly called *Kohleverflüssigung* ("liquefaction of coal") in German, is one of the most important tasks of science and industry, and the most able men in this field have worked on its solution.

As a matter of fact, the problem of *Kohleverflüssigung,* or *hydrogenation*, as it is called in English, has been solved in principle since the twenties of this century, but the industrial application of the process developed for synthesizing oil from coal is still rather restricted, on account of the costs involved. Fuller details about this matter are given in Chapter IX.

Underground Gasification

The use of gas, with its excellent qualities as a clean, smokeless, and handy fuel, would be more widespread if its price could be lowered. Gas, being a product of coal—gained in a very low-weight percentage from raw coal—is dependent as to price on the costs of coal production, which in their turn are rising rather than falling. For, in spite of technical improvements, coal production is still in a large measure dependent on man-power: no fewer than approximately 700,000 miners are working in British coalfields alone. Digging coal at the bottom of shafts several thousand feet in depth is hard labour, and workers in our age are no longer willing to toil for ridiculously low wages. We cannot expect, therefore, to lower the price of gas as long as it has to carry the costs of cutting, loading, and lifting enormous quantities of coal.

A method which is frequently used to save the transport costs of coal over long distances consists of erecting power-plants directly over the pit-heads of coalmines. This procedure is adopted especially in districts yielding coal of lower quality.

A further step in this direction is *underground gasification,* as practised in some plants in Russia. We quote here from Martin Ruhemann's excellent book *Power* (1946):

It has been found unnecessary to sink shafts for underground gasification. In the latest plants only bore-holes are drilled from the surface to the seam. Usually a number of concentric rings of holes are drilled some 30 feet apart, with one hole in the centre. At the beginning all the holes are sealed except the central hole and the innermost ring. Air and sometimes steam are driven down the central hole. The coal is ignited electrically at the bottom, and the gas is drawn off through the holes of the inner ring. When all the coal in the centre is consumed, the central hole is sealed up and the air admitted to the boreholes of the inner ring. Gas is then removed from the next ring of holes, and so on.

The gas made is generally a form of producer gas yielding about 180 B.T.U. per cubic foot, but various improvements have given gases up to 300 B.T.U. per cubic foot and even higher. It is usual to pump down air in which the oxygen content has been artificially increased. This makes the gas richer and also raises the gasification temperature.

Many difficulties had to be overcome before this process proved economical, and even to-day it is only in its infancy. It is probably not suited to every kind of seam, and it would be wrong to assume that it is going to do away with coal-mining altogether. But it is already producing gas at one-quarter its former cost, and many former miners are able to transfer to healthier and less hazardous occupations.

It is expected that gas produced by underground gasification of coal will be used extensively for generating electrical energy and thus for producing mechanical power, as well as for industrial heating in the chemical industry.

Tar and Other Coal Compounds in the Chemical Industry

In the gas-works, as well as in the coking industry, another by-product, coal-tar, is obtained whose importance as a raw material of the chemical industry is growing continually. It has been stressed at the beginning of this chapter that innumerable articles in daily use are made to-day from artificially synthesized carbon compounds using coal or coal products like tar as the starting substance. The giant chemical industries which grew rapidly after the First World War—for instance, I.G. Farben in Germany, I.C.I. in Britain, or Du Pont in the U.S.A.—are devoting the greater part of their activities to the production of organic substances made on the basis of coal products. It has been said frequently that mankind is on the way to leaving the Iron Age, passing on to the age of light metals and plastics.

What are commonly called ' plastics ' consist of a large number of different carbon compounds all belonging to the category of the so-called *high polymers*. We say that a compound B is a polymer of another compound A when B consists of the same elements in the same proportions, but grouped in larger molecules. If, for instance, A is a carbon compound of the formula $C_xH_yO_z$, then a substance B with molecules $C_{nx}H_{ny}O_{nz}$ (where n is an integer) is a polymer of A, and we speak of high polymers when n is of the order of magnitude of 100 or more.

The technique of polymerization cannot be described here, because it lies outside the scope of this book; it need be mentioned only that it has been possible to push it far enough to build up giant molecules with n surpassing 100,000 or so, and to synthesize artificial fibres from high polymers which excel by their extraordinary tensile strength, like nylon, perlon, and other new materials.

In the field of plastics and artificial-textile fibres a similar successful development takes place to-day to that which had started a century ago in the field of dyes. We mean the gradual transition from an initial stage of merely creating substitutes for natural material to the final stage of making artificial products which are superior to their natural prototypes.

The story of the coal-tar dye industry, as told by Hale in his book *Chemistry Triumphant*, began in 1856, when **W. H. Perkin** (later **Sir William Henry Perkin**) working under **A. W. Hofmann**, a German chemist, at the Royal College of Chemistry in London, happened to discover among the oxidation products of aniline ($C_6H_5NH_2$) a compound exhibiting tinctorial properties. When it was found out that this substance could be used for dyeing cloth as a substitute for certain costly plant juices a small factory was built by Perkin at Greenford Green in 1857, and the new dye named ' mauve ' entered production.

While the first step of making synthetic dyes was made in England, the successful following up of this development was soon transferred to Germany, whither Hofmann returned in 1865. From that date until 1874 not even one professorship of organic chemistry existed in England. As Hale critically remarks, " No instance of such extreme stupidity on the part of any two nations has ever been recorded in the history of the world as when France and England gave up the dye industry to Germany."

Under the leadership of Hofmann and, later **Bayer** German thoroughness and industrial skill started the rapid growth of a scientifically well-founded coal-tar industry, with its main branches of dye production and pharmaceutical production.

It was Bayer who in the eighties discovered the chemical constitution of indigotine, which is the effective agent of the natural dye indigo, and from 1897 synthetic indigotine, commercially produced on a large scale, was such a successful competitor to natural indigo that a total emancipation from the need of imports from India was possible. As early as 1868 two German chemists, **Graebe** and **Liebermann**, had synthesized alizarin, a red dye which had till then been prepared from madder grown in France. Shortly afterwards the French madder industry was practically ruined.

While the synthesis of alizarine and indigotine are outstanding milestones in the early history of synthetic dyes, the work of the coal-tar dye chemists went on incessantly, and extended to ever more sorts and colours of dyes, so that a modern catalogue of commercial dyes comprises several thousand items. Along with the increased variety of the assortment, the quality of the products has been improved. The characteristic steps from the poor substitute to the equal and finally superior products were taken in the dye industry within a period of hardly more than half a century, especially after **Bohn's** discovery of the indanthrene dyes in 1901, which in brilliance and fastness to light have few competitors.

The lead gained by thorough scientific research secured the German dye industry an absolutely dominant rôle until the First World War, and a good deal of the assets of the German export trade was due to its chemical industry. It was only after 1918 that by confiscating German patents large chemical industries like I.C.I. in England or Du Pont in America could take over a considerable part of the world's dye trade. Still, in spite of this considerable loss the German chemical industry soon recovered, and when at the beginning of the twenties the most important firms in this field were united in one big concern, the I.G. Farbenindustrie,[1] this undertaking grew within a few years into one of the mightiest and wealthiest industrial trusts in Europe which was accordingly split up in 1945.

Our digression from power economy to economic power was meant to impress on the reader the importance of that branch of industry which is utilizing coal products. Recent developments are extending the range of synthetic materials much farther. While up to the thirties of our century dyes, medicines, and fertilizers had been the most important products of that trade, an advance is now being made into the vast field of textiles. The first step—comparable, perhaps, with Perkin's mauve—was the production of artificial silk and cotton, originally an emergency affair caused by the shortage of raw materials during the First World War and the early post-war years. The first outputs of the production were cheaper, yet qualitatively inferior, substitutes for the natural material. But as soon as nylon entered the market it was recognized that here was a material of much greater durability for stockings and socks than any yarn from natural wool or cotton. Millions of housewives have been saved the labour of darning socks since the invention of nylon, and the problem of laundering has been vastly simplified by the use of nylon shirts.

Tensile strength and resistivity against attrition are, however, only two specific qualities among a great number of properties by which any textile thread is characterized, and therefore all the synthetic textiles on the market to-day are far from being able to compete in every respect with natural silk, cotton, or wool.

Science is, however, far ahead of practice, and a good many substances have been synthesized, yielding threads which will meet every demand. As in the case of the indigotine in the dye field, where between the unravelling of its constitution and its commercial production nearly two decades elapsed, future production of synthetic textiles will mark decisive progress in better and cheaper clothing for men and women. The salient fact is that the polymer chemist of to-day feels able to aim at a certain target by knowing how to build up molecules of material which complies with certain quite specific demands, as, for instance, elasticity, flexibility, porosity,

[1] I.G. is an abbreviation for *Interessengemeinschaft.*

thermal conductivity, waterproofness, etc., etc. With the obvious success of the synthesized threads like nylon, orlon, perlon, terylene, etc., which are on the market now, the transition has begun which leads from the poor-substitute stage to the advanced stage, in which the artificial material represents a far greater variety of high-class substances qualified for general use or for specific purposes.

To complete this short report on achievements of modern organic chemistry, it may be added that coal-tar is the most frequently used, but not the only, coal product which serves as the initial substance for making synthesized material. Another, much simpler carbon compound which on polymerization yields interesting products is acetylene, C_2H_2. Acetylene is a gas which, in contrast to ordinary high-quality gas, burns with a very brilliant flame, and at the beginning of our century it was widely used for headlights of cars and bicycles, as well as for portable lamps in general. Apart from the high intrinsic luminosity of its flame, acetylene has the advantage that it can be generated with very simple means by immersing calcium carbide in water, so that self-contained lights could be made, each fitted with a small generator fed by calcium carbide and water. Calcium carbide in its turn is made from limestone and coal by an electrothermic process which was developed on an industrial scale while there was a large demand for acetylene for illumination. Although at present most of the acetylene lamps have been replaced by electric headlights or torches, quite a different use for acetylene has proved to be practicable by turning the gas into solid substances through the polymerization process. Very long chains of molecules, each endlessly repeating the C_2H_2 sequence, can be used for producing synthetic rubber, among other useful substances.

Are We using Coal Economically?

In view of the function of coal as a raw material of ever-growing importance, one may rightly ask (*a*) whether it is wise at all to use this valuable substance as a fuel, and (*b*) whether in doing so we are proceeding with the necessary precautions of economy.

In answering the first question we have to bear in mind that at the present state of our technical development we have no other choice than to use coal as a fuel. Our oil reserves are many times smaller than those of coal, and water-power, though inexhaustible, will even after full development of all economically justifiable projects not suffice to satisfy the vast power demands of mankind. Atomic energy would give a limitless amount of power if ' fusion reactions ' or thermonuclear reactions (*cf.* Chapters XIV and XVI) could be made serviceable in controlled processes. As far, however, as our present knowledge reaches we do not see any possible method of realizing controlled nuclear fusion processes, and atomic energy can be derived to-day only from fissionable material gained from uranium and thorium, whose reserves, expressed in terms of energy, are not unlimited either. Besides, it will take some time until power production from atomic energy will be economic enough to replace conventional plants. Hence extensive use of coal as a fuel will be made during the next generations.

The answer to our second question is a decided *no.* It would be much too mild to say that we are not using coal economically: what is actually happening is that we are wasting it on an enormous scale! The amount of waste varies in the different uses of coal, but there is waste everywhere, and it is greatest, perhaps, in the most frequent and simple applications of coal as a fuel for domestic heating.

In starting a discussion on the wasteful use of coal with the problem of house-heating we must deal first with an objection frequently made in this connexion against economic considerations in general. The supporters of the traditional open fireplaces are trying to make a good case of their defence by stressing the fact that not everything must be viewed from the economic aspect alone, that there are other interests at stake than kilowatt-hours or calories, and that the comfort of a British house with its open fireplaces is worth the price one has to pay for it.

The advocates of such trains of thought should be reminded, however, that only a few centuries ago similar arguments were raised against the use of coal-fires at all. Grave economic reasons were raised against wood-fires for house-heating—" Save the British woods for timber! " may have been the slogan at that time—and, on the other hand, the argument that the comfort of the households would be endangered by the smoke and soot of these newly fashioned coal-fires was strong enough for Queen Elizabeth I to refuse to enter houses in which coal-fires were burning. And later she issued a proclamation forbidding anyone to burn sea-coal in London while Parliament was in session.

Although the quality of the coal and the construction of the fireplaces may have improved in the meantime, the amount of dirt and dust caused by the soot and ashes of the open coal-fires is quite large enough as every housewife knows from her experience during the heating period. This fact, together with the enormous waste of national wealth, ought to induce house-builders to abandon tradition and to reach a sensible compromise: one open fireplace per house or per flat in the sitting-room for a log-fire as a kind of embellishing auxiliary means of heating to other more economic contrivances for the rest of the house, as, for instance, central heating or stoves like the Continental coke-fired *Dauerbrandöfen* or the Swedish stoves built from Dutch tiles.

Another objection one may hear from people who have already changed over to central heating is that the net expenditure for fuel is the same as before, so that the talk of saving by improved heating is a bluff. If anything like that happens one may safely conclude that this is a case where the wasteful method of using a heating system of low efficiency was only replaced by the wasteful method of overheating the rooms. As a matter of fact, the existence of central heating causes a temptation to overheat, and while the typical British home is kept at a quite moderate temperature in winter, many American houses and offices have the atmosphere of a greenhouse. That is not only a waste, but also very unsound: a room temperature between 18°C. and 20°C. (or 64°F. and 68°F.) is quite sufficient, and appears to be the optimum from the standpoint of health. With rooms kept at the same temperatures as those attained by conventional fireplaces, the consumption of fuel will be considerably less in the case of central heating.

Having dealt first with the objections against improvements in our heating system, we can now make our charges against the traditional open fires. Taking, as usual, the ratio between useful delivery and total input as a yardstick of efficiency, the term 'efficiency of a heating device' will be used for the ratio between the heat conveyed to the object to be heated and the total heat consumed, or, expressed in an equation,

$$\text{Efficiency} = \eta = \frac{\text{useful heat}}{\text{total heat}}$$

where in the case of residential heating the 'useful heat' conveyed to the rooms is taken as the difference between total heat and the loss through the heat escaping up the chimneys. There are, of course, other losses too, as, for instance, from badly fitting windows and doors, or by the heat radiated away through the window-panes. Because these losses cannot be attributed to the heating system as such they are not considered in the figures given in Table 16—

TABLE 16. EFFICIENCIES OF HEATING SYSTEMS
(Approximate average values)

Open fireplace	10 to 20 per cent
Well constructed stove	50 to 60 ,, ,,
Central heating	Up to 70 ,, ,,
Electric fire	100 ,, ,,
Heat pump	300 to 500 ,, ,,

—in which, for instance, the efficiency of an electric fire is given as 100 per cent, because there is no flue at all. In spite of this fact, the use of electric fires ought to be avoided for economic reasons (except in countries with an abundance of very cheap water-power), because it is a waste to transform the highly valuable electric energy into low-grade heat energy.

As seen from Table 16, the efficiency of the heating system still widely used in Great Britain is the lowest in the list. More than five-sixths of the fuel consumed for residential heating is wasted by allowing 85 per cent of the heat to escape through the flues, thereby polluting the atmosphere with fumes, smoke, and soot. The reader is reminded that, according to Tables 13 and 14, about 22 per cent of Britain's coal consumption (178 megatons in 1938) is used for domestic heating. Taking a market price of £6 for a ton of coal, this means that of the £235 million spent for heating about 200 millions are wasted every year. By reducing this waste and exporting the coal saved Britain's trade balance could be very considerably improved.

It is very enlightening to read in this connexion what Professor F. E. Simon, of Clarendon Laboratory, Oxford, says in a report prepared for UNESCO.[1] In speaking about the much advertised projects of utilizing tidal power, he says:

> In particular a scheme in Great Britain, the so-called Severn Barrage Scheme, has been considered very carefully, and it has been estimated that it could replace a million tons of coal burnt each year in a power station. The present cost of such a scheme would be about 100 millions. When we consider that with half the capital investment 20 million tons of coal could be saved per year by improving heating appliances in Great Britain, it is obvious that such a scheme has no attractions.

A similar percentage of losses to that in the open fireplaces

[1] Our quotation, by kind permission of UNESCO, is taken from a summary of Simon's report printed in *Discovery*, May 1952, p. 150.

occurs in railway services. The efficiency of the steam locomotives, which originally, in Stephenson's time, was as low as 3 per cent approximately, could, in spite of all improvements, not be increased to more than about 15 per cent under optimum conditions. The average running service efficiency is, however, much lower, and, taking into account the losses caused by the energy consumed for cutting, hoisting, and transporting the coal, a energy-system efficiency of only 7 per cent results. This means that more than 90 per cent of all the coal consumed for railways is lost for other purposes than the proper work of traction, most of the lost energy being wasted for heating, and at the same time polluting the atmosphere. The cash value of the annual energy losses amounts to about £35 million for Great Britain alone.

Drastic measures against the waste of fuel in the railway system have been taken in several countries. The best method in countries with well-developed water-power plants is the full electrification of railway lines. Central Europe is in a fairly good position. You can go by electrically driven trains from Bâle, on the west frontier of Switzerland, through to Vienna, at the eastern corner of Austria, over a line 946 km, and in the north-south direction from Munich, through Bavaria, Austria, and half of Italy, down to Rome, over a line of 975 km. The completion of the electrification is slowly going on everywhere, being most advanced in Switzerland, with a 100-per-cent electric-railway system.

In countries devoid of water-power the transition to electric traction is on the march as well, especially when the reasons of power economy are strongly supported by considerations of the travellers' comfort and the avoidance of pollution of the atmosphere. The London transport system and the southern lines of British Railways are examples.

In less densely populated areas, with a lower passenger-mile rate on the lines, the interests of the installation costs of electrification may outweigh the saving in fuel. This holds especially for the very long transcontinental lines in America. As explained in Chapter IV, a different method is, therefore, adopted there by replacing the steam locomotives by Diesel electric engines.

The life expectancy of coal reserves will be discussed in Chapter X.

CHAPTER IX

Petroleum and Natural Gas

Treasures in the Earth

I N 1850 **Dr James Young**, a Scottish chemist, devised a process by which paraffin-wax, kerosene, and lubricating oil could be obtained from a natural oil which flowed into the workings of a Derbyshire coal-mine, and soon afterwards the technique of boiling off the volatile constituents of the crude oil was sufficiently developed to obtain petrol. A quarter of a century later, in 1875, **Otto** constructed in Germany his first petrol-motor; ten years later **Benz** inaugurated the era of motor-vehicles by fitting Otto's motor into a car. Another ten years later **Diesel** started the progress towards the better economy of internal-combustion engines, and on December 17, 1903, the first engine-driven flight with aircraft heavier than air was performed by the **Wright Brothers** at Kitty Hawk.

More surprising than the daily sight of the consequences of this development on our roads and in the air is its far-reaching effect on the welfare of people in rather remote corners of the world. Take as an example Kuwait, a small sheikdom on the Persian Gulf, mostly desert land, with a population of 200,000 souls, who from the oldest times till the beginning of the Second World War lived in the most primitive conditions. What happened to Kuwait within the last decade is like the spell of an Arabian Night's fairytale: forty-five completely new schools are being built under the present programme; the great modern Kuwait Secondary School, housing 640 boarders, has already been completed; a new town, Ahmadi, with modern spacious and comfortable houses, has risen from the desert near the capital, Kuwait; and the adjacent harbour is going to be one of the biggest oil ports in the world. It is easy to imagine what it means when a country of 200,000 people enjoys an annual revenue of £55 million from oil, especially when its sheik is generous enough to retain only £5000 for his personal needs and to spend all the rest on the development of his country.

Something similar, even on a larger scale, though not in such a sudden and concentrated form, has occurred to Venezuela, which is the world's largest oil-exporter to-day. All these countries like Kuwait, Qatar, and Iraq, in the Middle East, or Venezuela in the

Caribbean, owe the large sales of their natural riches to the invention of internal-combustion engines and to their widespread use for motor traffic. Certainly there had been a market for crude oil and its products before 1875, but the total consumption of petroleum at the time when paraffin for lamps was its most commonly used derivative was less than 1 per cent of the present figure. And even this small demand would have dropped considerably, owing to the competition of the electric light, if petroleum had not turned out in the meanwhile to be the most important source of energy next to coal. Thus the inventions of Otto and Diesel, together with those of their predecessors and successors, have created a market for petroleum products which justifies the vast amount of capital invested in prospecting for and drilling oil-wells. It was the ever-increasing demand for energy and its sources which acted as a stimulus for continued exploration of oil sources. Without it the treasures hidden deep below the desert soil of Kuwait would never have been lifted.

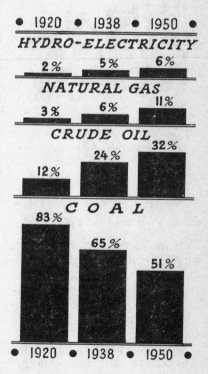

The increasing preference given to liquid and gaseous fuels can be seen from Fig. 54, in which the percentage of the global energy consumption supplied by the four major sources, including natural gas, are given.

A still more striking example of the downward trend in the use of coal is given in Fig. 55, showing the shift of fuel consumption in world merchant shipping between 1914 and 1949.

The reasons for the striking preference given to liquid fuels are obvious—greater ease in handling and transportation,

FIG. 54. MAJOR WORLD SOURCES OF ENERGY

*By courtesy of
the Shell Petroleum Company, Ltd.*

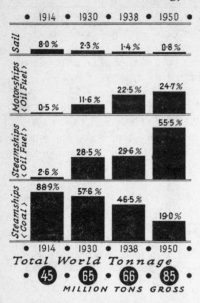

FIG. 55. WORLD MERCHANT
SHIPPING

By courtesy of
the Shell Petroleum Company, Ltd

better economy in the conversion to mechanical work. The motor-car would never have reached its popularity if its operation were as laborious as that of its predecessor, the steam locomobile. The greater ease in handling liquid fuel makes itself felt not only on the consumer's side, but on the producer's side as well. Seven hundred and seventy-four thousand miners in great Britain were doing hard work to produce approximately 225 million tons of coal in 1952. The efficiency of this kind of work can be estimated by calculating the electric power which could be obtained by burning all the coal in thermal power-plants. Assuming, as usual, a 20-per-cent overall efficiency and continuous operation throughout the 8760 hours of the year (100 per cent load factor), we obtain, according to Table 1, Chapter 1, a power of

$$1.68 \times 10^3 \times 225 \times 10^8 \text{ kWh} : 8760 \text{ h} = 43 \times 10^6 \text{ kW} = 43,000 \text{ MW}$$

This means that the work of 774,000 coal-miners is sufficient to supply the fuel required in a plant of about 43,000 MW, or 43 million kilowatts, operating with 20-per-cent overall efficiency and 100-per-cent load factor. In other words, eighteen coal-miners can produce the fuel for the current consumption of a 1000-kW plant—among them eight face-workers, operating under difficult conditions.

The conditions are quite different in the case of oil as a fuel. An exact comparison by simply giving the number of heavy manual labourers required to feed a 1000-kW plant with liquid fuel is hardly possible, because there is no proper counterpart to the miners in the oil industry. There is no pit-work, no underground face-cutting, hauling, prop-setting, or pack-wall building needed to extract the crude oil. Practically all the manual work is done on the surface, while it is only the drilling and pumping machinery which probes down to depths of several thousand feet. Of course, the extensive activities of the oil-producing undertakings require a good deal of heavy manual labour as well: the construction work of the derricks, tanks, and refineries, the building of roads, the clearing of forest and jungle, and the laying of pipe-lines are just a few examples. Still, only a small percentage of the entire personnel is doing really hard manual work, while the great majority are busy either in operating machines and driving transport vehicles or working in offices, laboratories, and the like.

Quite a rough estimate of the man-power needed to supply an all-the-year-round-operating 1-MW power station with fuel from different sources leads to the results shown in Fig. 56. Black figures indicate face-workers in the case of coal, and their equiva-lent in the case of petroleum. Improved mechanical equip-ment has greatly reduced the man-power needed in the American coal-mines, but the requirements for hard work per unit output of energy are still considerably lower in the case of petroleum extraction and refinery. As a result, there are fewer difficulties in recruiting the man-power for oil produc-tion. The fuel shortage in Great Britain in the forties was caused mainly by the lack of mining personnel, while in countries with sufficient petro-leum resources the fuel require-ments were met quite easily.

One of the main problems in oil production is to find the right place to drill a hole. If you are lucky enough to succeed at the first strike your

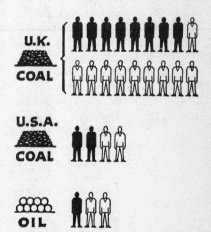

FIG. 56. MAN-POWER REQUIRED TO SUPPLY A 1000-KW POWER STATION OPERATING AT 20-PER CENT OVERALL EFFICIENCY AND 100-PER CENT LOAD FACTOR WITH FUEL FROM DIFFERENT SOURCES

Black figures, workers doing hard labour; white figures, other workers.

hole may turn out to become a flowing well with a yield of between 100 and 1000 barrels of crude oil daily representing an annual gross

income of between thirty thousand and several hundred thousand
pounds sterling. With an output value far in excess of the operating
and maintenance expenditure many oil-fields are better than gold-
mines, and the total value of the world's crude-oil production is
about a hundred times greater than that of its gold production. It
is not surprising, therefore, that the struggle for the possession of
oil-fields and for acquiring concessions creates problems of an intri-
cate nature in international economy and politics. It sounds quite
just when a nation, claiming its sole right to the natural resources in
its territory, opposes exploitation by foreign groups. On the other
hand, the value of financial aid and technical assistance in the
development of the resources must not be underrated. If a country
wishes to be left alone, after hundreds of oil-wells have been drilled
and one of the world's largest refineries has been built on its territory
with foreign capital, she should be reminded what would have
happened if she had been left alone from the beginning. Nations
with a lesser-developed technical civilization—whatever height their
culture in the field of arts, philosophy, or anything else may have
reached—have neither the scientific and technical skill nor the capital
and the enterprising spirit to do the job on an adequate scale. As
Ayres remarks in his book on energy sources, petroleum exploration has always been a gamble, and will always remain so. In
1949, 42,000 wells were completed, of which no fewer than 13,000
were dry. No state with a small budget, nor any small firm, could
take the risk of drilling holes at a cost of between ten thousand
and a hundred thousand pounds each, and failing, perhaps, a dozen
times before the first traces of oil were recovered. That is why
other states in the Middle East, like Iraq, Saudi Arabia, and Kuwait,
have not so far revolted against the exploitation of their natural
resources by foreigners. They know quite well that without efficient
help from abroad the heavy lid of the treasure-chest would never
have been lifted.

The Formation and Constitution of Petroleum

The story of petroleum begins, properly told, with its formation
millions of years ago, but we shall restrict this part of the report
to a few remarks which may help to an understanding of the
methods of crude-oil production. The most striking feature of
the technique of extracting petroleum from the depths of the earth
is the relative ease of the process compared with the burdensome
work of coal-mining. We are saved the trouble of digging wide

shafts, cutting haulage roadways, and hoisting innumerable tons of rock and fuel to the surface, because the oil flows either readily by itself or can be pumped by fully mechanized devices without anybody working underground. The result is, as mentioned in the preceding section, the far smaller amount of man-power required for supplying a given quantity of fuel. What combination of lucky circumstances is the reason for Nature providing us with sources of a most valuable fuel which, tapped at the right spot, jets out as if from a soda-siphon?

We know that petroleum deposits occur in sedimentary rocks, and we have reason to believe that they are fossil relics of micro-organisms inhabiting the oceans in former epochs, which after sedimentation formed layers at the bottom of the sea and were subsequently buried under the mud. By the combined action of pressure and temperature they were turned into the characteristic mixture of hydrocarbons which constitute the petroleum, and in certain favourable places the oil assembled in smaller or larger underground pools. The conditions for the assembly and conservation of such pools are (a) the existence of porous matter, as, for instance, layers of sand or limestone, and (b) the existence of non-porous surroundings, as, for instance, layers of clay, which prevent the dissipation of the oil.

The underground high-pressure decomposition process, lasting through millions of years, turns the organic matter, which is mainly composed of the four elements hydrogen, oxygen, carbon, and nitrogen, into compounds of hydrogen and carbon—shortly called hydrocarbons—while oxygen and nitrogen escape. The organic chemist has analysed the composition of many thousands of hydrocarbons, a few dozens of which are the more abundant constituents of petroleum. They fall into two main classes—the fatty and the aromatic hydrocarbons. The former are composed of chain-like molecules, among which again two groups can be discerned—the saturated fatty hydrocarbons, or *paraffins*, with the formula C_nH_{2n+2}, (with n being an integer), and the unsaturated fatty hydrocarbons, or *olefins,* with the formula C_nH_{2n}. The aromatic hydrocarbons, on the other hand, have ring-shaped molecules derived from the benzene ring C_6H_6. We have reason to believe that we do not only know the numbers of the single carbon and hydrogen atoms contained in the molecules of the different hydrocarbon compounds, but also their arrangement, so that along with the chemical formula the pictorial, or valency, formula can be given, as shown in the following examples. The dashes in the pictorial formulæ denote the valencies, or arms with which the atoms are holding each other to form a molecule.

A. EXAMPLES OF FATTY HYDROCARBONS (C_nH_{2n+2})

Name	Chemical Formula	Valency Formula

Methane CH_4

```
      H
      |
  H—C—H
      |
      H
```

Ethane C_2H_6

```
    H  H
    |  |
 H—C—C—H
    |  |
    H  H
```

Propane C_3H_8

```
    H  H  H
    |  |  |
 H—C—C—C—H
    |  |  |
    H  H  H
```

Butane C_4H_{10}

```
    H  H  H  H
    |  |  |  |
 H—C—C—C—C—H
    |  |  |  |
    H  H  H  H
```

Pentane C_5H_{12}

```
    H  H  H  H  H
    |  |  |  |  |
 H—C—C—C—C—C—H
    |  |  |  |  |
    H  H  H  H  H
```

Octane C_8H_{18}

```
    H  H  H  H  H  H  H  H
    |  |  |  |  |  |  |  |
 H—C—C—C—C—C—C—C—C—H
    |  |  |  |  |  |  |  |
    H  H  H  H  H  H  H  H
```

B. Examples of Aromatic Hydrocarbons

Name	Chemical Formula	Valency Formula

Benzene $\quad\quad C_6H_6$

Toluene $\quad\quad C_7H_8$

In the following valency formulæ the practice used in chemical literature is adopted, symbolizing the benzene ring by a simple hexagon, which represents the six carbon atoms and those of the six hydrogen atoms which have not been replaced by a radicle, such as the 'methyl group' (CH_3) or the 'ethyl group' (C_2H_5). The substituted radicles are expressedly denoted at the corner of the hexagon (one of the six C-atoms) to which they are attached. With these simplified symbols the valency formula of toluene is given here once more:

Name	Chemical Formula	Valency Formula
Toluene	C_7H_8	⬡ CH_3

Other aromatics are:

Name	Chemical Formula	Valency Formula
Ethyl benzene	C_8H_{10}	⬡ C_2H_5

Name	*Chemical Formula*	*Valency Formula*
o-xylene	C_8H_{10}	
m-xylene	C_8H_{10}	
p-xylene	C_8H_{10}	

Substances such as the last four compounds, which have the same composition (in this particular case eight carbon and ten hydrogen atoms) but different arrangements of the atoms, are called *isomers*. The letters *o, m, p,* used as prefixes to the three different xylenes, are abbreviations for ' ortho,' ' meta,' and ' para,' denoting the three kinds of positions in which the radicle CH_3 (' methyl group ') can be attached. In the ortho-position they are attached to adjacent C-atoms in the benzene ring, while in the meta- and para-position they are separated by one and two carbon atoms respectively.

All the hydrocarbons of a given series are the more volatile the smaller their molecular weight is. If we arrange, for instance, the first eight paraffins and a few aromatics according to the size of their molecules we obtain a series of substances with ever higher boiling-points, as shown in the graphic representation of Fig. 57. The list contains only the lighter hydrocarbons, which, depending on their boiling-points, are gaseous or fluid at room temperature. There are, however, hydrocarbons, and among them members of the paraffin series, with a higher molecular weight, the boiling-points of which are considerably higher still, while even their melting-points lie above room temperature, so that they are solid under normal conditions. Examples are the fatty hydrocarbons with formulæ ranging from $C_{18}H_{38}$ to about $C_{43}H_{88}$, a mixture of which is paraffin-wax.

As a result of the prolonged high-pressure process, decomposing the fossil micro-organisms, a whole gamut of gaseous, fluid, and solid hydrocarbons is formed, which, mixed in different combinations, constitutes crude oil. While mixtures with a relatively higher content of fluid and volatile components are assembled in the underground oil-fields, other mixtures enriched in hydrocarbons with a higher molecular weight are contained in the form of solid *bitumen* in the so-called *oil-shales*. By retorting the shales the liquid components can be separated from the mineral substance, and the *shale oil* thus produced can be used as a substitute for crude oil extracted from the petroleum wells.

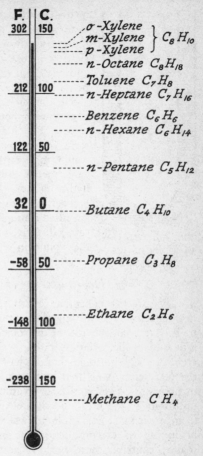

FIG. 57. BOILING POINTS (AT ATMOSPHERIC PRESSURE) OF SOME HYDROCARBONS

Reason for the Ready Flow of Petroleum

What is the reason for some wells turning into petroleum fountains as soon as a hole has been drilled down to the pool? The reason is simply the well-known principle of communicating vessels, illustrated in Fig. 58. If the two legs of a U-shaped tube are filled with fluids of different specific gravity the equilibrium heights of their free surfaces will be inversely proportional to their specific gravity. Petroleum is a fluid of lesser density than water, which in its turn is less dense than the mineral matter of the earth's crust. Whether or not the petroleum will rise by itself to the surface, tending to rise to even higher levels, depends on the conditions of the layers above the oil-pool. If they are rigid rock throughout, forming walls with a self-supporting ceiling around the pool, the pressure on the oil surface is only the atmospheric pressure corresponding

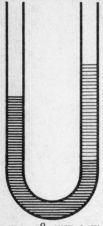

FIG. 58. THE LAW
OF COMMUNICAT-
ING VESSELS

to the depth of the pool, which is far too small to force the oil to rise through the bore-hole. In most cases, however, the soil, far from being a rigid structure of solid rock, is rather plastic, and yielding under its own weight, exerts a pressure sufficient to send the oil to some height above the surface level. By sitting on a water-filled rubber bottle and drilling a small hole in it you can obtain the same effect on a small scale. In some cases the upper part of the cavities containing petroleum is filled with its gaseous components, such as methane and other light hydro-carbons. These gases—known as *natural gas*—either escape first, if the bore-hole hits the gas-filled ceiling of the cavity (the so-called ' gas cap ') directly, or act like the carbon dioxide in the siphon in ejecting the fluid.

The Development of Oil Consumption

The present enormous consumption of petroleum is quite a new development of our civilization, but it would be a mistake to believe that the use of petroleum and its products was quite a new idea, like aviation, radio, or T.V. It has been deduced from archæological findings that asphalt (one of the solid residues of the petroleum) was used both as fuel and as construction material in prehistoric ages, and later on numerous historical references give testimony to the use of fossil fuel. Table 17 gives a few examples from ancient times, while in Table 18 data are collected on the beginning of oil production in some countries which nowadays make a contribution to world or home supply.

The dates given in the following tables show clearly that in a good many places scattered around the earth the use of petroleum as a fuel was known many centuries ago. Still, it was only the dawn of the era of motor-vehicles at the beginning of our century that caused the sudden increase in consumption. More petroleum was produced in the single year 1953 than in all the nineteen centuries from the beginning of the Christian era till 1900, and half the total oil and natural gas consumed by mankind has been burned since 1940. The global and regional increase in oil production since the First World War is illustrated in Fig. 59, which does not include, however, the latest data since 1950.

TABLE 17. USE OF PETROLEUM AND ITS PRODUCTS IN ANCIENT TIMES
(*Only roughly approximate dates are given*)

6000 B.C.	Mesopotamia. Asphalt used as fuel.
3000 B.C.	Mesopotamia. Flares of natural gas lit in temples.
1000 B.C.	China. Wells drilled down to depths of 3000 feet for producing natural gas, which was transported in bamboo pipe-lines for lighting and space-heating.
300 B.C.	Mesopotamia. Fires of fluid naphtha used for military purposes.
Before A.D. 500	North America. Hand-dug oil-wells drilled by Indians.
Before A.D. 1000	Mexico and Peru. Petroleum and asphalt used as fuel.
A.D. 1000	Burma. Wells drilled for petroleum production.

TABLE 18. START OF PETROLEUM PRODUCTION IN DIFFERENT COUNTRIES

About 1300. Baku (now in the Soviet Union).
1640. Modena, Italy.
1650. Rumania.
1692. Peru.
1750. Galicia (now Poland).
1859. Pennsylvania. (The famous Drake's well initiated large-scale oil production in the U.S.A.).
1876. California.
1887. Texas.
1893. Sumatra.
1893. Dutch Borneo.
1901. Mexico.
1908. Iran.
1909. Trinidad.
1913. Venezuela.
1913. British Borneo.
1927. Iraq.
1932. Bahrein.
1938. Austria.
1938. Saudi Arabia.
1938. Kuwait.
1940. Qatar.

Exploration of Petroleum Deposits

It has been mentioned before that the prospecting of oil-fields is a gamble. There are two circumstances acting in opposing directions which tend to increase and diminish the risk of failure in the exploration of new sources. One of them is the fact that

most of the easily accessible and detectable sources have been used
up already. The first oil-fields exploited in the early days of the
petroleum industry lay just a few dozen feet underground, and
revealed their existence by seepages of oil and/or asphalt, which

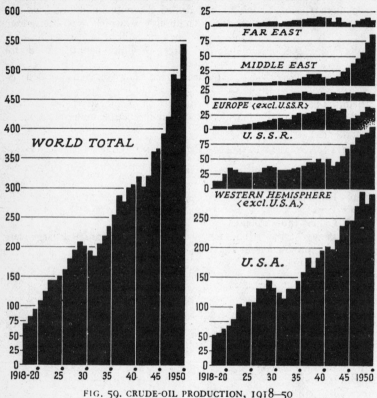

FIG. 59. CRUDE-OIL PRODUCTION, 1918–50
(Millions of metric tons)
By courtesy of the Shell Petroleum Company Ltd

could be seen and smelt by the explorers. Scarcely any fields of
that kind are left untouched in habitable areas to-day, and further
search either goes deeper down into the earth or extends to hitherto
unexplored areas of desert, forest, or jungle.

While, therefore, exploration continues to-day under less favourable external conditions than a century ago, modern scientific
methods provide more efficient tools for the discovery of oil deposits.
First of all, many empirical data have been collected on the con-

nexion between the existence of oil-bearing layers and the general geological make-up of a region. Secondly, geophysical methods have been developed for detecting and localizing underground layers. The boundaries between layers of different material act as reflecting surfaces for sound-waves; hence methods of echo-sounding similar to those used in determining sea-depths can be used. Small charges of dynamite are exploded on the surface, and the echoes of the shock-wave coming from the different underground layer-boundaries are received by recording instruments placed at suitable positions overhead. This so-called seismic method can provide valuable information on the sequence and localization of different geological layers. Another, more subtle, method requiring very exact measurements is the gravimetric method, based on high-precision measurements of local variations of the earth's gravitational field. From very tiny deviations of the acceleration constant from its average value, conclusions can be drawn as to the density of the material forming the vicinity underground. Combining data thus collected with those of the seismically determined localization of layer boundaries is helpful in building up a picture of what lies deep beneath the surface.

The widespread use of such methods, which are considered indispensable to-day, marks a radical change of attitude in the scientific approach to the exploration problem. Oil-prospecting in the early days relied mainly on a kind of scenting instinct of the explorers, and rather more confidence would have been put in the divining-rod than in a geologist's opinion. Modern exploration, however, listens to what geology has to say, and ample use is made of geophysical instruments before the expensive work of actual drilling is started.

Crude-oil Production

The separate steps in drilling a well are shown in Fig. 60, which is taken from a pictorial survey of the petroleum industry published by the Shell Company. The salient feature of the process is the combined action of a heavily loaded rotating-bit and a rinsing-fluid, removing the chips of rock which are being ground away. To perform this double task the drilling mechanism uses hollow pipes instead of massive spindles for rotating the bit. A mixture of water and clay, the so-called *mud*, is forced down the interior of the drill-pipes, and out through the bit, which it also keeps cool. The mud returns upward between the drill-pipe and the casing

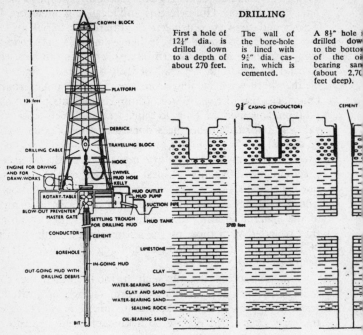

FIG. 60. DRILLING AN OIL-WELL

of the bore-hole, and passes over a vibrating screen for separating out the chippings, after which it is ready for use again. The drilling debris thus collected is thoroughly investigated, and conclusions can be drawn from their nature as to the types of strata the bit is passing through.

The final approach to the oil-bearing layer is a critical phase of the drilling process. It may happen that the pressure exerted by the soil and rocks on top of the oil-pool is strong enough to cause a ' blow-out ' of the well. A violent fountain of gas and/or oil rises from the bore-hole, carrying with it stones, drill-pipe, and tools, which are shot up hundreds of feet. Statistics show that in one case among four a ' wild-gone ' well catches fire from the sparks made by the impact of stone on steel. A giant torch rises to the sky, and the heat radiation makes any approach to the well-head a dangerous adventure. The taming of wild wells is, therefore,

PRODUCTION

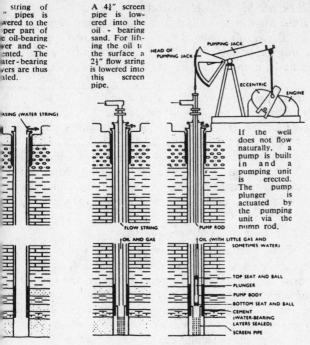

string of ″ pipes is wered to the per part of e oil-bearing ver and cemented. The ater-bearing vers are thus aled.

A 4¼″ screen pipe is lowered into the oil - bearing sand. For lifting the oil to the surface a 2½″ flow string is lowered into this screen pipe.

If the well does not flow naturally, a pump is built in and a pumping unit is erected. The pump plunger is actuated by the pumping unit via the pump rod.

ASING (WATER STRING)

FLOW STRING

OIL AND GAS

PUMP ROD

OIL (WITH LITTLE GAS AND SOMETIMES WATER)

TOP SEAT AND BALL
PLUNGER
PUMP BODY
BOTTOM SEAT AND BALL
CEMENT (WATER-BEARING LAYERS SEALED)
SCREEN PIPE

HEAD OF PUMPING JACK
PUMPING JACK
ECCENTRIC
ENGINE

By courtesy of the Shell Petroleum Company Ltd

a difficult task. A very famous case is that of Moreni No. 160, in Rumania, which, after having been drilled down to 4798 feet, blew wild on May 28, 1929, and burned for 890 days, consuming several million cubic feet of gas daily, until a special expert called from California was able to subdue the flames on November 4, 1931. The final taming of the well took three more months, and only by February 1932 was the well-head capped by a valve strong enough to withstand the gas-pressure.

The normal procedure is to prevent the blow-out of the well by using a weighted drilling mud—that is, one of higher average density—to keep down the high underground gas-pressure as soon as oil is struck. The top of the well is closed with a system of high-pressure control-pipes and valves, known as a ' Christmas-tree.' Depending on where the drill-hole strikes the oil-pool, the well produces either gas or oil, and in many cases a mixture of both emerges. Most wells drilled into a newly discovered pool are flowing, or 'flush,' wells. The period of flush production varies with the size of the pool and the number of wells drilled. In course of time, after a good deal of the pool's contents has been consumed, the pressure decreases, and the period of 'settled' production begins, in which the oil has to be pumped to the surface, as shown in Part 6 of Fig. 60.

Instead of pumping the oil directly, as shown in Fig. 60, an indirect method is sometimes used, called the 'air-lift,' or 'gas-lift.' By pumping air or gas (which in the form of natural gas is abundant near oil-wells) down into the oil-pool a soda-water-like mixture of oil and gas bubbles is made, the average density of which is less than that of the pure oil. Hence the residual gas-pressure in the pool, no longer sufficient to lift the oil to the surface, may still be strong enough to lift the artificially made gas-oil mixture.

The Use of Natural Gas

The separation of the natural gas from the fluid, which is necessary before storing the crude oil, is performed in the ' degassing stations,' which consist of a series of large pipes in which the gas escapes from the slowly flowing oil. The crude petroleum is then fed to large containers, before being transported to the refineries, while the natural gas is either collected also and fed through pipe-lines to suitable places of consumption, or simply wasted by being left to escape into the atmosphere. In the latter case the precaution is generally taken of igniting it, in order to prevent unforeseen explosions or poisoning of the surrounding air. Thus, for instance, the feeding station of the big pipe-line at Kirkuk, in Iraq, is encircled by fires which burn night and day, wasting several thousand million calories daily. It may seem strange that a waste of such magnitude is allowed to continue by undertakings which are proud of the economic management of their business. The reason lies in the difficulties of economic transport of natural gas over wide distances. Pipe-lines—which alone can be taken into account—require larger diameters, and therefore higher costs, to transport amounts of energy in the form of gaseous fuel equal to those of liquid fuel, because of the much smaller density of the gas. While, therefore, the problem with oil is to continue to produce as rapidly as we consume, the inverted problem turns up with natural gas: to develop useful methods of consuming it as rapidly as we produce it. About two-thirds of the production of natural gas comes from proper gas-wells, which can be ' throttled ' to conform as closely as possible with demand. One-third, however—and probably a larger percentage in the future—stems from oil-wells, as described above, and is produced with the same increasing rate as crude oil. Since it cannot be stored indefinitely, it must be either consumed on the spot by suitable industrial establishments or transported in costly pipe-lines, or wasted.

This brings us back to one of the basic principles of world power economy which can be applied to the conversion of heat into work, as well as to the use of fossil fuels or to the distribution of electric power or to the exploitation of water-power resources. *Capital investment for efficiently producing or saving energy pays well in the long run.*

According to this principle, over 100,000 miles of pipe-lines for natural gas have been built in the United States within a relatively short period, and three and a half million horse-power are installed in the pumping-stations. The result of this development is that natural gas leads among all energy sources used to-day in respect of the increase of consumption rate. While coal consumption has shown no radical change, or even decreased, within the last fifteen years, American oil consumption has doubled in the same period, and the consumption of natural gas is about three times greater to-day than it was in 1939.

The method of transporting natural gas in pipe-lines to the consumers is not yet feasible in the Middle East, with its less dense and less industrialized population and its far smaller demand for domestic heating. Schemes are therefore made for the foundation of local plants in or near Kirkuk, in which the natural gas associated with the oil production could be utilized partly as fuel and partly as a raw material for the manufacture of ammonium sulphate, sulphur, carbon-black, and cement.

Refining

Modern equipment does not consume crude oil in its natural state, but uses those of its fractions best suited for the specific purpose. Therefore the oil extracted from the wells is first treated in a refinery, where groups of the lighter and heavier hydrocarbons —a very complex mixture of which constitutes the petroleum—are separated from one another. Two different processes play the main rôle in the work of a refinery. One operation is the fractioned distillation, performed by simply heating the crude oil in a large still. The lighter fractions of the mixture distil off first, then the medium ones, and, last, the heaviest fractions, while a residue is left in the form of a viscous fluid which solidifies at low temperatures. The still itself no longer resembles the glass vessel for distilling spirits, but is a cylindrical steel tower—in larger plants as high as an eight-storey house—in which the crude oil, heated from below, partly evaporates. The temperature within the tower decreases towards the top, so that the vapours of the hydrocarbons of different boiling-points condense at higher or lower levels, the lighter fractions rising to the top before they become fluid, and the heavier ones liquefying lower down. The condensing vapours of the single fractions are caught in special trays at the different levels of the tower, and then removed through pipes.

The distilling process can do no more than single out the different fractions contained in the crude oil, without being able to influence the percentage in which they occur in the raw material. This is a grave disadvantage, because the demand for the different sorts of hydrocarbons, though having undergone several radical changes in the last hundred years, never corresponds to the ratio of their natural mixture. As a result of this disproportion between demand and natural occurence, appreciable quantities of residual fuels were left over which the market could not absorb. At the beginning of the petroleum era, the demand for paraffin-oil, used in lamps and stoves, was greatest, while the highly volatile and easily inflammable fractions like petrol were almost useless. In our days of motor traffic it is the petrol which is coveted more than any other of the components of the crude oil, and a much larger percentage of it is required than that which ' straight-run ' distillation yields.

Cracking

Decisive progress was achieved, therefore, by the introduction of the second main process into the art of refinery—cracking. Since hydrocarbons with lighter molecules have lower boiling-points (*cf.* Fig. 57), we can get more volatile components if we succeed in splitting up the heavier molecules contained in the mixture. This is exactly what the cracking process does: by exposing hydrocarbons to high temperatures and high pressures some of the larger molecules are broken up into smaller ones, so that more fractions of low boiling-points are obtained. Consider, for instance, a medium-weight paraffin $C_{12}H_{26}$, which by cracking can be split into C_8H_{18} + C_4H_8. The former is octane, a lighter paraffin which can be used as a motor-spirit of high anti-knock qualities, while the other fragment, C_4H_8, butylene, does not belong to the paraffin series (which are saturated fatty hydrocarbons), but to the olefins, which are unsaturated hydrocarbons. Although the latter have excellent burning qualities, they are unstable, since their unsaturated valencies have a tendency to polymerization—that is, unison of several equal molecules to a heavier compound. In this way cracking not only produces lighter, but also heavier, molecules than those originally present in the charge, and therefore the process has to be completed by fractional distillation.

Thermal cracking, which, as explained, consists essentially of heating under high pressure (temperatures up to about 900°F.—1000°F. or 482°C.—538°C. and pressures up to 90 ata, or 1300 psia), was introduced in 1912, and proved to be an efficient means of obtaining a higher yield of gasoline from crude oil, as well as of improving the quality of motor-spirit.

Catalytic Cracking

Still further progress has been made since 1936, and more particularly during the Second World War, when *catalytic cracking* came into commercial use, and turned out to be superior to the purely thermal process in quantity and quality of its products. A catalyst is a substance which in a chemical reaction plays the rôle of a kind of best man in assisting the process by its mere presence, without being involved in it as one of the actively reacting substances. Catalysis plays a fundamental rôle in all organic life, the enzymes and ferments produced in the living body being the catalysts for the complicated processes of metabolism. Much simpler substances than the very complex enzymes can be used in industrial processes as catalyst: metals or metal salts in contact with the reacting materials can accelerate or inhibit chemical reactions. Since 1868, when commercial production of chlorine from hydrochloric acid and air, using cupric chloride as a catalyst, was taken up, the technique of catalytic processes has been developed more and more, and large-scale production of widely used chemicals—for instance, ammonia, nitric acid, and sulphuric acid—is based on the use of catalysts. Research done in the laboratories of the petroleum industry revealed that a combination of silicon and aluminium oxides is an efficient catalyst for cracking hydrocarbons, and within the last two decades an increasing number of large plants for catalytic cracking have been erected. Among the different methods used in these new plants, the Fluid Catalyst Cracking Process, as shown in Fig. 61, seems to be the most successful one. The idea underlying this method is to make as close as possible a contact between catalyst and gaseous hydrocarbons. Since the melting-point, and still more the boiling-point, of the silicon and aluminium oxides is too high to achieve

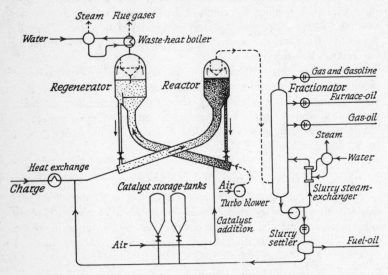

FIG. 61. FLOW DIAGRAM OF FLUID CATALYST CRACKING PLANT

a real gas mixture between catalyst and oil-vapour, an artificial 'fluidization' is obtained by using the catalyst in the form of a very fine powder with grain diameters between 100 and 200 microns, which circulates through the plant by the action of rapidly blown oil-vapours, like dust in a windstorm. A thorough mixture of re-acting material and catalyst is obtained in this way, and the catalytic action is efficient enough to provide better cracking outputs, with pressures of only 1–2 atmospheres, than those of purely thermal cracking, which requires pressures up to 90 atmospheres.

Fig. 61 gives the flow diagram of fluid catalytic cracking. The charge of material to be cracked—the so-called cracking stock—is first heated to a temperature of about 500°C. either by heat-exchange or in a direct fired furnace (not shown in the diagram), and then fed into the reactor, a vertical cylinder approximately 25 feet in diameter and 50 feet high. Before it enters the reactor it is mixed

with a stream of hot powdered catalyst, which completes the vaporization of the oil. The blast of hot vapour flowing with high
velocity through the relatively narrow intake-pipes carries the dust
along like a whirlwind, but in the wide space of the reactor the
velocity of the upward-moving vapour decreases considerably, so
that time is left for cracking the hydrocarbon molecules while the
catalyst powder is partly settling by its own gravity and partly
prevented from escaping with the oil-vapour by the action of suitable filtering means. During the cracking process a fine coating
of carbon is deposited on the grains of the catalyst, which makes
them ineffective, and therefore a regeneration process is necessary,
which simply consists of burning off the carbon coating. As shown
in Fig. 61, the dust particles, which grow more and more sooty
during the process, are removed from the reactor and join a blast
of hot air, which takes them to the regenerator, where the coating
is burnt off. This latter process releases heat, which can be used
to cover a part of the energy expense involved in the process.
Stripped of its coatings, the catalyst is fed back to the intake charge.
The cracked gaseous hydrocarbons, however, are fed to a fractionary
column, where the separation in gaseous and more or less volatile
oil fractions is performed in the usual way.

Octane-number and Anti-knock Value of Fuels

The development of the modern catalytic cracking plants, which
caused very considerable capital investments, was accelerated during
the Second World War by the demand for aviation fuel with high
anti-knock qualities.

A small digression may be inserted here to explain the technical
term 'octane number.' It has been found that pure iso-octane,
which, according to the valency formula

$$CH_3\text{---}C\text{---}CH_2\text{---}CH\text{---}CH_3 \qquad \text{iso-octane}$$

is a branched chain isomer of normal octane, C_8H_{18}, has a particularly good anti-knock value. Normal straight-chain heptane, on the
other hand,

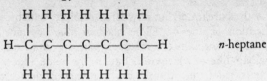

n-heptane

is liable to detonation in the motor. Mixtures of these two substances will, therefore, according to the ratio octane-heptane, represent a whole gamut of fuels, ranging from very poor to high anti-knock values. The octane number of any fuel is thus defined as the percentage by volume of iso-octane in the octane-heptane mixture having the same resistance to detonation as the fuel under consideration.

Apart from increased gasoline yields, catalytic cracking improved the anti-knock value of motor-spirits by supplying motor gasoline of about 80 octane number, compared with about 70 by the thermal method, which in itself constituted an improvement over straight-run gasoline of scarcely more than 60 octane number, produced by simple distillation. It was quite a lucky chance for motorists that the large demand for aviation fuel at the beginning of the forties started before the production of military aircraft was switched over to turbo-props and turbo-jets. For it is only in carburettor piston engines that high octane numbers are essential, while gas turbines can operate quite well with cheaper fuel. It is rather doubtful whether the investment of nine hundred million dollars, leading up to the erection of modern cracking plants, would ever have been made if as early as 1940 most of the planes used in the War had already been equipped with turbines. With the now existing plants production of high anti-knock fuel will be continued, of course, and those who profit indirectly from this development are the motorists. For with fuel of higher octane number available motors can be built with a higher compression ratio, and, therefore, better efficiency, and, consequently, more miles to the gallon. On the other hand, little is gained by these improvements as regards economy of the world's petroleum reserves, because the greater efficiency of car-motors is as good as compensated for by lower efficiency of the production process. We can define the efficiency of a refinery as the heating value of the combined output products divided by the sum of the heating values of the crude input and the fuels consumed in the process. While a simple, well-designed ' topping ' plant, in which nothing is done to the crude oil except distillation, has an efficiency of about 98 per cent, the efficiency of a thermal cracking plant is about 87 per cent, and that of a catalytic cracking plant 82 per cent. Thus most of the savings which can be made in car-motors by the use of anti-knock fuel have been spent already in the course of its production. This remark does not imply that we should go back to the old methods, for having knock-free fuel is a distinct advantage ensuring smoother and safer running of both engine and car. It would merely be a mistake to believe that this progress could help to delay the final exhaustion of fossil liquid fuels.

Synthetic Gasoline

Widely diverging opinions exist as to the time when all reserves of crude oil will be exhausted. Statements of competent experts predicting that the end of the petroleum era will come within the next generation are strongly contradicted by optimists pointing out that similar predictions made half a century ago proved to be erroneous, and that the fast-growing petroleum production is accompanied by an almost faster exploration and discovery of new oil-fields. American oilmen, obsessed with the idea of prosperity, say, " We won't run out of oil until we run out of ideas."

As a matter of fact, the optimists may be right in so far as it will, perhaps, not be our children, but our grandchildren, who live to see the world running out of oil. But there can be no doubt that within a period which is very small compared with historical epochs—and even infinitesimally small compared with the many million years of the future of mankind—the production from oil-fields will be far from meeting the demand for liquid fuel. Fig. 62, taken from the book *Energy in the Future*, by P. C. Putnam, shows the prospective deficiency of the American domestic supply during the next decade. The subsequent global development will make imports from foreign oil-fields ever more difficult.

Petrol production from oil shale, which is not economic to-day, might cover a small part of future demand, while conversion of coal to hydrocarbons will probably be the solution during the short period of a few centuries between the exhaustion of oil-fields and coal-mines.

The problem of making synthetic gasoline by hydrogenation of coal has been technically solved, but the methods used so far are not economic enough to compete with present fuel production from crude oil. Two different methods have been developed in Germany,

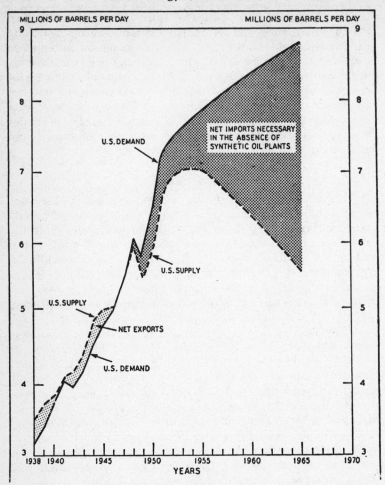

FIG. 62. PETROLEUM SUPPLY AND DEMAND OF THE U.S.A., 1938–65

From P. C. Putnam, " Energy in the Future," reproduced by permission of Van Nostrand Co., Inc.

and no decisive progress beyond them has been made elsewhere.

Direct hydrogenation of coal is made in the Bergius process by treating a coal-tar paste with hydrogen under a pressure of several hundred atmospheres after having added small amounts of tin chloride or iron oxide as a catalyst. The product of this process is a mixture of hydrocarbons which after being fractionated yields gas, gasoline, and tar. The latter is recycled to the feed-charge of powdered coal, in order to make the coal-tar paste.

The second process, which, after the inventors, is called the Fischer-Tropsch synthesis, might be characterized as an indirect method of hydrogenation, because it does not start from coal directly, but from 'synthesis gas,' a mixture of carbon monoxide and hydrogen which is a product of incomplete combustion of coal in the presence of steam. The preliminary process is, therefore, the reaction

$$3C + O_2 + H_2O \rightarrow 3CO + H_2$$

The product of this process is called 'water-gas,' and by enriching it with hydrogen the Fischer-Tropsch 'synthesis gas' is obtained. The hydrogenation is made by bringing the gas into contact with a suitable catalyst. The stuff leaving the catalyst chamber contains gaseous as well as liquid hydrocarbons, and in addition to it some oxygen-containing chemicals, such as ethyl alcohol.

Both processes are wasteful, the Bergius process even more than the Fischer-Tropsch; in both cases the liquid-fuel product contains less energy than is consumed for running the process. Defining as conversion efficiency the ratio of the energy content of liquid-fuel product to that of the coal input, the efficiency of the Bergius process is only 40 per cent, that of the Fischer-Tropsch 45 per cent. Taking into account the far greater demand for man-power in coal production than in oil production, it will easily be understood that there is little chance of commercial use of either of the two processes in present economic circumstances. The case is different, however, as soon as for any reason a larger industrial area is cut off from crude-oil supply. During the Second World War a good deal of the vast demand for aviation and motor fuel in Germany was covered by synthetic gasoline produced by the Fischer-Tropsch method. This fact shows clearly that it is basically possible to depend on coal alone.

Under the assumption that technical improvements will result in achieving greater efficiency in the conversion from coal to liquid fuel, it seems likely that the use of coal will later be more and more concentrated on the production of liquid fuel for motor-cars and aircraft, while stationary power-plants will be driven either by atomic energy or by water-power.

And still later the use of liquid fuel might be restricted to aviation, while motor-cars might be replaced by electromobiles, supplied with power from overhead lines.

CHAPTER X

Life Expectancy of Fossil Fuel Reserves

The Unit Q

BECAUSE of the large amounts of energy involved in global consumption the unit

$$Q = 10^{18} \text{ B.Th.U.}$$

is used throughout this chapter for the energy content (heat value, not electricity equivalent) of fuel reserves. To convert this into the other usual units the thumb-rule may be remembered that one Q is approximately as much as 300×10^{12} kilowatt-hours, or 400×10^{12} horse-power-hours, or one-quarter of 10^{18} kilocalories. The exact conversion formulæ are:

$$1 \text{ Q} = 293 \cdot 1 \times 10^{12} \text{ kWh} = 393 \cdot 0 \times 10^{12} \text{ H.P.h} =$$
$$0 \cdot 251996 \times 10^{18} \text{ kcal} = 1 \cdot 05518 \times 10^{21} \text{ joule} = 1 \cdot 07599 \times 10^{20} \text{ kgm} =$$
$$7 \cdot 7826 \times 10^{20} \text{ ft-lb}$$

Coal Reserves

Optimistic expectations regarding coal reserves were backed by statements made at the beginning of this century. In 1909 M. R. Campbell published an estimate that there were in the United States $3 \cdot 1 \times 10^{12}$ tons of coal of all kinds within 3000 feet of the surface, and at the Twelfth International Geological Congress at Toronto in 1913 a world review of coal estimated the total amount capable of being mined to be about 8×10^{12} metric tons. A comparison of these assumed reserves with the current annual consumption of coal—which was then of the order of about 5×10^8 tons in the U.S.A. and about 15×10^8 tons in the whole world—appeared to prove that the present rate of burning could be continued for another five thousand years. It took some time before the full scope of the error underlying such optimistic forecasts was realized. As recently as 1947 the U.S. Bureau of Mines and the U.S. Geological Survey jointly submitted a statement to the Senate Public Lands Committee, saying:

> Reserves of bituminous coal, sub-bituminous coal, and lignite in place in continental U.S. are estimated at about 3.1 trillion tons as of

January 1944, sufficient for nearly 3400 years at a yearly rate of production approximating the maximum rate in the past—600 million tons—and with current mining losses.

It is only within the last few years that drastic reductions in the estimated life expectancy of fuel reserves from over thirty centuries to one or two centuries were made. These radical cuts result from reductions in estimates of reserves on the one hand and considerations of the future increase of consumption on the other. In connexion with the world-wide search for oil, more knowledge about the make-up of the earth's crust has been gained, and this has revealed, among other things, that Campbell's geological inferences erred on the high side. The error extends both to his estimate of the total amount present and to the percentage of this amount which is economically recoverable. The assumed $3 \cdot 1 \times 10^{12}$ tons included bituminous coal existing in seams with a thickness down to 14 inches, while on sober consideration 24 inches thickness appears to be the limit for economic mining. Besides, we must not overlook the fact that for obvious reasons past coal-mining has skimmed off the cream of high-grade bituminous coal easily accessible in thick seams relatively near the surface. The tonnage left is greater than that consumed so far, but it is of inferior quality and of lower average heat content. For this reason the above-quoted 1947 report to the U.S. Senate Committee assigned a total heat content of about 67 Q to the then assumed reserve of $3 \cdot 1 \times 10^{12}$ tons, instead of the product of this quantity with an officially accepted average heat value of $26 \cdot 2 \times 10^{6}$ B.Th.U. per ton, which would have given

$$3 \cdot 1 \times 10^{12} \times 26 \cdot 2 \times 10^{6} = 81 \times 10^{18} \text{ B.Th.U.} = 81 \text{ Q}$$

Further critical analysis made by Crichton (1948) and Fieldner (1950) resulted in far more drastic cuts. A thorough discussion of all the independent correction factors which must be taken into account is given in Putnam's book (1954); they are summarized in Table 19, taken from Putnam:

TABLE 19. FACTORS TO APPLY TO THE 1947 ESTIMATE OF UNITED STATES COAL RESERVES TO CONVERT IT TO AN ESTIMATE OF COAL " RECOVERABLE BY PRESENT METHODS AT SUBSTANTIALLY PRESENT COSTS "

Correction	Factor
1. To correct for the coal inferred in 1909 to exist, but thought in 1950 not to exist	0·50

Energy for Man

Correction	Factor
2. To correct for upgrading the seam-thickness specification from 14 to 24 in.	0·50
3. To correct for the coal that will be left in the ground	0·50
4. To correct for high-ash coal	0·95
5. To correct for the 50-per-cent thermal losses in converting the presently non-transportable low-rank Western coals to transportable fuels	0·83
Composite factor (product of all factors)	0·099

The net result is a reduction in the estimated U.S.A. coal reserves to one-tenth of the formerly assumed value—that is, to about 6 or 7 Q. Applying similar reduction factors to reserve estimates made in other areas from which considerable coal-supplies can be expected, Putnam obtains the global summary given in Table 20. The word ' economically ' as used in Table 20 is defined as recoverable at unit costs not greater than twice 1950 costs.

TABLE 20. ESTIMATES OF HEAT CONTENT OF ECONOMICALLY RECOVERABLE COAL RESERVES OF THE WORLD, 1950

Country	Heat Content (Q)
United States	6
Canada	2
United Kingdom	1
Other (Western) Countries	7
China	6
U.S.S.R.	10
Total	32

Further adjustments will have to be made in these revised estimates for the losses involved in the conversion from coal to liquid fuels. This will be discussed later in this chapter.

Petroleum Reserves

The proved petroleum reserves in the United States were estimated at 28×10^9 barrels in 1950. Comparison of this figure with the U.S. annual consumption, which amounted to $2·5 \times 10^9$ barrels and $2·6 \times 10^9$ barrels in 1952 and 1953 respectively, shows that the proved reserves will hardly cover the demand for another ten years unless a considerable part of it can be supplied from undiscovered sources or from imports and synthetic fluid fuels. In spite of all the uncertainty of forecasts of the future, there is good reason to assume that the amount from newly discovered sources will not greatly surpass the proved reserves. The reason for this assumption lies in the fact that the rate of new discoveries in the thoroughly searched country is sharply declining. Since 1913 1·3 million wells have been drilled in the U.S.A.; the reserves discovered per exploratory well have decreased in twelve years from about 600,000 barrels to about 200,000 barrels. It has become necessary to drill down to ever-increasing depths, and the daily average yield per well has dropped to 12·3 barrels in the U.S.A., as compared with 219·7 barrels in Venezuela and 5754·7 barrels in the virgin soil of the Middle East!

In estimating the heat content of oil reserves the fact is taken into account that on the average every barrel of crude petroleum produced has associated with it 3000 cubic feet of natural gas of a specific heat content of about 1000 B.Th.U./cu ft. Hence the composite oil-gas heat content can be assumed to be $(5·8 + 3) \times 10^6$ B.Th.U. = 8·8 million B.Th.U./barrel, and the 28 thousand million barrels proved reserves in the U.S. represent a total heat value of about 0·25 Q. Undiscovered crude oil in the same area, which has been estimated to total between 46 and 60 thousand million barrels, may yield another 0·50 Q to the U.S. reserves.

Little account has been taken in these estimates of submarine oil-fields lying off-shore on the continental shelves. They may contain bigger reserves, but, except for a comparatively narrow strip along the coastline, they do not seem to be recoverable at costs competitive with production from oil-shale or synthesis from tarsands. Restricting the survey to economically recoverable reserves (both proved and undiscovered), Putnam summarizes the petroleum world situation as in Table 21, given below.

Reserves of Oil-shale and Tar-sands

Liquid fuels are produced not only from crude petroleum, but also by distillation from certain kinds of solid fuels. The so-called oilshales are rocks containing about 10 per cent or a little more of a bitumen called ' kerogen,' which is a mixture of different, mostly heavy, hydrocarbons with impurities like sulphur, nitrogen, oxygen, and other elements. After being mined oil-shales are crushed and then heated, yielding liquid shale-oil, which, like crude petroleum, can be used as a raw material for refining processes. It differs from crude petroleum in its higher content of ill-smelling impurities.

TABLE 21. ESTIMATES OF THE HEAT CONTENT IN THE WORLD RESERVES
OF OIL-GAS THOUGHT TO BE RECOVERABLE AT REAL COSTS NO HIGHER
THAN ABOUT 1·3 TIMES 1950 COSTS

	Proved Q (Jan. 1, 1950)	Undiscovered Q (based on Weeks)	Total Q
United States	0·25	0·25	0·5
Rest of Western Hemisphere	0·11	0·9	1·0
Soviet Union	0·04	1·3	1·3
Middle East	0·57	1·1	1·7
All other regions	0·02	0·5	0·5
TOTAL	0·99	4·05	5·0

The yield varies from 60 gallons of crude oil per ton down to 10
gallons per ton. As shown in Table 22, taken from Ayres and
Scarlott, considerable oil-shale reserves exist, according to our
present knowledge, only in the U.S.A. and Brazil.

TABLE 22. POTENTIAL OIL-SHALE RESERVES OF THE WORLD

Country	10^9 Tons of Oil-Shale	10^9 Barrels of Recoverable Oil
United States	700	365
Brazil	550	300
Sweden	5	1
Estonia	1·5	1·5
Manchuria	0·5	0·1
France	0·06	0·03
Australia	0·04	0·06
Tasmania	0·009	0·002
Approximate Total	1257	668

Assigning to shale-oil the same specific heat content as crude
petroleum—namely, 5,800,000 B.Th.U. per barrel—we could expect
an energy reserve of 3·7 Q in the oil-shales. King Hubbert (1949)

estimates the heat content of the world's oil-shales to be about 6 Q, while Putnam's estimate is only 1 Q. This lower value is reached by discarding as economically not recoverable all shales with an oil-content below 30 gallons per ton.

Tar-sands are deposits containing a semi-solid sticky substance mainly composed of higher hydrocarbons which is similar to the tars we know as by-products of the distillation of coal and petroleum. Liquid fuels can be produced from this naturally occurring tar, but by a process which is still more expensive than retorting oil-shale. Therefore it would be only in a situation of emergency that recourse would be had to the production of a further supply of fuel from tar-sands. The deposits along the Athabasca river, in North Alberta, Canada, are said to contain up to 300,000 million barrels of crude oil, but if, according to Putnam, the same economic proviso (30 gallons per ton as a minimum) is applied to these sands as has been applied to oil-shale, the heat content of the economically recoverable reserves in the Athabasca area does not exceed 0·02 Q. All other tar-sand deposits are much smaller.

Total Reserves of Fossil Fuels

The aggregate reserves of fossil fuels enumerated in the foregoing sections are summarized in Table 23, taken from Putnam. According to Paley (1952), the expression ' economically recoverable '

TABLE 23. HEAT CONTENT IN THE WORLD RESERVES OF ECONOMICALLY RECOVERABLE COAL, OIL AND GAS, OIL-SHALE, AND TAR-SAND

Coal	32·0 Q
Oil + Gas	5·0 Q
Oil-shale	1·0 Q
Tar-sand	0·02 Q
Total	38·02 Q

is used for recoverable at costs no higher than 1·3 times 1950 costs in the case of oil-gas, and at costs no higher than twice 1950 costs for all other fuels.

A further reduction of the total amount of 38 Q results from the fact that with progressive depletion of the oil reserves the necessity will arise to convert coal into fluid fuels, a process which involves considerable fuel losses. As mentioned in Chapter IX, the thermal efficiency of the Bergius process is 40 per cent, and that of the Fischer-Tropsch 45 per cent. Assuming that future technology succeeds in raising the efficiency to 50 per cent, we must take into account the fact that for every million B.Th.U. derived from synthesized liquid fuel two million B.Th.U. of coal will be consumed. With the present trend to substitute liquid fuels for solid ones, the demand for the former can be expected to rise steeply. Putnam estimates that the percentage of the global energy load carried by liquid fuels, which is 30 at present, might reach more than 55 by A.D. 2000. Taking into account the foreseeable rise in global energy consumption (as explained in the next section of this chapter), the total conversion losses are estimated to amount to 11 Q. In this way the net assured 1950 world reserve of coal will be only $32 - 11 = 21$ Q, reducing the total heat content of the world's economically recoverable fossil fuels to 27 Q.

Ayres and Scarlott, who are too cautious to give exact figures of world reserves, restrict their estimates to an upper and a lower limit, as shown in Table 24, taken from their book *Energy Sources*.

TABLE 24. ULTIMATE RESERVES OF ENERGY FROM ALL FOSSIL-FUEL DEPOSITS OF THE WORLD

(In horse-power-hours $\times 10^{14}$)

	Maximum	Minimum
Coal	750	75
Petroleum (liquid)	5	3
Natural gas	3	2
Oil-shale	12	2
Tar	20	—
Peat	5	5
Total	795	87

Expressed in Q, the maximum and minimum, according to Ayres and Scarlott, is 202 Q and 22 Q respectively, with Putnam's estimate of 27 Q lying nearer to the minimum.

A graphic representation of the contribution of the different sorts of fossil fuels to the total reserves is given in Fig. 63.

Future Rise in Demand for Energy

Life expectancy of fuel reserves is determined by their recoverable amount and by the rate of demand. Fuel consumption is rising at a tremendous rate, which can be appreciated best by considering

the following facts put forward by King Hubbert in a symposium
Energy from Fossil Fuels in Washington during the Centennial

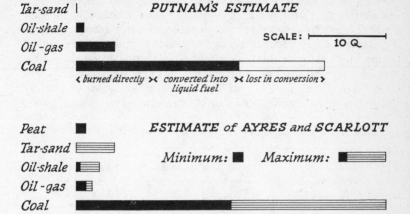

FIG. 63. CONTRIBUTION OF DIFFERENT FUELS TO TOTAL WORLD RESERVES
ACCORDING TO ESTIMATES OF PUTNAM AND OF AYRES AND SCARLOTT

Celebration of the American Association for the Advancement of
Science:

> By the end of 1947 the cumulative production of coal during all
> past human history amounted to approximately 81 billion metric
> tons.[1] Of this, 40 billion, or approximately one-half, has been mined
> and consumed since 1920. Sixty-two billion, or more than three-
> quarters, has been produced since 1900. . . . The cumulative production
> of petroleum by the end of 1947 was 9·17 billion cubic meters (57·7
> billion U.S. barrels). Of this one-half has been produced and con-
> sumed since 1937, or during the last ten years, and 97 per cent since
> 1900.

Production figures of coal and petroleum are practically identical
with consumption figures, because no appreciable amount of such
fuels is stocked over a long period. When Hubbert gave his report
in 1948 the latest annual production record of petroleum was 3
billion barrels in 1947; in the meantime the record has risen to

[1] The reader is reminded that in America 1 billion = 10^9.

5·02 billion (= 5·02 × 10^9) barrels in 1953. Naturally the trend in increase of population and energy consumption cannot continue for a long time. The well-known example of one penny deposited in a bank in A.D. 1 at 4 per cent compound interest annually through all the centuries shows that an undisturbed rise in a geometrical progression would lead to figures of impossible magnitude even within time intervals quite short compared with historical periods. Table 25 shows how any figure increasing continuously at a compound rate of 0·5, 1·0, 2·0 . . . per cent annually will grow excessively within a few centuries. As can be seen from the table, a steady

TABLE 25. MULTIPLICATION FACTORS OF GEOMETRICAL PROGRESSIONS

Annual Increase	Multiplication Factor in Years			
	50	100	500	1000
½ per cent	1·284	1·648	12·16	148
1 ,,	1·644	2·704	144·5	20,900
2 ,,	2·692	7·244	20,000	4 x 10^8
3 ,,	4·386	19·24	2·63 x 10^6	6·92 x 10^{12}
4 ,,	7·104	50·47	3·27 x 10^8	1·08 x 10^{17}

increase of 4 per cent *per annum* would result in a multiplication by a little over 50 in a hundred years, with 327 millions in five hundred years, and with 1·07 × 10^{17} in a thousand years.

There is no doubt that the rapidly rising curve of oil consumption will suffer a break sooner or later through depletion of the resources. When the new prospect of deriving electrical power from nuclear fuels turned up the United States Atomic Energy Commission requested Mr Palmer Putnam to make a study of the maximum plausible world demands for energy over the next fifty to a hundred years. An answer to this question is given in Putnam's book *Energy in the Future* (1954), which we have quoted repeatedly, and from which the following considerations are also taken. His analysis deals fully with the two factors determining the global energy demand—namely, rise in population and average *per capita* consumption. The effort is—to put it in the author's own words—" to identify and describe those maximum plausible contingencies with which a hypothetical Trustee of the world's energy should reckon." It is obvious that, seen from this viewpoint, cautious predictions should rather overestimate the demand and underestimate the reserves than *vice versa*. No wonder, therefore, that Putnam's estimates of the life expectancy of fossil fuels turn out to be considerably lower than those of Ayres and Scarlott.

According to our present knowledge, world population seems to have been stable at a figure between 200 and 300 millions from the days of Christ till about A.D. 700. A period of slow increase lasting till about 1650 followed, during which the population doubled. For unknown reasons the middle of the seventeenth century is a milestone in the history of mankind. From about A.D. 1650 the slow and steady population growth (which during a millennium averaged about a 0·07-per-cent increase *per annum*) turned abruptly into an accelerated rate of growth. This remarkable event appears to have occurred almost simultaneously both in Europe and in other parts of the world, particularly in the areas of large populations in the Far East. Table 26 shows clearly the

TABLE 26. ESTIMATED ANNUAL INCREASE OF WORLD POPULATION FROM 1650
(Approximate figures)

A.D.						Persons per thousand per annum
1650 ...	...	...	...	...	...	2
1750 ...	...	...	...	...	...	4
1850 ...	...	...	...	...	...	6
1950 ...	...	...	...	...	...	8·7

explosive character of the population growth during the last three centuries.

A purely mathematical extrapolation of the present trend of growth would lead to grotesque results within a few centuries. There are, however, sociological evolutions foreseeable which sooner or later will cause a change in the rate of growth. What we have to take into account is the contingency that a large part of the world's population now living in a state of illiterate subsistence farming will ultimately move into literate industrial-urbanfarm patterns of life. This evolution, called *demographic transition,* leads, according to past experience, to declining mortality, and, later, also to declining fertility. With the second consequence lagging behind the first, the combined effect can be expected to be a temporary increase of the rate of population growth, followed by a decline, because the decrease of the birth-rate will outweigh in the long run the smaller death-rate.

Considering all these factors, the conclusion is reached by experts that the world population, which is about 2400 millions to-day, may be between five and eight thousand millions in A.D. 2050.

Another consequence of the demographic transition will be a still further increase in the *per capita* demand for energy. Since 1860 not only has the global and national consumption of energy risen considerably, but also its quotient by the population figure. However, the considerable increase of the *per capita* ' output ' of usable energy was not accompanied by a proportional increase in energy ' input ' (that is, fuel consumption), because a part of the growing output could be provided by improved efficiency of the engines and heating devices. The improvements in engine efficiencies have been shown in Fig. 16, Chapter III; the corresponding improvements in heating systems may be illustrated by the fact that the efficiency of domestic heating in the United States has risen from that of the colonial fireplaces—about 8 per cent—to the present central-heating efficiency of nearly 50 per cent. This development resulted in compensating for a while for part of the increased demand for energy output. The rising efficiencies will, however, reach a ceiling sooner or later, and from that time on the increase in energy input will have to keep step with the rising output.

As a result of all these considerations, the maximum plausible curves of cumulative fuel consumption reproduced in Fig. 64 have been drawn by Putnam. They are based on the expectation (derived from the premises explained above) that (*a*) the world population will rise to between 5000 and 8000 millions in A.D. 2050; (*b*) the annual rate of growth of the demand for usable energy *output* will be between 3 and 5 per cent; (*c*) the average efficiency of fuel uses will reach its ceiling along a certain curve which is described in detail by Putnam.

The ordinates of Fig. 64 do not represent the world's annual consumption, but its cumulative consumption—that is, the total heat content of all fuel used from the beginning of civilization until the time represented by the abscissa of the diagram. It is assumed that by 1860 the cumulative consumption of all fuel (mainly wood-fuel by that time) had been 9 Q, and had risen to 13 Q by 1947. This figure, added to Putnam's 27 Q of economically recoverable resources, gives altogether 40 Q as the ultimate reserves, past and future—neglecting income from recurring sources, which will be dealt with later.

The steeply ascending curved areas of the diagram (vertically shaded) include the curves representing the rise in cumulative consumption corresponding to the assumed annual rates of growth of 3, 4, and 5 per cent respectively. The lower border of each of these areas corresponds to the smallest assumed population growth, ending with 5000 millions by A.D. 2050; the upper border corresponds to the largest likely growth, ending with 8000 millions.

A depletion of the economically recoverable fossil fuel reserves

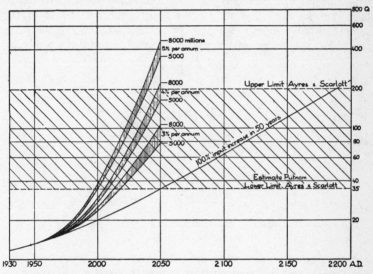

FIG. 64. PROBABLE TRENDS OF CUMULATIVE FUEL CONSUMPTION

would occur, according to Putnam, between A.D. 2000 and 2025, unless other plentiful energy sources were tapped in the meantime.

A considerably longer life expectancy results from estimates made by Ayres and Scarlott of future energy demands. Apart from larger estimates of the total fuel reserves—the upper limit being over 200 Q, as compared with Putnam's mere 40 Q—their expectations concerning the growth of world demand are considerably lower than those of Putnam. According to their assumption, the demand for usable energy output will grow at a rate of 50 per cent in fifty years (equivalent to a compound rate of growth of 0·8 per cent *per annum*). Considering the conversion losses caused by the growing demand for synthetic liquid fuels (which are not taken into account in Ayres' estimates of the reserves), the assumed rate of growth of the output will involve roughly an input rate of growth of 100 per cent in fifty years. The curve of cumulative consumption corresponding to this assumption is drawn in Fig. 64, and can be seen to cut the line of Ayres' lower reserves limit at about A.D. 2050, and the upper limit at A.D. 2190.

The rather wide divergence between the estimates of Putnam and Ayres and Scarlott results from their different viewpoints. Putnam, in the rôle of the prudent Trustee feeling responsible for well-timed development of new methods for power production, has to take into account all contingencies causing early depletion of the hitherto known energy sources. Ayres and Scarlott, on the other hand, are not under an obligation to provide for the worst case,

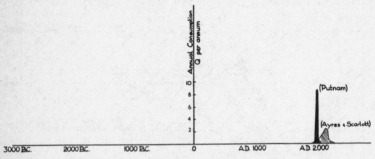

FIG. 65. WORLD'S ANNUAL CONSUMPTION OF FOSSIL FUELS VIEWED FROM A LARGER PERSPECTIVE

but are free to predict what they feel most likely to occur. Still, even the longer life expectancy of about two and a half centuries is much shorter than what had been believed only recently, and is extremely small compared with the past and the future life-span of mankind. This is shown in Fig. 65, in which the world's annual consumption of fossil fuels is represented by the ordinates of the curves and the total reserves by the enclosed area. The black and the shaded areas are equal, and correspond to the upper limit (200 Q) of Ayres' estimate of recoverable world reserves. The curve enclosing the black area refers to Putnam's assumption of a 4-percent annual rate of growth of usable energy output; the curve enclosing the shaded area refers to Ayres' estimate of a 50-percent output growth rate in fifty years.

Man's reliance on fossil fuels for his supply of energy can but be a short episode in his history—indeed, merely a " pip," to use Hubbert's term! If we want to preserve our civilization under the pressure of the fast-growing world population we must find ways of tapping other energy sources than the quickly vanishing fossil fuels.

CHAPTER XI

Fuels from Vegetation

WITH the prospect in view of having spent the rich heritage of fossil fuels within the next few generations, it might appear reasonable either to look for other energy sources or to make preparations for living off income. The growing general disproportion between rising energy expenditure and current income has been illustrated in Fig. 9, Chapter III. More details about the possible rôle of fuels from recurrent sources are given in this chapter.

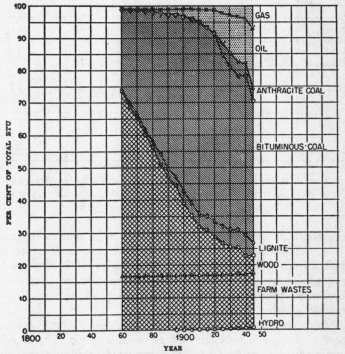

FIG. 66. THE PERCENTAGE CONTRIBUTION TO THE ENERGY SYSTEM OF THE WORLD FROM EACH MAJOR ENERGY SOURCE, A.D. 1860–1945

From P. C. Putnam, " Energy in the Future "

Wood-fuel

From the beginning of civilization, many millennia back in prehistoric times, until about A.D. 1800 more than 95 per cent of the global energy load had been carried by fuels from recurrent sources, with wood-fuel leading and farm wastes second. It was only in a very few countries like Great Britain and Germany that coal began to dominate earlier; in the world's total energy system wood was not overtaken by coal until 1880. The global trend of

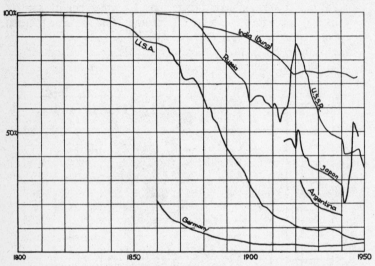

FIG. 67. THE PERCENTAGE CONTRIBUTION OF WOOD-FUEL TO THE ENERGY SYSTEMS OF ARGENTINA, GERMANY, JAPAN, THE U.S.A., AND THE U.S.S.R. AND OF FARM WASTES TO THAT OF INDIA

the percentage contribution of the main energy sources is shown in Fig. 66, taken from Putnam. While the percentage of energy supplied from wood has dropped from about 57 in 1860 to about 5 in 1945 (and has been decreasing still further since), the relative contribution from farm wastes—which we shall discuss later—is remarkably stable.

Particulars about the use of wood in a few selected countries where statistics are available are given in Fig. 67. The general downward trend can be seen to be interrupted only in times of war or revolution, when there is a grave shortage of coal and liquid fuels. The question one may feel tempted to put is: Imagine a general state of emergency caused by full depletion of the world's reserves of fossil fuels. Would it be possible then to repeat on a world-wide scale what happened in the Soviet Union in 1920 and in Japan in 1945—namely, to shift over to wood more than half of the energy load?

The answer is decidedly *no*. The peaks of the curves in Fig. 67 must not be understood to imply that in Russia in 1920 or in Japan in 1945 far more wood was produced than normally. The curves do not represent production figures, but merely the percentage of wood contribution to the energy systems of the countries, and the sharp peaks are simply due to extreme shortage of coal and oil, causing a severe decline both in domestic heating and production. In those times of national emergency neither Russia nor Japan was in an economic condition which could be upheld for longer periods without seriously endangering the health and welfare of the nation. The pattern of modern civilization, together with rising population figures, has changed the demand for energy to such a degree that relying mainly on wood as a fuel is no longer possible to-day.

The world's annual energy income derived from wood was 1.5×10^{15} kcal in 1936, according to the Guyol Report, while the corresponding figures for the present time are 1.9×10^{15} kcal, according to Ayres and Scarlott, and 1.0×10^{15} kcal, according to Putnam. In the total energy budget of about 23×10^{15} kcal per annum the present contribution from wood lies between 4 and 7 per cent, as compared with more than 90 per cent in A.D. 1800. Unless appropriate measures are taken a further decrease can be expected, because the total energy consumption is fast rising, while, on the other hand, in countries with formerly great wood reserves the forests are dwindling rapidly.

In addition to the insatiable demand for fuel there is an ever-growing demand for timber-wood and pulp-wood, so that the annual cuttings in many areas are considerably greater than the annual growth. According to Ayres and Scarlott, soft wood is being used up in the United States at eight times the rate of growth, and hard wood at four times the rate of replacement. If this sort of exploitation is tolerated without limit the American forests will, in the opinion of some experts, be wiped out before the end of the century. The situation is similar in Canada and in some other countries.

It is true, too, that the depletion of forests in some countries is man's own fault, and is not the unavoidable consequence of the

ratio between world population and forest area. Nearly two-thirds
of the forest area of the globe are virgin forests, with no net output
at all, and the rest is far from being used in the most economic way.
If the total forest area could be brought under proper management
and made to yield two tons per acre, and if one-half of this were
used for fuel, the annual input to the world energy system would
be nearly equal to the present total income from all energy sources.
Greater details about actual and potential yields according to
different estimates are given in Table 27.

TABLE 27.   ACTUAL AND POTENTIAL ENERGY PRODUCTION FROM
WOOD AS FUEL

	"Encyclopaedia Britannica"	Putnam
Total forest area, acres	6·2 x 10⁹	8 x 10⁹
„ „ sq km	25 x 10⁶	32 x 10⁶
Annual output ⎰ cu ft	26 x 10⁹	20 x 10⁹
of wood fuel ⎱ m³	740 x 10⁶	570 x 10⁶
Heat value ⎰ B.Th.U./lb.	5820	
of wood fuel ⎱ kcal/kg	3230	
Energy content of ⎰ B.Th.U.	5·2 x 10¹⁵	4 x 10¹⁵
present output ⎱ kcal	1·3 x 10¹⁵	1 x 10¹⁵
Potential output, tons per acre	0·8	2
Potential global energy ⎰ B.Th.U.	32 x 10¹⁵	100 x 10¹⁵
income from wood-fuel ⎱ kcal	8 x 10¹⁵	25 x 10¹⁵

It would be wrong, however, to expect that the problem of
future global energy supply could be solved in this way. The reclamation and adequate management of all the virgin forests of the
earth—most of them lying in underdeveloped areas—would be a
task of global dimensions for which neither the political nor the
economic situation is ripe. To perform it would take nearly a
century, because trees cannot be persuaded to grow faster than they
are used to do, and by the time of its completion the world's consumption of energy would have risen to several times the present
demand, which, as stated above, might just be covered by full
exploitation of the maximum potential growth of wood. Quite
apart from this fact, fire-wood is a much too bulky sort of fuel to

be suitable for world-wide distribution. As can be seen from Fig. 49, Chapter VII, the transport costs of wood-fuel over long distances are prohibitive.

While, therefore, better cultivation of wood would be no solution of the global energy problem, it might be of great importance for certain single countries gifted by nature with particularly rich resources. Among them are Canada, Finland, French Equatorial Africa, and Sweden, where, after full development of all water-power sites and proper management of all forests, the energy income from recurrent sources might transcend the home demand. This does not mean, however, that these countries would no longer depend on fuel imports unless better economic methods than those used to-day were developed for converting wood into liquid or gaseous fuels. The primitive and rather uneconomic method of generating gas as a motor-fuel from wood in single tiny plants on each car was used in Germany as an emergency measure during the Second World War, but was abandoned as soon as petrol became available again. Future conversion method would employ large plants in which a well-devised combined production of gaseous and fluid hydrocarbons, together with charcoal, can be achieved.

Farm Wastes

It can be seen from Fig. 67 that about three-quarters of the energy load of India is carried by farm wastes. This is possible in a country with much agriculture and little industry, where most of the energy input is used as heat for cooking and as space-heating. Cakes of dried dung are a very commonly used form of fuel in those countries of the Middle East and of Asia where wood is rare and relatively expensive. Straw, stalks, and corn-cobs are also burned as fuel. On account of the large population in the whole zone of rather woodless countries between Casablanca and Rangoon the total amount of energy supplied from farm wastes is quite considerable. According to Putnam, it is nearly 20 per cent of the total world consumption, or about four times the contribution from wood. If all the present hydro-electric power were converted into heat in electric stoves it would supply only one-sixteenth of the heat gained from burning farm wastes.

Although every kind of fuel from recurrent sources is a most welcome income item of the energy budget of mankind, the use of dung-cakes as a fuel should be abandoned in order to allow the return of dung to the soil. The depletion of soil may become an even more grave problem in the long run than that of fuel shortage, and particularly the gradual and irresistible shift of the phosphates away from the lands of the earth to the bottom of the oceans may lead ultimately to serious consequences. Fertilizing material should, therefore, not be destroyed, and for this reason the progressive Government of India has inaugurated the production of cheap solar cookers, described in Chapter XIII, which are meant to replace the dung-cake fires for cooking.

There would be no objection, however, against burning the large quantities of cellulosic wastes, such as corn-stalks, cobs, straw, husks, and shells, which are associated with food production. According to Ayres and Scarlott, the heat content of these wastes in the U.S.A. is about 3·5 times that of the wood consumed. However, the costs of collecting, transporting, and converting wastes into marketable fuel would be many times too high for competition with petroleum products at their present prices.

Alcohol

The kind of fluid fuel which has been produced from recurring sources since early times is alcohol, made by fermentation of plants. Table 28, reproduced from Ayres and Scarlott, gives the yield of alcohol expected from different plants.

TABLE 28. ALCOHOL YIELDS FROM CERTAIN MATERIALS

Material	*Gallons per Ton*	*Gallons per Acre per Annum*
Wood	70	70
Corn	84	89
Potatoes	23	178
Sugar-cane	15	268
Sugar-beets	22	287

The yields are not sufficient to cover the demand of highly industrialized countries. The present oil consumption of the U.S.A. is nearly three thousand million barrels annually, with an energy content corresponding to about four thousand million barrels of alcohol. (The heat value of alcohol is about 70 per cent that of gasoline.) Not fewer than 2300 million acres of woodland would be required to produce this quantity of alcohol from wood, or 560 million acres of cropland to produce it from sugar-beets. Comparing these figures with the woodland area in the U.S.A., totalling approximately 700,000 square miles (= 1·8 million square kilometres), of which only one-third—that is, about 160 million acres —is marketable, and with the practically usable cropland area of approximately 350 million acres, we must reach the conclusion that the fuel demand of a country like the U.S.A. cannot be covered from fermentation alcohol. This result is based on chemical and geographical facts, and holds, therefore, quite independently of the additional economic difficulties.

Thus alcohol, in the same way as wood, cannot be used as a universal substitute for petroleum or coal products, but may make valuable contributions to the energy system in areas where other fuels are expensive or scarcely available, and where sugar crops are easily raised. New power-alcohol plants have lately been built, therefore, in Brazil, Jamaica, India, Pakistan, and in the Dominican Republic.

With wood as the raw material for synthesizing fluid fuels, the conversion into gasoline is likely to become more economic than the conversion into alcohol.

The Chlorella Experiment

All fuels used so far consist of carbon and/or hydrogen, or of compounds of these two elements. The building up of hydrocarbons from the inorganic substances—water, H_2O, and carbon dioxide, CO_2—originally present on the earth's surface has been done, and is done continuously, by the process known as photosynthesis. Under the action of sunlight in the presence of pigments like chlorophyll the carbon dioxide in the atmosphere is reduced, and the freed carbon is compounded in various ways with hydrogen taken from the water molecules.

Methods of large-scale production of fuels from vegetation will have practical value only if more annual energy output per acre can be obtained than at present. This output again depends on the tonnage of vegetation yield per acre, on the heat value of the crop, and on the efficiency of its conversion into marketable fuel. As shown in Table 28, the ultimate energy output is better, for instance, via sugar-beets and alcohol than via wood and alcohol. Since all kinds of fuel production from vegetation will be a most unwelcome rival, as to required acreage, of food production, the best method of obtaining fuel from plants will be that which not only requires as small an acreage as possible, but can also be performed in non-fertile areas.

The whole problem boils down, therefore, to the more general question: How can we best use photosynthesis with a minimum of arable land? Biochemical experiments have shown that the solution may be expected from certain types of marine micro-organism. Among the many different varieties of algæ, which range from rather complex structures ('seaweed') down to microscopically small particles, it is just the smallest ones, the single-celled algæ, which may be best suitable for large-scale culture on account of their high reproduction rate and certain other advantageous properties. More than 10,000 different species exist, and one of them, the *Chlorella pyrenoidosa*, has been carefully studied, and has been the object of experiments in a pilot plant which proved that the yields per acre—both concerning fat and protein content, and energy content—are much higher in the case of Chlorella than with the most economic land crops, such as corn, sugar-beets, or soya-beans. The reason can easily be understood. All land crops use solar energy for photosynthesis at a maximum rate during a short time just before harvest. And even this rate can be kept only if all weather conditions, such as wind and rain, sunshine and temperature, are at their optimum value, and if the right amount of fertilizer in the soil and of carbon dioxide in the atmosphere is supplied. The latter is, by the way, always below the optimum value.

All these factors, however, determining the rate of reproduction can be kept continuously at their optimum value, and can be mutually best adapted in artificial cultures like those used for growing algæ in a pilot plant constructed and operated for the Carnegie Institution by **Arthur D. Little.** The plant belongs to the category of hydroponics cultures, in which the vegetation does not grow on soil, but within an aqueous solution of nutrients exposed to daylight. In the pilot plant of Little the aqueous medium, enriched with nitrogen-compounds, is permanently circulated in an endless flat tube of transparent plastic material, which in addition to the fluid contains an atmosphere artificially enriched with carbon dioxide. Since Chlorella is not anchored by roots to a soil, but is freely floating in its nutrient solution—with some twenty thousand million single cells in one quart of the fluid—harvesting need not be performed batch by batch, but can be done continuously by passing the solution through a high-speed centrifuge which separates the grown-up cells. By keeping the right temperature and illumination the conditions can be made such that " the algæ are always at the height of their growing season."

The flow diagram of a possible cycle process of producing and burning fuel from Chlorella is given in Fig. 68. The carbon dioxide escaping from the furnace is fed to the tubes containing the nutrient solution, and, after being reduced there by photosynthesis, is used for building up more hydrocarbons in the Chlorella.

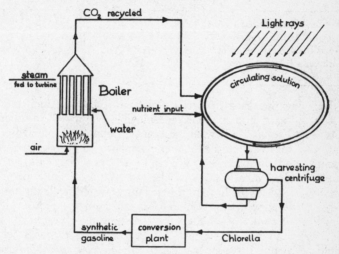

FIG. 68. FLOW DIAGRAM OF CHLORELLA POWER-PLANT

A plant like that sketched in Fig. 68, in which the energy-supplying element, carbon, is freed from CO_2 by the action of daylight, can be considered as indirect solar-power work. Direct solar-power plants could be made with greater efficiency. The advantages and disadvantages of both systems will be discussed in Chapter XIII, dealing with solar energy.

Under laboratory conditions a reproduction rate has been achieved corresponding to an annual acre-yield of some 75 tons dry weight of Chlorella. With a heat value of between 9000 and 13,600 B.Th.U. per lb—that is, between 5000 and 7500 kcal/kg— the energy income per unit area is incomparably higher than with any other fuels produced from plants. Taking a heat value of 11,300 B.Th.U./lb as the arithmetical mean of the above-given data, and assuming that the dried Chlorella—a fine yellowish-green

powder—can be converted into gasoline by the Fischer-Tropsch
process with a thermal efficiency of 45 per cent, we obtain the
energy yields given in Table 29.

TABLE 29. ANNUAL ENERGY YIELDS OF FUEL PRODUCED FROM
VEGETATION (APPROXIMATE FIGURES)

Fuel	Million B.Th.U. per Acre	Million kcal per Hectare
Alcohol from wood	5	3·2
Alcohol from sugar-beets	22	14
Gasoline from Chlorella	832	526
Chlorella burned directly	1850	1170

By using the dried Chlorella powder directly as a fuel, instead
of converting it into gasoline, considerable savings of installation
costs would result, and twice the energy output could be achieved.
Assuming that half of the Chlorella yield would be converted into
gasoline, and the other fed directly to a power plant, the energy
content of the annual yield would be 1341 million B.Th.U. per acre,
or 848 million kcal per hectare. These figures would make the nation-
wide energy supply from Chlorella plants even for industrialized
countries a technically—though not economically—feasible proposi-
tion. This can be seen from Table 30, showing the area which
would be required to produce fuel from *Chlorella pyrenoidosa* at a
rate sufficient to cover the total present demand of energy.

Recent research has shown that another species of Chlorella
has a much better reproduction rate at temperatures lying
above normal room temperature. Considering that the nutrient solution, exposed to strong sunlight under a transparent cover, will be
warmed to well above the surrounding temperature, we can reasonably expect still better results than those given in Tables 29 and 30.

Any waste land might be used as the site of algæ plants, assuming only that the necessary, not too large amount of water could
be supplied, and, therefore, the rather moderate percentage of area
required would not in itself exclude the possibility of energy supply
from photosynthesis, as it did in the case of wood.

Still, the realization of such projects will have to be postponed
until the time—perhaps some centuries ahead of us—when not
only fossil fuels, but also the uranium reserves, are exhausted. For
as long as energy can be produced at the present relatively low

TABLE 30. TOTAL AREA OF THE U.K. AND U.S.A. AND PERCENTAGE
REQUIRED FOR ENERGY PRODUCTION FROM CHLORELLA

| | Country | |
	U.K.	U.S.A.
Total land area (millions of sq mi)	0·0943	3·2
„ „ (millions of km²)	0·244	7·8
Total yearly energy demand		
(10¹⁵ B.Th.U.)	6·8	34
„ „ „ „ (10¹⁵ kcal)	1·7	8·5
Area required for Chlorella cultures		
(sq mi)	5,550	28,300
„ „ „ „ (km²)	14,600	73,000
Percentage „ of total area	6%	1%

costs the gigantic expenditure needed for installing tens of thousands of square miles of Chlorella farms, requiring, among other
things, many million miles of plastic tubes, could not be afforded.

Summarizing, it may be stated that fuel production from growing vegetation, although being within the range of technical feasibility, cannot be taken into consideration as long as reserves of fuels
exist which are as cheaply available as those we are using to-day.
This statement should not, however, be understood to diminish the
importance of the algæ experiments altogether. For they have been
made with a different aim in view. Dried Chlorella contains 50 per
cent protein, 20 per cent fats, 25 per cent carbohydrates, and 5 per
cent ash. The large output per acre, together with the possibility of
using non-arable land, may show the way to increase food production in the future, when conventional agriculture no longer suffices
to meet the demands of the ever-growing world population.

CHAPTER XII

Water-power

The Units MW and TWh

THE reader may be reminded once more that the strict meaning of power is energy per second, and that the internationally accepted unit is the kilowatt (kW), while the unit of energy is the kilowatt-hour (kWh). The formulæ for conversion to other units are

$$1 \text{ kW} = 1 \cdot 341 \text{ H.P.}$$
$$1 \text{ kWh} = 860 \text{ kcal}$$

In order to avoid unnecessary use of very large numbers for giving the power and the energy production of large plants, larger units will be used in this and the next chapter. (It may be mentioned incidentally that the distance between Liverpool and New York is usually expressed in miles or kilometres, but not in inches or feet.) By using M = mega for a million, G = giga for 10^9, and T = tera for 10^{12}, we get

$$1 \text{ MW} = 10^6 \text{ W} = 1000 \text{ kW}$$
$$1 \text{ GW} = 10^9 \text{ W} = 10^6 \text{ kW}$$
$$1 \text{ TW} = 10^{12} \text{ W} = 10^9 \text{ kW}$$

The power of large hydro-electric plants will generally be given in MW (Megawatts), the energy produced annually in TWh.

$$1 \text{ TWh} = \text{one thousand million kWh}$$
$$1 \text{ Q} = 293{,}000 \text{ TWh}$$

The use of water-wheels, though one of the earliest-known methods of power production, is still to-day the best of all methods as regards economy, efficiency, cleanliness, reliability, and inexhaustibility. The short episode in human history of ample energy production from fossil fuels (see Fig. 65), and even the possibly somewhat longer period of a high standard of living from atomic energy, will long have been over when the giant water-power plants we have built to-day, and more and larger ones built in the future, will indefatigably still be doing their beneficial work for the comfort and welfare of mankind. As long as climatic conditions permit the existence of life on our planet there will be rainfall and rivers ready to supply power.

Developing water-power sites is very often a multi-purpose enterprise which, in addition to the supply of valuable electric power, offers the possibility of installations for flood-control, irrigation, and inland waterways. Very large combined projects of this kind have been completed already—for instance, the gigantic enterprise of the Tennessee Valley Authority (TVA)—and others, still larger, are under construction in the U.S.A. and the Soviet Union.

Viewed from the standpoint of world power-supply, however, a disadvantage lies in the fact that water-power, though inexhaustible, is far from being available in sufficient quantities to cover the global demand for energy. Hydro-electricity accounts to-day for about 1·5 per cent of the world's total energy requirements, and full development of all economically justifiable sites is expected to increase the amount of water-power to about twenty times the present output. By the time this task has been performed the total consumption will have risen so far that the twentyfold hydro-electric power may amount to perhaps 5 per cent of the total demand. Another disadvantage lies in the fact that the geographical distribution of abundant water-power resources differs widely from the distribution of regions with a large power demand, and that the transmission of electric power over distances of more than about 1000 miles, or 1600 km, is scarcely possible without either great transmission losses or enormous installation costs of the lines. (See Chapter VII.)

Types of Water-wheels

The technique of using water-wheels for doing mechanical work is perhaps three thousand years old. From the time of the Babylonian Empire, through all antiquity and the Middle Ages, flour-mills and, later, forges and sawmills, have been driven by water-power, and many thousand installations of this kind are still in use all over the globe. As explained in Chapter VII, the first successful electrical power transmission from Lauffen to Frankfurt-am-Main in Germany opened in 1891 the way for making use of distant water-power in places hundreds of miles away from the river. Since the beginning of the twentieth century the use of water-power for generating electricity has made rapid progress, so that to-day the ratio of hydro-electric power to water-power used directly in mills, etc., may be more than ten thousand to one.

Along with the amount of total output some progress has been made in quality of equipment. The primitive mill-wheels shown in Fig. 69a and 69b, with flat blades, have a rather poor efficiency, but this defect was not felt seriously in view of the modest power requirements of the mills. In larger plants, however, where a small percentage of loss in power would involve serious losses in income,

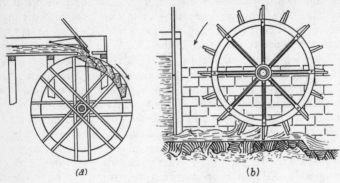

FIG. 69. MILL-WHEELS
(a) Overshot wheel; (b) undershot wheel

the necessity arose for making improvements, and thus the three types of water-turbines shown in Figs. 70 to 72 were developed, each of them serving best within a certain range of power-head. (The difference in height between the water-levels above and below the power-plant is called its ' head.')

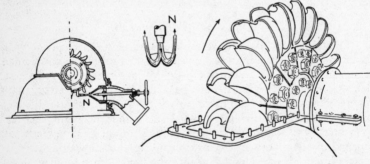

FIG. 70. PELTON WHEEL
N = nozzle.

As explained in Chapter 1, the potential energy of water rushing down the penstocks of a power-plant is converted into kinetic energy, and this, again, is converted into useful work when the water by its impact on the blades drives the water-wheel. A hundred-per-cent efficiency of the latter conversion process is achieved by consuming all the kinetic energy—that is to say, if the water after hitting the blades comes to rest. This would not be the case if a flat blade of the old mill-wheel type were exposed to a speedy jet of water coming down from a considerable height, because a good deal of kinetic energy would be left in the turbulent motion of the water sputtering to all sides after having hit the blade. To avoid these losses the Pelton wheel is fitted, therefore, with blades in the shape of a twin cup, as shown in Fig. 70. The jet, pointed symmetrically to the middle ridge of the twin cups, is split into two halves, each of which runs smoothly along the walls of the cup, and is ultimately turned round 180 degrees from its original direction without appreciable loss of its velocity relative to the blade. If v is the 'absolute' velocity of the water—that is to say, its velocity relative to the power-house—and if the peripheral velocity of the turbine-blades is kept at $v/2$, then the velocity of the water relative to the receding blade is $v - v/2 = v/2$, and the absolute velocity of the deflected jet will be $v/2 - v/2 = 0$. Even if the latter condition is not exactly fulfilled, so that a small percentage of the initial velocity is left over in the escape-water, the loss may

not be large, because the kinetic energy is, according to equation (6), Chapter I, proportional to the square of the velocity. If, therefore, 5 per cent of the initial velocity remains in the water outflow the relative loss is not 0·05 times the total energy, but $0·05^2$ = 0·0025 times the energy, or a quarter of 1 per cent.

Pelton wheels are used in water-power plants with relatively small flow and high heads. A free jet flowing with high velocity from the nozzle (N in Fig. 70) strikes the blades of the wheel, setting it into motion.

In plants with greater water-flow and medium head no free jet is used. In the Francis turbine, shown in Fig. 71, the

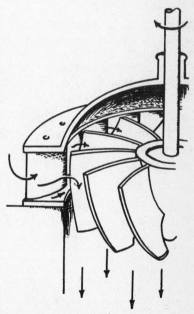

FIG. 71. FRANCIS TURBINE

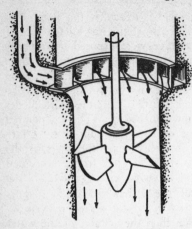

FIG. 72. KAPLAN TURBINE

water flows in a ring-shaped canal, fitted with directional vanes, obliquely from the circumference towards the axis of the turbine. The streaming water is then deflected by the suitably shaped wheel-blades, and leaves the turbine flowing in a direction parallel to the axis of the wheel.

In river plants with low heads and very great water-flow the Kaplan turbine (Fig. 72) is preferred, which is a kind of inversion of the ship's screw or the plane propeller.[1] In the same way as a rotating screw exerts a force in an axial direction on the medium through which it moves, a moving fluid will cause the rotation of a propeller which is placed with its axis parallel to the direction of the flow. The characteristic feature of the Kaplan turbine is the adjustable angle of inclination of the blades, so that by altering the pitch of the screws according to different velocities of the water-flow a permanently good efficiency can be achieved.

The ranges of application of the three different types of water-wheels are shown in Fig. 73 at p. 237.

Power Capacity

According to equation (5), Chapter I, the potential energy E_p of a body with a mass of m grammes at a height of H centimetres above ground-level is given by

$$E_p = mgH \text{ ergs} \dots\dots\dots\dots\dots\dots\dots(34)$$

with $g = 981$ cm/sec^2, denoting the gravitational acceleration. With the metric ton $= 10^6$ grammes as the unit of mass, the

---

[1] It may be mentioned incidentally that both the ship's screw, by which more than 90 per cent of marine tonnage on all waters of the globe is driven, and the Kaplan turbine, which is installed in power-plants with a total power output of many millions of horse-power, have been invented by Austrians. Josef Ressel was the inventor of the ship's screw, and his screw-driven steamer *Civetta* made her first cruise in the harbour of Trieste in July 1829, when paddle-wheel-driven steamboats were to be in general use for another generation. Victor Kaplan, an Austrian engineer, invented his turbine at the beginning of this century.

metre = 10^2 cm as the unit of length, and the kilojoule = 10^{10} erg as the unit of energy, equation (34) reads

$$E_p = 9.81mH \text{ kilojoules} \quad\text{...........................(35)}$$

Denoting the mass of water flowing down per second as the water-flow Q, and assuming that the potential energy is fully converted into useful work, the power—that is, the work done per second—is

$$P = 9.81QH \text{ kilojoules/sec} (= \text{kilowatt}) \quad\text{...............(36)}$$

Since a cubic metre of water weighs one ton—the small deviations from the standard value 1·000 of the density of water at temperatures differing from 4 °C. can be neglected—the flow Q can be given also in cubic metres per second. Neglecting further the small difference between 9·81 and 10, we can bear in mind the thumb-rule that the gross power of a water-power plant is ten times the product of water-flow and head, the power being expressed in kW, the water-flow in m^3, and the head in metres. The net power delivered to the consumer is less on account of the losses in the turbines, the dynamos, and the transmission lines. The overall efficiency of the conversion from potential energy of the water to electric power delivered from the plant can be estimated at about 75—80 per cent, and therefore a (rather optimistic) thumb-rule for net power is

$$\text{Net power} = 8 \ QH \text{ kW} \quad\text{.........................(37)}$$

The velocity of the free jet impinging on the blades of a Pelton wheel can be computed by equalizing the potential energy of the water and the kinetic energy into which it is converted in rushing down the penstocks. From

$$\frac{m}{2} v^2 = mgH \quad\text{..................................(38)}$$

the equation for the velocity v is derived

$$v = \sqrt{2 gH} \quad\text{......................................(39)}$$

Table 31 gives the variation of flow velocity with height of the fall.

The head of a power-plant is a quantity determined by the local conditions, and is fixed within certain limits. Small variations occur in storage plants fed from a reservoir according to the state of filling. The flow, however, is subject to wide variations, and therefore the power supplied from the plant to the grid also varies

TABLE 31

Head (metres)									Velocity (m/sec)
1	...	...	...	...	...	...	...	...	4·43
2	...	...	...	...	...	...	...	...	6·26
5	...	...	...	...	...	...	...	...	9·9
10	...	...	...	...	...	...	...	...	13·9
20	...	...	...	...	...	...	...	...	19·8
50	...	...	...	...	...	...	...	...	31·3
100	...	...	...	...	...	...	...	...	44·3
200	...	...	...	...	...	...	...	...	62·6
500	...	...	...	...	...	...	...	...	99
1000	...	...	...	...	...	...	...	...	139

within wide limits. The maximum power output for which the water-wheels and generators of a station are designed is called the *installed capacity* of the plant, and is one of its most significant data. Fig. 73 is a chart in which the heads of a group of power stations of different sizes, together with the flows used at maximum power capacity, are represented graphically. On the left side of the chart are shown the ranges of heads within which the three main types of water-wheel are used.

Annual Production

Although water-power plants have definite advantages over thermal-power plants, there is one serious drawback. In contrast to the thermal plants, which can be run at full capacity for long periods if required, the power output of a hydro-electric station may be limited by the available water-supply, quite irrespective of larger or smaller demand from the consumer side. A second characteristic datum of a plant besides its capacity is, therefore, the annual energy production to be expected under normal conditions from the seasonally varying flow. Thus, for instance, the Kaprun main stage (No. 8 in Fig. 73), with a capacity of 220 MW, would produce 1900 million kWh annually if the water-supply were sufficient to run the turbines at full capacity all the year round. Actually its production averages only 460 million kWh per year.

In surveying the water-power resources of a country or of a continent the relevant datum is, therefore, not the sum of the installed capacities of all stations, but the total annual production. Denoting by F the annual water-flow, expressed in cubic kilometres per year, and assuming an overall efficiency of about 74 per cent,

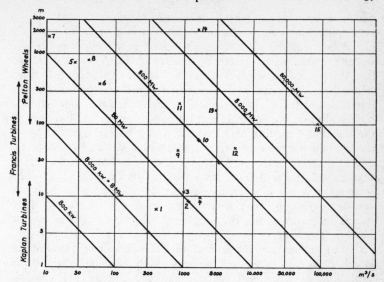

FIG. 73. HEADS AND WATER-FLOW AT MAXIMUM CAPACITY OF DIFFERENT
WATER-POWER PLANTS

Installed capacity given in MW: (1) Oberaudorf, river Inn, Austria, 20 MW;
(2) Birsfelden, Switzerland, 82 MW; (3) Braunau, Inn, 90 MW; (4) Jochenstein,
Danube, 140 MW; (5) Vermunt, Ill, Austrian Alps, 140 MW; (6) Rodund, Ill,
170 MW; (7) Chandoline, Switzerland, 190 MW; (8) Kaprun, Austria, 220 MW;
(9) Aura, Norway, 282 MW; (10) Dnjeproges, Russia, 650 MW; (11) Hoover Dam,
Colorado, U.S.A., 1300 MW; (12) Kujbyshev, Volga, U.S.S.R., 2100 MW; (13)
Grand Coulee, Columbia river, U.S.A., 2370 MW; (14) Pe, Tsangpo, Eastern Tibet
(undeveloped site, estimated capacity >17,000 MW); (15) Gibraltar Dam (Utopian
project, estimated capacity 63,000 MW).

we obtain for the annual production **E** expressed in gigawatt-hours
(1 GWh = 1,000,000 kWh), the thumb-rule

$$\mathbf{E} = 2FH \text{ GWh} \dots\dots\dots\dots\dots\dots\dots\dots(40)$$

This formula can, however, be applied only on the assumption
that the whole water-flow can be utilized without discharging surplus water over the spillways at times when the flow exceeds the
installed power capacity of the water-wheels, or when the power
consumption is too low.

The Problem of Balancing Supply and Demand

Even with equal capacities and equal annual production two
hydro-electric plants may differ considerably in their economic value,

which is very much dependent on the seasonal variation of its
availability. The trouble with hydro-electric power lies not only
in the fact that the most abundant resources are not situated con-
veniently near places where the energy demand is greatest, but
that the maximum water-flow is very often not available at times
when the power consumption reaches its peak. Large water-power
resources are found, for instance, in mountainous regions and in
higher altitudes where in late spring and summer, at the time of
the maximum melting of snowfields and glaciers, plenty of water
flows down the mountain rivers, while in winter and early spring,
when the whole landscape is covered by snow and ice, the water-flow
may be reduced to a small percentage of its summer value. This
seasonal variation of water-flow is just out of phase with the variation of demand, because the consumption of electric power reaches
its maximum in winter on account of the increased requirements
for light and heat.

There is also a wide diurnal variation in power consumption
which does not correspond, anyway, with a similar variation in
water-flow. As can be seen from Fig. 48, Chapter VII, the load
curve of a power station shows several sharp peaks during day-time
and a very deep low in the small hours of the morning. While in
a thermal plant the steam production, and therefore the fuel consumption, can be adapted to the power level of the hour, the water-
flow of a big river hardly varies in twenty-four hours, and therefore
in run-of-river plants with no pondage basin for daily balance the
surplus water is discharged through the spillways during the hours
of small load, so that valuable energy is wasted.

Several measures are taken in order to balance supply and
demand. Attempts are made to smooth out the daily load curve by
shifting a part of the load to night power. Water-boilers and
specially adapted stoves for space-heating can be constructed heavily
enough for their thermal inertia to keep them warm during the
daytime after they have been heated by night current. The connexion to this kind of load is automatically switched on at night
and off in the daytime, and the kilowatt-hours consumed by these
devices are recorded on a separate electricity meter, so that a special
account can be made for night power, which is sold at reduced
rates.

No perfect smoothing out of the diurnal load curves, and still
less of the seasonal fluctuations of power consumption, is possible,
however, and some other methods have to be used in order to
reduce energy waste caused by the remaining unevenness of the

curves. An important step is the transition from the system of local power supply to a compounded power economy by feeding high-tension current into a national or regional grid of transmission lines which distributes power over a greater area.

If the grid is fed from a combination of thermal and hydro-electric plants the run-of-river stations can carry the base load operating most of the time at nearly full power capacity, while the peak load is carried by thermal plants which can be run only as long as additional power is required.

Reservoirs

Among the hydro-electric stations there are also plants whose power output can be adjusted without spilling the excess flow— namely, the *storage plants,* which are fed from a large reservoir. It is particularly the stations with large head (some of them being represented in the upper half of Fig. 73) which are best suited to be storage plants, because the product of the head H and the volume of the reservoir is proportional to the stored energy. This is the reason why within the last decades gigantic dams have been built in several places in the world for reservoirs in which enough hydraulic energy can be stored for the supply of power at times when the demand is greatest. On account of their flexibility of power output, storage plants fed from large high-level reservoirs are the most valuable kind of power station, but the necessary capital invest-ment for their construction is very high too. Table 32 gives some data of the dimensions of large dams and their reservoirs. Lake Mead, in the Boulder Canyon of the Colorado, dammed up by the Hoover Dam, exceeds all other reservoirs in its energy-storing capacity, which is greater than the annual production of the plant. Hence, even if the Colorado river ceased to flow one day the water stored in Lake Mead would suffice to keep the station working at full power for more than a year. With a reservoir like this, the seasonal variations of water-flow can be fully compensated.

The ratio of stored energy to annual output will be still greater in the case of some of the reservoirs to be built in the Soviet Union (see below under multi-purpose plants). Even with relatively small reservoirs, however, a sufficient balance between supply and demand can be achieved by suitably compounding the power from storage plants and run-of-river stations. An example is given by the Austrian federal grid, to which power is supplied from thermal stations as well as from hydro-electric plants of both kinds. In some of the storage plants, like that of Kaprun, for instance, two reservoirs

TABLE 32. DAMS AND RESERVOIRS

NAME	DAM			RIVER	RESERVOIR			
	Height (m)	Length (m)	Volume (million m³)		Area (km²)	Capacity (km³)	Head (m)	Stored Energy (TWh)
Hoover	220	380	3·36	Colorado	683	36	210	10
Oahe	73	2340	60	Missouri	1450	29	70	3
Garrison	63	3660	53	Missouri	1800	28	60	2·4
Fort Peck	76	6400	96	Missouri	993	24	70	2·3
Grand Coulee	167	1270	8·1	Columbia	329	11·7	160	2·6
Limburg	120	350	0·446	Kaprun	1·2	0·084	850	0·16

lying at different levels, each with a power-house, are provided. At times of large demand both stations are running and supply power to the federal grid. At times of very low consumption, however—for instance, during summer nights, when the power required is less than that supplied from the run-of-river stations alone—the lower (main) stage of the Kaprun plants is stopped, while in the upper stage the dynamos are used as electric motors driving water-pumps which are pumping water from the lower basin to the upper one. The power for the motors is drawn from the grid, which itself is fed with dump energy from run-of-river plants. In compounded supply systems with sufficient capacity of pumping installations it will no longer be necessary to spill the surplus water of run-of-river stations, and thus to waste energy. Instead the whole power of the rivers can be utilized, partly for supplying the current demand, partly for storing energy in the higher reservoir, which is tapped again at the time of peak consumption.

International Power Trade

Trading power across state frontiers is another method of avoiding waste of surplus water-flow, since a well-balanced supply may be the easier obtained the larger the market. There are neighbouring countries with complementary power resources which can well profit by bartering energy gained from different sources. There is abundant developed water-power in Vorarlberg, in the western corner of Austria, where transmission to the capital, Vienna, over a 600-km line would involve rather considerable losses. It is more economic, therefore, to export hydro-electric power to the neighbouring parts of Germany, which, though rather poor in water-power, are rich in coal, and can, therefore, easily recompensate by supplying peak power from their thermal plants.

Objections have been raised against the export of electric power. One argument stresses the fact that electric energy produced from inland water-power plays an analogous rôle to that of valuable raw materials, which should, if possible, not be exported as such, but employed for the manufacture of finished products that can be exported with higher profits for the country. Another target of the critics is the considerably lower price obtained for the kilowatt-hour of exported power from hydro-electric plants, compared with the price paid for imported power. It may seem at first glance a very bad bargain to export any goods at prices definitely lower than those for which the same goods must be re-imported some time afterwards.

The answer to the first objection is that in certain cases geographical and demographic conditions of a country are such that the development of abundant water-power resources may be far ahead of its industrial production capacity, so that the available hydro-electric energy will surpass the domestic requirements for years to come. In reply to the second objection it should be stated that electric energy—though valuable at any rate—has no fixed value irrespective of the time of purchase. An ounce of gold costs pretty much the same in winter as in summer, while the use value, and in many cases also the price, of the kilowatt-hour depends on the season of the year, and even on the hour of the day. The reason lies in the peculiar character of production and in the difficulties of storing electric energy. Although it is possible in principle to store electric energy, the installation costs for storage plants are so enormous that in general only a small fraction of the energy produced can be stored. It is therefore quite justifiable for hydro-electric energy to be sold and exported at cheap prices in summer while in winter higher rates are paid for thermally produced power. It can turn out to be much more economic for a country to import 100 million kWh of peak power from its neighbour in exchange for, say, 150 million kWh of exported summer energy than to make large capital investments for plants which are needed only for a short time of the year.

Multi-purpose Water-power Plants
(A) Tennessee and Missouri

Modern economy demands that man's natural resources, and also the installations for using them, should be fully exploited without appreciable waste of material and energy. While this rule is fairly well obeyed in the chemical and metal industries, its general application to power production is slow to follow. In the majority of existing fuel-operated power stations about three-quarters of the fuel energy is run to waste with the lukewarm cooling water, and only now the system of utilizing reject heat of steam plants for district heating or as process heat is going to be adopted more generally in designing new stations. There will be ever more dual-purpose thermal plants in future, supplying both electricity and heat, electricity being the main purpose and heat the by-product.

One of the most decisive advantages of hydro-electricity over all other known methods of power production lies in the fact that the developments of water-power sites can serve quite a number of other important purposes besides the generation of electricity, as, for instance, flood-control, irrigation, navigation, and recreation. In designing such multi-purpose projects one cannot confine the work to the planning and building of a single dam plus its power station, but must carefully consider the totality of the river, its tributaries, and the whole country through which it flows. As D. E. Lilienthal (1944) emphasizes in his book on TVA (Tennessee Valley Authority), it is not only the head and the water-flow of the plant to be built that matters, but " the oneness of men and natural resources, the unity that binds together land, streams, forests, minerals, farming, industry, mankind." By properly taking into account all these issues, and by extending the planning of hydro-electric plants to the entire region of interdependent areas, work can be performed the power production of which is only the smaller part of the total benefits afforded to the country.

The necessity and utility of flood-control are not so strikingly obvious under normal conditions as, for instance, the availability of cheap electric power. Still, viewed from a wider perspective, and taking into account the statistics of damage, material losses, and casualties caused by the ever-recurrent floods in many parts of the world, it turns out that the prevention of such disasters is well worth the capital invested. People are quick to forget events which did not harm them personally; it may be remembered, therefore, that among the innumerable floods that have plagued mankind since Noah's days there have been, within the last two decades alone, two really devastating ones—namely, the Ohio and Mississippi Valley disaster of 1937, which made over a million persons homeless and killed 466 people, and the big Indian flood of August 1943, which drowned 10,000 and caused indirectly the death of more than a million through the subsequent famine. In contrast to earthquakes or volcanic eruptions, the force of a river can be kept under control by constructing dams and reservoirs, not, however, in one place alone, but along the whole river system, and thus harnessing the entire network of the watershed. It does not suffice to turn the energy of the falling water into electricity; what is needed is, rather, collective action from all the members concerned.

The group of the TVA plants consists of twenty-eight dams and as many power stations, in which the operators of all the hundreds of outlet-tubes and spillway-gates of the dams receive orders from the office of central control. The water-dispatcher on duty is continuously kept informed by the head of the TVA forecasting division, who himself receives reports from three hundred stations gauging rainfall and flow of the streams, so that the height of water can be predicted for days in advance. As soon as warning messages come in from the headwaters of the river the water-dispatcher begins to play on his giant instrument of dams (twelve times as heavy as the seven great pyramids in Egypt, taken together), like a master organist drawing and stopping his registers. The operators of the downstream reservoirs are ordered to release water to make room for the water from above; the operators of the plants on the tributaries outside the area of heavy rainfall are instructed to hold back the water and keep it from adding to the main river; and thus the heavy output of rainfall, which, uncontrolled, would have been discharged in a devastating wave of flood, can be wisely released during a longer period without rising to a fatal peak. The beneficial effect of measures taken within the region controlled by the TVA is not restricted to that territory, but extends still farther downstream by preventing the flood-crest from adding what might be fatal inches to top the levels and spread desolation on the lower Ohio and Mississippi.

According to Lilienthal's report, excessive rainfall in Tennessee and Virginia, which occurred in the winter of 1942, would have caused floods seriously threatening the machinery of vital industries down the river at Chattanooga without the aid of the then completed dams operated in the manner described. It is regrettable that human achievements like these are scarcely recorded at all in history, while catastrophes, and still more the results of human follies like wars, fill pages and pages of articles and books. The saviour gets less publicity than the killer.

Even greater economic importance than flood-control can be attributed to the task of irrigation and rebuilding soil. Not only the arid zones of the globe, but also areas with quite a large amount of annual rainfall, cannot contribute to food-supply because the amount of rain is too unequally distributed over the year. If for three or four months hot sunshine beats down daily from a cloudless sky the vegetation starves, and the soil, robbed of its protecting cover, is mercilessly exposed to the eroding action of wind and water. During the next rainy season torrents raging down the gullies carry away innumerable tons of valuable soil, and almost irreplaceable fertile material is lost to the country for ever. Quite apart from the vast deserts, such as the Sahara, Gobi, etc., the areas within the temperate zone of the earth which could be turned into fertile land by first levelling out the rainfall and subsequent proper reclamation are in their totality a region which could feed millions of people. A striking example of what can be done in this respect within a particular area is given by the results achieved in the states profiting from the work of TVA. According to Lilienthal, the increases in production levels in the Tennessee Valley within the decade from 1933 to 1943 were: small grains increased 13 per cent, corn maintained at the same level, but using fewer acres; hay increased 33 per cent. But on the demonstration farms, where new farming methods taught by the TVA-sponsored Farm Improvement Associations have been put to use, the rate of increase was three times greater. Under the new method of farming the same acreage and the same man-power produced from 30 per cent to 60 per cent more meat, eggs, milk, and dairy products.

Considering that these results have been achieved shortly after the first tentative beginnings, and that, moreover, other areas with a total acreage a hundred times larger need rebuilding soil far more urgently, one can easily realize what an enormous task lies before us, and what benefit could be done by extending to other countries the kind of work TVA did in the United States.

The construction of inland waterways is another kind of work which can be combined with the development of hydro-electric power sites. Transporting any goods adds to their costs without improving their quality, and is, therefore, a loss for a trade. Since shipping is still the cheapest method of transporting over long distances, the possibility of shifting the transport from roads or railways to navigation will cause decreased losses, and accordingly saving in production and distribution expenses. It is, therefore, an appreciable advantage for inland towns when they are turned into ports on one of a system of interconnected lakes between which an extensive shipping traffic is kept going. Many towns in the seven states belonging to the TVA area—as, for instance, Decatur, in Alabama—were lucky enough to become ports, and the slow but steady process is beginning to work which makes prospering cities out of these formerly small provincial towns.

Shorelines for public recreation, with parks, playgrounds, and bathing-beaches, are assets of a country which cannot be expressed in cash value. Anybody who is aware of the enormous flow of men and vehicles pouring down from London to the coast resorts in summer days will easily realize what it means for people living far from the seashore when beautiful lakes are created in the immediate neighbourhood. Nine thousand miles of shoreline, says Lilienthal, are available for the recreation of the people within the area developed by the TVA.

The development of all the sites of a watershed as an entity is advisable even if viewed from the standpoint of power production alone. The reason lies in the danger that single reservoirs will lose their capacity by being filled with silt in the course of years. Water-power itself is certainly inexhaustible, because rain will not cease to fall on earth for ages. But the most valuable and also most expensive part of the installations for using water-power—namely, the reservoirs—may be made useless by sedimentation of mud in a relatively near future unless the upstream parts of the river are harnessed by a series of other dams. Thus Lake Mead, behind the Hoover Dam, in the Boulder Canyon of the Colorado, which is No. 1 of all reservoirs as regards energy storage, has already lost 4·5 per cent of its capacity in the first thirteen years of its existence, and would, therefore, not be able to fulfil its function of levelling out the variations of water-flow for an unlimited period if steps were not taken to stop the influx of sediments. Under present conditions a whole cubic kilometre of sand and silt is deposited in the lake every seven years. The reason for this vast influx lies in the fact that Lake Mead is not only the first on our list of great storage basins, but also the first reservoir which the waters of the big Colorado river meet after a thousand-mile journey from its source. With a chain of upstream dams and lakes cascading the drop of the water-flow, and by regulating the torrents which cause the erosion of the valley's walls, the sedimentation can be reduced enough to extend the lifetime of large reservoirs to more than a thousand years. In periods of that length enough capital can be accumulated from the revenues of power production to pay for the installations and the labour required to dredge sediments from the bottom of the reservoirs and carry them away. Thus a fully developed water-power site can be made to be a capital investment of incomparably longer working life than any installation for recovery of fossil fuels.

What has been done by TVA in the south-eastern states of the U.S.A. is going to be repeated on a still larger scale in the Missouri river-basin, where reservoirs are being built with the primary purpose of flood-control, irrigation, and navigation, with power in some of the plants only the by-product. Table 33 gives a comparison between the water capacities of the five greatest reservoirs of the TVA sites and those partly completed and partly under construction in the Missouri river-basin.

Owing to the larger drainage area and to the greater differences in level in the Missouri basin a considerably greater power

TABLE 33. CAPACITIES OF TVA AND MISSOURI RESERVOIRS GIVEN IN CUBIC KILOMETRES

($1 km^3 = 8.11 \times 10^5$ acre-feet)

TVA		Missouri River-basin	
Kentucky	7·4	Oahe	29·1
Norris	3·2	Garrison	28·3
Cherokee	2·0	Port Peck	23·9
Douglas	1·8	Fort Randall	7·8
Guntersville	1·2	Canyon Ferry	2·5

production can be expected there than from the TVA plants. The full development of the Missouri power sites will, however, take some time.

Multi-purpose Water-power Plants
(B) The Soviet Projects

In the field of hydro-electricity the U.S.A. is leading as regards total annual production, while Norway is far ahead of all other countries in *per capita* production. In respect, however, of total utilization of water resources for combined purposes the Soviet Union will become first if she succeeds in completing the giant works which are partly under construction and which are partly very bold plans needing decades of uninterrupted peace for their execution.

The power-plant of Dniepropetrovsk-Zaporoshie, on the lower Dnieper (originally completed in 1932, destroyed by the Germans during the War, and reconstructed since with an installed capacity of 650 MW), is so far the greatest single hydro-electric station in Europe, but it will be surpassed by the two big Volga plants at Kuibyshev (formerly Samara) and Stalingrad, which are designed for a capacity of 2100 MW and 2300 MW respectively. When completed in 1956 they are expected to contribute together 22 TWh = 22,000 million kilowatt-hours annually to the energy system of the Soviet Union. This is about one-fifth of the U.S.A. hydro-electric production in 1954, but the area of the reservoirs on the Volga, and particularly the amount of reclamation and irrigation to be done according to the Soviet plans, will be considerably greater than what has been achieved so far in the U.S.A. One of the main tasks of the newly developed sites will be the reclamation of an area of about 13 million hectares on the left bank of the Volga, reaching from the 55th Parallel down to the mouth of the river on the Caspian Sea at 47° latitude North.

The whole project, called the " Great Volga Plan," will provide for altogether twenty power plants on the Volga and its tributaries. The first big reservoir, completed in 1951, is the " Rybinskaja Sea " north of Moscow, with a surface area of 4750 square kilometres, or 1835 square miles, which is about seven times the area of Lake Constance, the largest Central European lake. The reservoirs dammed up at Kuibyshev and Stalingrad will be of similar dimensions, while the biggest of the artificial lakes within the Volga river basin will be that of the Solikamsk plant, on the upper Kama (one of the eastern tributaries of the Volga), with an area of 10,000 square kilometres, or 3860 square miles. The usable capacity of all the reservoirs of the Volga system will be 80 cubic kilometres, or about 65 million acre-feet, which is about one-third of its annual water flow.

While the large hydro-electric plants in Central Europe serve mainly for power production, with navigation and irrigation a minor by-product, the reservoirs and a dense network of irrigation canals of the Volga Plan are mainly intended to increase the food production, with power as a by-product. The irregular and greatly varying water-supply of the country in its natural state causes inundations and floods carrying away valuable fertile soil, while during the hot season severe droughts occur, causing clefts in the dry soil and killing mercilessly all vegetation. It has been estimated that the fertility of the Volga region might be increased enough to produce food for about 50 million people as soon as it is released from the ever-recurrent droughts by adequate regular supply of water.

The main arteries of the widespread network of irrigation canals will be two large canals which will be used as waterways. One of them has been built between Kalatch, on the Don, and the Volga bend near Stalingrad, so that naval traffic between the Black Sea and the Caspian Sea has already been opened.

Another canal of similar capacity is being built from the Zaporoshje reservoir, on the Dnieper (feeding the Dniepropetrowsk power-plant), to Nogaisk, on the Sea of Azov. This will furnish a short cut between the Don-Volga waterways and the Dnieper river, which in its turn will be connected to the Central European Bug-Vistula-Netze-Warthe-Oder network by the newly built Pripet canal.

According to the sixth Five Years Plan (1956–60), published in January 1956, the construction will be started of a series of plants in Siberia whose power will surpass that of the Grand Coulee, in the north-west of the U.S.A., which is at present the world's most powerful single hydro-electric plant. The reason lies in the particu-

larly favourable hydrographic conditions of the big Siberian rivers.
Lake Baikal, with its length of 400 miles, situated at 1540 ft above
sea-level, and fed from 330 rivers and rivulets, represents one of
the world's largest usable reservoirs. It discharges its waters through
the lower Angara into the upper Tunguska, which finally empties
into the big Yenisei. The Angara seems to be one of the rivers best
suited for hydro-power, and within a decade the district north-west
of Lake Baikal will become one of the most important industrial
centres of Asia. A first power-plant will be built near the influx of
the Baikal waters into the Angara, at Irkutsk. Its capacity of 660
MW will be just a little more than that of Dniepropetrovsk. But
farther down the Angara, at Bratsk, the high rocks on either side
of the river reduce the width of the valley to about half a mile,
so that a natural giant door is made which after being closed
by a dam of 300 ft height will dam up a reservoir of 340 miles
length and 180 cubic kilometres capacity—which is more than twice
the sum of all the reservoirs of the Volga system, or five times
that of Lake Mead. The Bratsk plant, with sixteen turbine generator
pairs of 200 MW power each, making a total power capacity of 3200
MW (or 4·3 million horse-power) will on completion take the rôle
of Grand Coulee as the world's biggest hydro-plant, but it will soon
be equalled by another 3200 MW station near Krasnoyarsk, on the
Yenisei. Two plants on the river Ob, one generating 400 MW near
Novosibirsk and another with 500 MW at Kamensk, are included
in the project of hydro-plants which will feed the great Central
Asian grid. As in the case of the Volga grid, the distribution will
be made by 400 kV transmission lines—that is, with voltages exceeding the highest ever used in Great Britain and Western Europe.

All the Siberian projects. enumerated in the sixth Five Years
Plan refer to the upper course of the rivers Ob and Yenisei and
their tributaries. They can be executed with relatively low costs
per MW installed power capacity, and are scheduled to be completed in the early sixties. The utilization of the lower Ob and
Yenisei is the aim of the so-called Davidov plan, which is a very
bold project with far-reaching consequences. The plan is shortly
sketched here, although its realization may be only a matter of the
tenth or eleventh Five Years Plan.

The idea is that the big Siberian streams Ob and Yenisei, with an
aggregate annual flow of about 900 km³, are wasting their water
by pouring it into the Arctic Sea, where nobody needs it, causing
at the same time frequent inundations and floods in Northern
Siberia. According to Davidov's project, the fact could be used

that the watershed between the Arctic Sea and Lake Aral runs in the wide flatland of Western Siberia across a very small elevation, the so-called Turgai Gate, at a point about 51° 35′ N. and 65° 30′ E. 23 km north-west of the town of Turgai so that a dam of only 78 m = 258 ft in height at Bielogorie (61°8′ N. 68° 30′ E.), near the confluence of the rivers Ob and Irtish, together with a deeply cut canal through the Turgai Gate, would suffice to send the water of the Ob over a distance of more than 1000 miles across a third of the width of the Asian continent into Lake Aral, and from there down through another canal to the Caspian Sea.

By further damming up the Yenisei where it crosses the 60th Parallel, and by enlarging the canal between the Yenisei tributary Kass and the Ob tributary Ket, additional water could be gained from the Yenisei, thus making available a total flow of about 300 km^3 annually for all the purposes to be achieved by enterprises of this kind. With a difference in level of 90 m = 300 ft between the planned Siberian reservoir and the Caspian Sea, together with the power-heads of the dams on the Ob and Yenisei, an annual production of 83 TWh = 83,000 million kWh is expected. This is little more than 80 per cent of the present total hydro-electric production of the U.S.A., and would, taken alone, scarcely suffice to justify the expenditure of capital and labour involved in a project of such exceptional dimensions. But the main purpose of the work would be the radical change of the map of Siberia, causing improvements in climatic, agricultural, and economic conditions of a considerable part of Asia. The swampy lowland on either side of the rivers Tobol, Irtish, and Ob will be converted into an inland sea 2000 km = 1250 miles in length, with a water area approximately equal to the land area of the United Kingdom and Northern Ireland, connected by navigable canals to the open sea both on the north and south of the continent. After completion of the work it will be possible to go by ship from the Mediterranean through the Dardanelles to the Black Sea and the Azov Sea up the lower Don, through the Don-Volga Canal and the lower Volga to the Caspian Sea; from there through the Turkmenian Main Canal to Lake Aral, and through the Turgai Gate Canal to the new Siberian Sea and down the lower Ob to the Arctic Sea. What has been said in connexion with TVA about the benefits of inland waterways and of turning provincial towns into harbour cities holds just as well for the development of traffic and trade in inner Asia. The main benefit for the country, however, would be the reclamation of about 25 million hectares of particularly fertile land with record harvests, yielding food for another 100 million people.

A plan like this sounds very tempting. Will the Soviet Union be able to mobilize the enormous amount of labour needed to realize the giant project? If, as planned for the ultimate stage of the work, the canal through the Turgai Gate is made wide enough for a waterflow of 11,000 m³ per second, the necessary soft earth dug-out would amount to 18 km³, or 18,000,000,000 m³ = 23,500,000,000 cu yds. Even with the efficient equipment of modern Russian machinery, which uses land-dredgers handling 300 m³ earth per hour, an army of 1000 machines working simultaneously would need 60,000 hours uninterrupted work to dig the canal. Apart, however, from the work on the canal two dams on the Ob and Yenisei, with a considerably greater volume than any existing dam, will have to be built, and also quite a migration of peoples must be organized if an area equal to that of Great Britain is set under water.

The realization of the Davidov plan will mainly depend on whether or not the world powers succeed in keeping peace for the next two or three decades. If we live to see the year 1980 without any major war the way will be open for a luckier humanity to spend its energy on productive instead of destructive work, because by that time the folly of destroying a world on account of ephemeral political quarrels will have become obvious enough.

World's Greatest Power-plant on the Tsangpo

There is another idea which to my knowledge has nowhere been widely publicized, but will appear as very tempting to all who study attentively a detailed map of Asia. It differs from the Davidov plan in the following points.

The project of creating an inland sea in Siberia has been made in a country particularly keen on progressive industrialization and electrification. It has, therefore, some chance of being realized, although the returns of capital investment expressed in terms of annual kilowatt-hours per dollar invested in construction work are relatively poor. The Tsangpo project, on the other hand, refers to a water-power site which lies in one of the most backward countries, far from any industrial centre, and has, therefore, little chance of being realized in our generation. Still, it deserves attention, because the Tsangpo bend in Eastern Tibet is of all places on the earth exactly the one which has by far the greatest concentration of water-power in one spot, and therefore the world's most powerful single hydro-electric plant is likely to be built there some day, perhaps about A.D. 2000.

The Tsangpo originates east of one of Tibet's holy places, Lake Manosarovar, from where it flows due east along the Tibetan highland between Himalaya and Trans-Himalaya, until, after a 1000-mile journey, it takes quite a sharp hairpin bend at latitude 29° 50′ N., longitude 95° 10′ E., surrounding the 25,445 ft peak Namcha Barwa, and breaking in a series of deep-cut gorges through the Himalayas. It flows then down to the plains of Assam, first under the name of Dihang, and finally as the Brahmaputra, which, crossing Eastern Pakistan and joining the Ganges, empties into the Bay of Bengal. The identity of the Tsangpo and Dihang-Brahmaputra had long been a subject of discussion, because as late as the beginning of this century no white man had succeeded in penetrating the Tsangpo gorges on the bend of the river, and it seemed doubtful whether the west-east bound Tsangpo, high up in the Himalayas, and the Dihang, flowing at a level more than 7000 ft lower in the opposite direction, were really identical. This geographical problem has been definitely solved in the meantime (see Bailey, 1914; Kingdon Ward, 1926), and it is just the combination of a sharp bend and an enormous drop which causes the remarkable concentration of water-power.

In Fig. 74 a sketch map of the general run of the Tsangpo-Brahmaputra is given, together with an enlarged map of the bend. By building a dam near Pe, above the gorges, and making a tunnel about ten miles in length under the pass Doshong La, the Tsangpo might be diverted into one of the side-valleys running into the Dihang at Yortong below Kapu, where the enormous power of the river could be utilized in a series of appropriately staged plants. Yortong, where the lowest power-house could be erected, lies at a distance of not more than twenty-five miles from the dam, but not less than 2160 m = 7140 ft below Pe, and the combination of an exceptionally high head with the water-flow of a great river makes the opportunity unique. The estimates about the water-flow of the Tsangpo at Pe vary between 34 km³ and 77 km³ annually. Inserting these values, together with $H = 2160$ m, into equation (41), we get an annual production of

147 TWh = 147,000 million kilowatt-hours (low estimate)
333 TWh = 333,000 million kilowatt-hours (high estimate)
240 TWh = 240,000 million kilowatt-hours (arithmetical mean).

Even according to the lowest estimate the annual energy would be about three times the total electricity produced in the United Kingdom, or one and a half times the total hydro-electric production of the U.S.A. The high estimate of 333 TWh would be approximately equal to the present total hydro-electric world production. But it must be borne in mind that equation (41) cannot be applied unless the storage capacity of the reservoir is great enough to use the total flow without spilling surplus water. Since the Tsangpo belongs to the rivers with strong seasonal variations

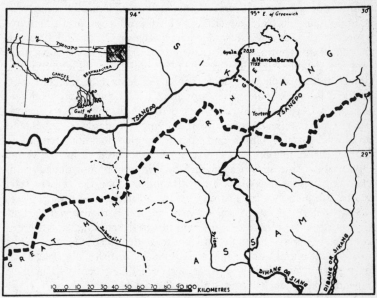

FIG. 74. MAP OF TSANGPO AND ITS BEND
The site of the projected tunnel is indicated by the broken line crossing meridian 95° E. *Scale (excluding inset) approximately* 1:3,000,000.

of water-flow, it would be necessary to build a reservoir with a storage capacity of between 50 and 80 per cent of the annual flow. This seems to be quite possible with one or two dams of not unusual dimensions.

The installed power capacity of the Tsangpo plant, adapted to the annual average flow, would be:

$P = 17,000$ MW $= 23$ million horse-power (low estimate)
$P = 38,000$ MW $= 51$ million horse-power (high estimate)
$P = 27,000$ MW $= 37$ million horse-power (arithmetical mean).

Nowhere else on the globe would it be possible to tap a practically inexhaustible energy source yielding something near forty million horse-power with such a small expenditure for construction as just there at the far end of Tibet, where nobody is able at present to use it properly.   Still, what seems Utopian to-day may be a reality within two or three generations. With the decline of fossil-fuel reserves the value of energy from recurrent sources will rise to such a degree that the migration to focal places of hydro-electric power will become a natural demographic phenomenon. After the termination of the cold war it might be economically quite wise to build giant electrolytic works for cheap production of aluminium, copper, and heavy water somewhere between Dacca and Gauhati, on the Brahmaputra, at a place where the river is easily navigable, and at a distance from the Tsangpo plant which still allows economic power transmission.   A joint Indian, Pakistanian, Tibetan, and Chinese interstate enterprise, supported by technical and economic assistance from the United Nations, might complete here a work which could raise enormously the living standard of Tibet and adjoining parts of China and Assam by electrification, irrigation, and land reclamation.   At the same time, it could supply power to chemical and metallurgical plants down in Assam and Bengal suffi-cient to produce five times more aluminium than the present total world production.   In a similar way, as in former times, places of particular fertility of soil or with natural harbours were attrac-tive centres of migration, future industrial settlements will con-centrate in areas of ample and cheap hydro-electric power.   The newly-built aluminium plant at Kitimat, in British Columbia, is an example of this kind, and what has been done there will later be repeated on a much larger scale in the Tsangpo region.

The Gibraltar Project

Another, still larger, project, which was designed by the German **Hermann Sörgel** about a quarter of a century ago, would probably require between one and two hundred times more capital expendi-ture than the Tsangpo project, and is, therefore, not likely to be realized until the total exhaustion of coal and oil, as well as of uranium and thorium, necessitates drastic measures.   It is based on the fact that the evaporation in the Mediterranean basin consumes far more water than all the tributary rivers, like the Rhône, Po, Danube, or Nile, can supply.   The consequence is that a perpetual stream of water from the Atlantic enters the basin through the Straits of Gibraltar and covers the deficit.   With an annual flow of approximately 2700 km^3/y, the Gibraltar stream is second to the Amazon among the biggest rivers of the earth.   From the stand-point of hydro-electric power this big flow is valueless as long as there is no head. A difference in level between the Atlantic and the Mediterranean could, however, be created artificially by locking the Straits of Gibraltar with a dam. As soon as the west-east flow is stopped there, the deficit due to evaporation exceeding the income from rivers and direct rainfall will cause the surface of the Mediterranean to fall by about one metre a year, so that in a century a difference in level of 100 metres will have been established, which can be used as the head of big power-plants near Gibraltar, on the Bosporus, and on the mouths of all big rivers emptying into the basin. It is estimated that the Gibraltar stream contributes about two-thirds of the total income of the Mediterranean, which amounts to 4100 km³/y. The annual yield from the Bosporus and all the rivers is approximately 1400 km³/y, corresponding to an average water-flow of

$$Q = 45{,}000 \text{ m}^3/\text{sec}$$

With this water-flow and a head of $H = 100$ m, the aggregate power of all the plants, excluding Gibraltar itself, would be

$$P = 8QH = 36{,}000 \text{ MW} = 48 \text{ million horse-power.}$$

The Gibraltar plant itself would be in a unique position by being fed from the Atlantic, which represents a reservoir of practically infinite capacity. Hence there would be no limit for the installed power capacity of the plant, and if it should be desirable to build a station to supply peak power of, say, 500 million kilowatts one might do so without risking the exhaustion of the Atlantic. While, however, such a plant could run at full load for hours and days, it could not do so for longer periods without again raising the level of the Mediterranean, decreasing thereby the head H and submerging the station. The primary source of the energy is the evaporation of the water in the Mediterranean basin, which is about 4100 km³ per year under present conditions. After lowering the sea-level, however, the water area—and correspondingly also the evaporation—would be decreased by about 20 per cent, and since the inflow from the rivers would remain unchanged the contribution through Gibraltar would have to be reduced from its present value of 2700 km³/y to 1900 km³/y. According to equation (41), the annual production would be:

Gibraltar plant: $2 \times 1900 \times 100 = 3.8 \times 10^5$ MWh = 380 TWh
All other plants: $2 \times 1400 \times 100 = 2.8 \times 10^5$ MWh = 280 TWh

Total production of Mediterranean basin = 660 TWh

In addition to the power made available, a very considerable increase in the land area of most of the adjoining countries would be achieved. Broad belts of very fertile land would emerge from the receding waters along the whole Mediterranean coast, and new countries would be born by drying out the Gulfs of Gabes and of Sidra, on the African coast, and the northern half of the Adriatic Sea. The total gains of new land are estimated at half a million square kilometres, or approximately twice the land area of the United Kingdom and Northern Ireland.

In spite of all these tempting prospects, neither we nor our grandchildren will live to see the realization of the Gibraltar project, because the time is far from being ripe for it. Above all, the energy which can be gained by the Gibraltar dam is too small to justify the costs. To obtain between two and four times the energy of the Tsangpo plant with more than a hundred times the capital expenditure would be a poor bargain from the standpoint of power economy. The trouble with the Gibraltar project is the unfavourable distribution of shallow and deep waters in the Mediterranean. If the coast profile descended as steeply off Marseilles, Genoa, and other Mediterranean ports as it does near Gibraltar these cities could be preserved with less cost as harbours even after lowering the sea-level. As it is, most of the great Mediterranean ports like Marseilles or Trieste, and also the picturesque Venice, would be turned into inland cities, while, on the other hand, the Straits of Gibraltar, with their depth of 400 m = 1300 ft, could be dammed up only by a structure with a volume of about 400 times that of the Hoover Dam, in the Boulder Canyon of the Colorado, which was in process of construction for five years, and took the combined efforts of six of the largest American private building-contractor firms.

Taking into account that the whole energy output of 660 TWh annually is less than 10 per cent of the present total energy consumption of Europe, and would at the earliest possible date of operation of the plants (even if we started building the dam to-morrow) be less than 3 per cent, considering further that more than half of this energy would be produced near Gibraltar, at the far end of Europe, whence economic transmission of electric power would be restricted to Spain and Morocco, we can easily understand that the Sörgel plan will scarcely ever be realized for reasons of power production alone. It is conceivable, however, that the need for more arable land will become so urgent that the prospective gain of 500,000 km², or about one hundred and twenty million acres of productive area, between Europe and Africa might one day justify the undertaking. The 660 TWh obtained annually, along with the other benefits of the enterprise, will then be quite a welcome premium.

The Rôle of Water-power in Future Energy Production

The world's present hydro-electric power production amounts to about 370 TWh per annum, or 0·0012 Q/y, which is between 1 and 2 per cent of the global energy consumption, and about 7 per cent of its electricity equivalent. The widely held belief that full development of all economically justifiable water-power sites will increase the hydro-electric production by a factor of not more than ten is based on present fuel prices. In view of these relatively low prices, large capital investments may seem to-day unprofitable for certain projects which will prove to be quite economic in times of declining production of fossil fuels. Thus Putnam (*loc. cit.*, p. 174) estimates the total available hydro-electric capacity, harnessed and unharnessed, for the whole area, which includes the Eastern U.S.S.R., India, China, Japan, Australia, and New Zealand, at 100 million kilowatts, while, according to the medium estimate given above for the Tsangpo power, the installed capacity on that small spot alone would amount to 27,000 MW = 27 million kilowatts, not to speak of the other great streams in the Himalayas, like the Indus, as well as the still mightier streams in China and other sources with potentialities together of ten times as much power. There is certainly a category of giant projects like the Davidov plan or the Tsangpo project which require for their successful execution not only large capital investments but also radical demographic changes involving migration, resettlement, and creation of big industrial centres in hitherto undeveloped regions. Whether or not these great plans will ever be realized depends partly on future world politics and also to a high degree on the costs of producing electrical power from nuclear fuels. The latter point will be discussed in Chapter XV.

Even assuming forced development of hydro power, including such sites as the Tsangpo, a ceiling would be reached finally with a global production not exceeding twenty or thirty times the present level of 370 TWh annually. By the time when this is reached the total energy demand will have risen so far that hydro-electricity will again account for only a small percentage of the total supply.

Although, therefore, increased development of water-power will not be a solution to the global problem of energy production, it will be of vital importance for a series of single countries. In Fig. 75 the present consumption of sixteen countries is compared with the potential hydro-electric production. All graphs give *per capita* values; the lines referring to present consumption and production denote the ratio of total consumption to present population figures, while the empty bars referring to hitherto unharnessed hydro-electricity denote the ratio between additional production after full development of all sites and the population figure of the country to be expected in A.D. 2000. In computing these numbers the very different rate of growth of population in the individual countries has been taken into account. The solid line in the first row of the graph of each country gives the present *per capita* consumption of energy, the shaded line in the second row gives its electricity equivalent (fuels taken at 20 per cent efficiency), and the third row gives present hydro-electric production (black) and potential future production (white). The disproportion between demand and supply can be seen from the figure. While in all the great countries only quite a small fraction of the energy demand could be covered by hydro-electricity, there are undeveloped countries with abundant potential supply. This abundance is so great that with really full development of all hydro-power sites of the world—irrespective of the present local requirements—the disproportion between demand and supply would be considerably less on world average than in the countries of the present Great Powers. Although, however, the last graph of Fig. 75 shows that almost 80 per cent of the present electricity equivalent of world consumption could be obtained by harnessing all the water-power resources of the globe, it would be wrong to conclude that mankind, after having spent its heritage of fossil fuels, could rely (with more strict economy) on hydro-electric energy alone. Three reasons make it impossible. First of all the graphs for consumption refer to the present state. By the time when all hydro-power is really exploited the demand will be a multiple of the present figure. Secondly, the transition from total energy input (as represented by the solid lines in Fig. 75) to its electricity equivalent gives a fair comparison only as regards energy in the form of mechanical work or of specific uses of electricity, but not in the form of heat. Even, therefore, if the hydro-electric supply were equal to the electricity equivalent of the consumption, a balance between supply and total energy demand, including heat, without the aid of fuels could be established only by the universal use of heat-pumps with a performance

energy ratio of at least P.E.R. = 5 (see Chapter VI). This, however, would be practically possible only in a small percentage of
devices of comfort heating, and not at all for process heat.

Thirdly, global reliance on hydro-power is not possible, because

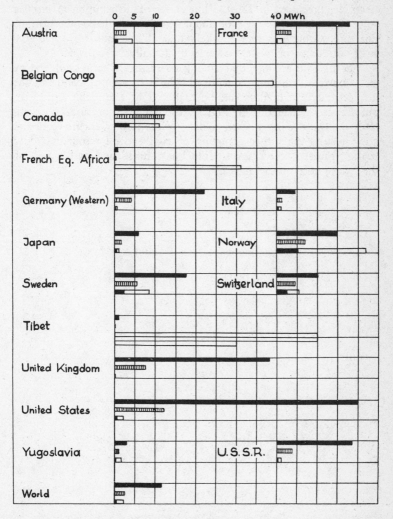

FIG. 75. COMPARISON BETWEEN PRESENT PER CAPITA CONSUMPTION OF
ENERGY AND POTENTIAL HYDRO-ELECTRIC POWER PRODUCTION OF SIX-
TEEN COUNTRIES AND THE WORLD

world trade is not a communicating vessel for electric power. While it is possible to supply motor-cars running in Australia with petrol produced from Iranian or Venezuelan crude oil, it would not be possible to feed a grid of Australian electric railway-lines from electricity produced in Tibet. The range of distribution of electric power is restricted to areas of about 600 miles radius at the utmost. For this reason it is very unlikely that all the existing, even very large, resources of water-power will actually be tapped.

Taking into account that among all recurrent sources of energy water-power is the only one from which significant contributions to electricity production can be expected at reasonable costs, we can see quite clearly from Fig. 75 the necessity of finding substitutes for the declining fossil fuels. Without very extensive use of nuclear fuels the future depletion of coal and oil resources would change most of the ' haves ' into ' have-nots,' and just a few of the ' have-nots ' into ' haves.'

CHAPTER XIII

Solar Energy and Other Sources

The Available Amount of Energy

THE sun's rays are the ultimate source of all life on earth by generating vegetation through photo-synthesis; we are dependent on them for all our food and all our fuels, fossil and others, and in addition for rain and water-power and the heat which keeps the atmosphere at a tolerable temperature. A wide spectrum of rays of different wave-lengths falls on the earth, and is partly reflected, but mainly converted into heat and re-radiated into space in the form of heat-rays of longer wave-lengths.

The gross energy income from sun-rays is very large. The energy falling on our globe every fifteen minutes is considerably greater than the annual energy consumption of mankind. The reason why direct use of solar energy has not yet made larger contributions to our power economy lies in the fact that the earth's total energy income is much diluted in the vast ocean of the atmosphere, and locally subject to seasonal, climatic, and meteorologic variations. It has been more economic so far, therefore, to make indirect use of solar energy by tapping fossil fuels and water-power resources in which the offshoots of solar energy are presented in a more concentrated form.

Table 34 shows some of the uses Nature makes of the energy falling on the earth. Quite a considerable part of it, (e), which is spent in the evaporation of water, is stored in the form of the latent heat of water-vapour. It may be transported and distributed by air-drifts, and then released again in the form of heat when the vapour is condensed and clouds are formed. The mild oceanic climate, with its relatively warm winters, is a result of this global kind of steam-heating. A much smaller part, (f), is spent in lifting the vapour into the atmosphere, and a tiny fraction of the potential energy obtained in this way is used for water-power plants. The bulk of our sources of energy—namely, fossil fuels— is derived from the cumulative action of photo-synthesis represented by parts (i) and (j) of Table 34.

TABLE 34. THE EARTH'S ANNUAL ENERGY INCOME FROM THE SUN

(*Compiled from data given by Ayres and Scarlott*)

		Approximate Figures	
		Q	10^{12}kWh
(*a*)	Gross radiation energy striking the earth ...	5140	1,500,000
(*b*)	Part of (*a*) reaching the surface of the earth ...	3200	940,000
(*c*)	Part of (*a*) converted into wind	90	26,000
(*d*)	Part of (*b*) reaching land areas	900	260,000
(*e*)	Part of (*b*) spent in evaporation of water ...	1000	290,000
(*f*)	Part of (*e*) spent in lifting water-vapour ...	17	5,000
(*g*)	Part of (*f*) recovered in the form of rivers ...	0·17	50
(*h*)	Part of (*g*) used for power plants	—	0·4
(*i*)	Part of (*d*) photo-synthesizing land vegetation	0·15	45
(*j*)	Part of (*b*) photo-synthesizing marine vegetation	1·25	375
Present annual energy consumption of mankind ...		0·09	26

Trapping Sunshine

The method of trapping sunshine for heating purposes is an old and well-tried system. It is based on the physical fact that the transparency of glass windows is greater for visible light than for wave-lengths in the infra-red part of the spectrum. About 90 per cent of the incident radiation—with its broad spectrum extending from the ultra-violet over the visible light to longer wave-lengths—is transmitted through window panes. Only a small fraction of the entering rays is reflected and leaves the room again, while the bulk is converted into heat. The point of the story is that glass windows are nearly opaque to the infra-red thermal radiation which is emitted from the walls of the room, so that a large percentage of the incident radiation is trapped.

This fact is widely used for heating greenhouses, and is partly employed also as an aid to domestic heating. Large windows on south-facing walls, provided with suitable shutters that can be closed at night, can help to make considerable savings in fuel. Modern architecture is applying this principle extensively to homes as well as to offices and factories. A symptom of this trend is the rise in production of window-glass in the United States, which from 3 sq ft per capita per annum in 1930 rose to 6 sq ft in 1940, and reached 10 sq ft by 1950. In latitudes between 40° and 43° fuel bills can be reduced by 30 per cent, and in some cases even by 50 per cent, when large double-paned southern windows help to introduce additional heat to the rooms.

Solar House-heating

A large and bold step had to be taken from the trapping of sunshine for auxiliary heating to a complete solar house-heating installation, without any fuel at all. But this problem can be regarded as technically solved to-day for areas with a certain minimum amount of sunshine in winter. The borderline of feasibility of solar heating does not everywhere run parallel to the latitude, but, as far as present experience goes, the range of economic performance seems to be restricted to latitudes well below 45°.

The salient point in the problem of solar house-heating is the storage of heat. The influx of sunshine is intermittent not only on account of the change of day and night, but also because of meteorological factors. Sequences of four, five, or even more dull days may occur, and to secure the heat supply during such periods a certain stock of heat must be kept as reserve. The heating system must therefore consist of the following items: (*a*) a collector of sunshine; (*b*) a storage device; (*c*) a distribution system.

A successful solution of the above problem is achieved by the famous Dover House, which was designed by Dr Maria Telkes, Research Associate of New York University (College of Engineering), with Eleanor Raymond as architect. Fig. 76 gives a diagram of its heating system. The collector is a black sheet of thin-gauge iron mounted vertically behind double glass panes along the whole south wall on the first floor of the house. A narrow air space of a few inches' thickness behind the collector sheet is covered on the rear side by a well-insulated panel, and an air-circulating stream conveys the heat from the iron sheet through ducts to heat-storage bins formed by double walls between some of the rooms of the house. The walls of the bins are constructed as radiant heating panels, supplying a background of heating effect sufficient to keep the rooms at a comfortable temperature when the outdoor temperature is milder. When more heat is required, thermostatically operated fans circulate the air from the bins through the rooms to supply additional heat, while the colder air of the rooms returns to the bins through louvers.

The heat storage itself, which is the essential part of the system, might be performed in principle by placing into the storage bins any material with a sufficiently large heat capacity, as, for instance, great masses of substances with high specific heat. Large water-tanks constructed like boilers or heat exchangers with bundles of vertical tubes through which the heating air flows would be among the possible solutions. During the hours of sunshine, while heat is supplied from the collector plate, the water is warmed to a degree well above room temperature, and during the night or on dull days the heat stored in it is given off to the rooms.

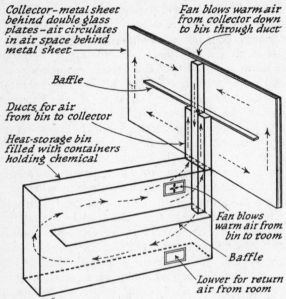

Collector-metal sheet behind double glass plates – air circulates in air space behind metal sheet

Fan blows warm air from collector down to bin through duct

Baffle

Ducts for air from bin to collector

Heat-storage bin filled with containers holding chemical

Fan blows warm air from bin to room

Baffle

Louver for return air from room

FIG. 76. HEATING SYSTEM OF THE DOVER HOUSE

The disadvantage of storage by using specific heat lies in the vast amount of water, and consequently also of available bin space, which would be required. More heat within a given space can be stored by using the heat of fusion. As is well known, heat is required to melt a solid substance, and, *vice versa,* heat is given up when a fluid solidifies. The latter process is not so familiar to us from daily experience, because the heat from water solidifying to ice is released at such a low temperature that it does not make us feel hot. Still, the heat released by any quantity of solidifying water is nearly as great as the heat required to raise its temperature from room-level to boiling-point. There are other substances with considerably higher melting-points which, after being melted and then solidified again, release their latent heat of fusion at temperatures well suited for space heating. The requirements for heat-storage materials appear to be best met by a series of hydrated salt compounds of sodium and calcium in which water molecules are included in the crystal lattice on solidification. Table 35, taken from Telkes (1949), gives melting-points, heat of fusion, and approximate prices in dollars of some of these salts. The cheapest of them, sodium sulphate decahydrate, is used as the heat-storing material in the Dover House. The salt is filled into standard five-gallon drums, which are then hermetically sealed, so that on heating no gas or odour can be produced. The material does not wear out in use, and if care is taken that corrosion of the container drums is avoided they can last indefinitely. The drums are stacked to form vertical columns in the heat-storage bins through which the air from the collector is circulated. During hours of heat collection the salt within the drum melts, thereby absorbing fusion heat, and on solidification the latent heat is returned to the surrounding air, and conveyed by the circulation to the rooms.

TABLE 35. CHEMICALS FOR HEAT STORAGE

Chemical Compound	Melting-point		Heat of Fusion		Cost (dollars per ton)
	F.	C.	B.Th.U./ lb.	kcal/kg	
$CaCl_2.6H_2O$	84–102	29–39	75	42	13
$Na_2CO_3.10H_2O$	90–97	31–36	115	64	24
$Na_2HPO_4.12H_2O$	97–118	36–48	114	63	70
$Ca(NO_3)_2.4H_2O$	104–108	40–42	90	50	*
$Na_2SO_4.10H_2O$	88–90	31–32	104	58	9
$Na_2S_2O_3.5H_2O$	120–126	49–52	90	50	58

* Cost not certain.

No manipulation at all is necessary to operate the heating system. By pushing a button the fans and thermostats are switched on, all the rest being done by the sun. The data of the installation in the Dover House are: collector surface 720 sq ft = 67 m²; heat-storage volume 470 cu ft = 13·3 m³; amount of sodium sulphate 21 tons; total heat-storage capacity 4,700,000 B.Th.U. = 1,200,000 kcal, sufficient for twelve average days' heat requirements; total cost of material, including fans, electrical equipment, and thermostats, 1865 dollars.

There will be many people who may feel that the large wall area occupied by the collectors and the extra costs for material and construction work are too high for the heating system of a house with only one living-room, two bedrooms, kitchen, and bathroom. As a matter of fact, solar heating in a climate like that of Massachusetts cannot yet compete economically with coal- or oil-fired conventional heating. In places of lower latitudes, however, both the collector surface and the amount of storage material can be kept smaller, and within a few decades, when fuel costs are considerably higher than to-day, solar heating, as well as heat pumps, will play a great rôle in countries where the conditions are favourable. The decisive advantage of solar heating lies in the fact that no running costs arise, except the electricity bill for the power driving the fans, which is very small. Thus the one single investment for the installation pays once and for all the heating costs for the lifetime of the house. In addition, the system works automatically without smoke, soot, and fume production, and saves all trouble in stoking, refuelling, cleaning, repair, and other work. Adding solar heat to the energy system of a country helps to increase the wealth of the nation, and if all houses in areas with favourable conditions were equipped with solar heating systems, fuel saving worth millions of pounds yearly could be achieved. The work of Telkes, Hottel, Löf, Bliss, and other scientists who are paving the way for solar heating is real pioneer work, the full significance of which will emerge more clearly in the future.

Solar Cookers and Water-heaters

It has been mentioned in Chapter XI that the global amount of heat obtained from burning farm wastes represents about sixteen times the energy supplied from all hydro-electric power plants in the world. While burning dung-cakes makes use of a recurrent source of fuel, it is still a wasteful method, because it deprives the soil of fertile material, and a wise economy should therefore try to find suitable substitutes. Solar energy is abundant in areas where burning dung-cakes is customary for heating and cooking. But on account of the very low living standard in these countries there is no possibility of using solar house-heating, with its expensive installations for heat collection and storage. Still, the replacement of kitchen fires by solar cookers may be a practicable solution. Putnam (1954) quotes a Press release of the Indian Embassy in Washington, stating that a 5000-a-month production-rate of low-cost solar cookers was inaugurated in Bombay in May 1953. The cooker consists of an aluminium reflector with a diameter of about four feet, mounted on a turnable shaft, and a collector-plate near the focus of the mirror on which a cooking-pot is placed. It may be used nearly every day in the year, in most parts of India, from mid-morning to mid-afternoon. It takes about twenty minutes to prepare a standard Indian meal of lentils and rice. The cooker sells for about five pounds, and the market was estimated at over a hundred million.

Unfortunately, the present price is beyond the cash resources of the poor people, especially in India and Egypt, who could theoretically benefit most by the use of solar cookers. For this reason the experiment of saving dung material by the introduction of solar cooking has failed so far.

While solar cookers will be used mainly in regions where coal, oil, and electricity are relatively expensive, the method of obtaining hot water from solar heat may find a more extended application. The installation of a solar water-heater, dispensing with long-term storage, is much less costly than that of solar

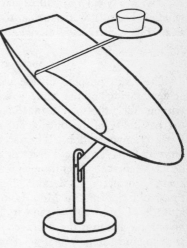

FIG. 77. SOLAR COOKER

house-heating. In regions with suitable climatic conditions a cost-free supply of hot water can be obtained during a large part of the year with a system of pipes behind glass panes, backed by a black surface and mounted on a tilted south-facing roof, together with a storage-tank under the roof and a small pump for circulating the water. In latitudes not higher than 35° every two square metres of collector area will suffice for a hot bath daily if the weather is fine.

Putnam estimates that the minimum likely cumulative contributions in the next 180 years to the world energy system by solar heat collectors at costs not greater than twice the present costs will be:

Solar house-heating	4 Q
Solar water-heating	0.05 Q
Solar cooking	0.02 Q

Smaller contributions can also be made from solar distilling plants and solar refrigerators. In countries with lack of fresh water distilleries are used for the water-supply. Wherever enough sunshine is available the evaporation can be done by solar heat, instead of burning fuel. A solar water-distilling plant was built as long ago as 1883 at Salinas, in Chile, with a daily yield of 28 tons of distilled water from a collecting area of 5000 m² = 1·24 acres. Other plants of the same kind have been built since 1914 with a yield of 5 litres daily per square metre of collector, which is just a little more than that of the Chilean plant. Improved solar stills are expected to yield almost twice this output.

Solar Furnaces

The main purpose of the solar heaters described above is to achieve savings of fuel and/or electricity. In addition to operation without fuel, the solar furnace has two other advantages—namely, the possibility of heating bodies to extreme temperatures and perfectly clean operation. The largest solar furnace so far has been built by Felix Trombe, Directeur de Recherches au Centre National de la Recherche Scientifique, in France. His laboratory of solar energy, formerly in Meudon, near Paris, was transferred in 1950 to an old *citadelle* on Mont-Louis, in the Eastern Pyrenees, at an altitude of 1600 m = 5200 ft, latitude 42° 30′ N. The place is favourable because the average annual number of sunshine hours is 2750, and the power of solar radiation striking a surface perpendicular to the sun's rays reaches, under good weather conditions, a maximum of 1·0 kW per square metre (see Trombe, 1953).

The essential parts of the furnace are a plane mirror and a parabolic concave mirror. The plane mirror acts as a heliostat moving with the sun and deflecting the sun's rays so that they strike the parabolic mirror in the axial direction whatever the position of the sun may be. Two smaller installations and a large one have been built. The heliostat mirror of the large furnace is a rectangle of 13 m × 10·5 = 43 ft × 34 ft, and the parabolic mirror is circular, with a 10·5 m = 34 ft diameter. Since it would not be possible to grind mirrors of that size in one piece perfectly plane or perfectly parabolic, both mirrors are mosaic-like, composed of many smaller properly adjusted mirrors. This arrangement serves the purpose well enough, because, instead of a perfect photographic image of the sun, only a series of overlapping images have to be projected into the focal plane of the parabolic mirror. The large furnace of Mont-Louis collects a power of about 75 kW;

temperatures of over 3000°C. have been obtained in the focus, and higher ones are expected after further improvements. According to a new French project, the next step will be a 1000-kW furnace.

Although higher temperatures have been reached in the electric arc, and still higher ones in electric sparks, the solar furnace is superior as regards the spatial extension of the hottest zone, so that larger bodies can be heated to extreme temperatures. Besides, there is no contact with, or interference from, any electrodes, so that chemical processes with high purity standards can be performed at temperatures of 3000°C., or even more. One of the industrial applications will be the melting of highly refractory substances. The solar furnace is an example of one of the uses of solar energy in which its specific merits are most important. In contrast to the application for house-heating or cooking, however, the fuel-saving as a result of the use of solar energy for furnaces is scarcely worth mentioning in proportion to world consumption.

Power Production from Solar Energy

An area of one square metre (approximately 1·2 square yards) placed perpendicular to the sun's rays at the edge of the earth's atmosphere receives a solar irradiation with an energy of 0·32 kilocalories per second. Conversion into electric units gives 1·34 kilowatts per square metre as the power of solar radiation just outside the atmosphere. About one-third of this energy is absorbed and scattered in the atmosphere before it reaches the earth's surface, and therefore the power striking one square metre of a horizontal plane at sea-level is 0·9 kilowatt with the sun in the zenith, and correspondingly less with oblique incidence. At an altitude of 1600 m the solar power is 1·0 kW/m². Considering that in the arid zones of the world many million millions of square metres of desert land are free for power production, we find that by utilizing only 1 per cent of the available ground for solar power plants a capacity could be reached far higher than the present installed capacity of all fuel-operated and hydro-electric power plants in the world, which is about 200 million kilowatts.

There is, therefore, a theoretical possibility of the survival of civilization after the depletion of all fossil and nuclear fuels. But as long as these fuels are available at reasonable costs solar power plants can hardly compete economically. A serious difficulty is the intermittent supply of energy. No power during the night and little or no power at all, according to the latitude, during the winter months in places beyond the 40th parallel. Solar power-supply would therefore require the construction of plants not only for power generation, but also for storage of energy. Within the tropical zone storage for the daily demand would suffice; in higher latitudes storage for a half-year's supply of energy would be necessary.

The diagrams in Fig. 78 show the seasonal variation of solar energy at different latitudes. The ordinates of the diagram represent the energy of full sunshine falling on one square centimetre (= 0·155 square inches) of a horizontal surface at sea-level per day. The abscissæ represent the time of the year, from January to December, and the different diagrams refer to northern' latitudes 15°, 30°, etc. The same diagrams can be used for the southern hemisphere simply by shifting the abscissæ by six months. The numbers in the top centre of the diagram (for instance, 6115 at 0° lat.) give the sums of the daily energies on the 15th of all twelve months. Taking 30·4 days as the average length of the months, and neglecting the difference between the sunshine in the middle of a month and its mean value for that month, we obtain an approximate value of the annual income of solar energy per square centimetre by multiplying these numbers by 30·4. Thus, for instance, the annual energy on the equator is approximately 6100 × 30·4 = 186,000 gcal/cm², or 1·86 million kilocalories per square metre. Two facts emerge from Fig. 78 which might not be expected by the layman: the maximum amount of annual energy is not received at the equator, but at a latitude near 15°, and the receivers of the maximum daily energy are, of all places on the globe, the poles, with 667 gcal per (sq cm) (day) at the North Pole on June 21, and even more at the South Pole on December 21, if the day happens to be cloudless. Nevertheless, the Arctic regions are much less suitable places for solar power plants because (a) the total annual income is much less (about 0·67 million kcal/m² year), and (b) the long winter pause would necessitate the installation of large storage plants. Even at latitudes as low as 30° the winter supply drops to about 43 per cent of the maximum rate, and at 45° lat. the contribution in the winter months is so small that a continuous power supply could be upheld only at the cost of building storage plants with a capacity holding at least a three or four months' load.

We are therefore faced with the unfortunate situation that it is just in the tropical zone of the earth that (a) most of the water-power is available, (b) the conditions are most favourable for solar power plants, and (c) the demand for power is smallest.

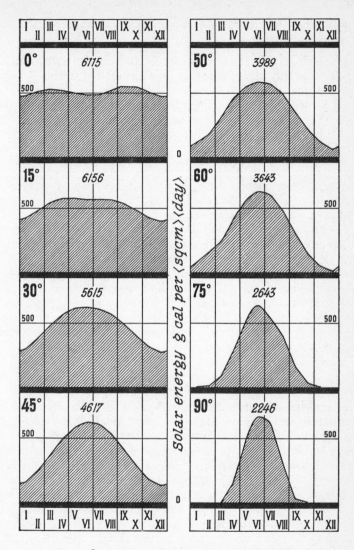

FIG. 78. SEASONAL VARIATION OF SOLAR ENERGY
ON CLOUDLESS DAYS

Production of electricity from solar energy can be performed in several different ways:

(1) Indirectly, by using photo-synthesis for growing vegetation (especially algæ), which is used as fuel for thermal power plants.

(2) Semi-indirectly, by heating boilers or air-heaters (see Chapter V) of conventional thermal power plants with sunshine concentrated by collecting mirrors.

(3) Directly, by converting radiation into electric voltage by means of thermocouples or photo-voltaic cells.

The indirect method through photo-synthesis has already been discussed in Chapter XI. Another indirect method based on the production of fuel from solar energy could use the so-called photolysis—that is, the photo-chemical breakdown of water under exposure to sunlight. Experiments are going on at the Stanford Research Institute with various organic and inorganic photocatalysts, which will absorb sunlight and then transmit the absorbed energy to a second reactant which will initiate the decomposition of water. If these experiments succeed, so that hydrogen can be cheaply produced from sea-water and sunlight, then the combination of photolysis with hydrogen fuel cells, as described in Chapter VII, could achieve an overall conversion efficiency from sunlight to electricity of 26 per cent.

Thermal Solar Power Plants

Steam-engines with boilers heated by focused sun-rays have been constructed, and were operating in France about ninety years ago. The names of the French scientists Bernière and Mouchot are mentioned as the pioneers of solar steam-engines. While Bernière's drawing shows a very large convex lens as the focusing means, another engine constructed in 1884 used a paraboloid reflector with a cylindrical boiler vessel in the axis near the focus. Both constructions are technically wrong, because they require a movable boiler to adjust the system to the changing position of the sun. Considering that modern boilers or air-heaters use bundles of long, thin tubes for the water or the air to be heated, the proper solution of the optical part of the problem is a system of straight, long boiler tubes, each placed along the focal line of a *cylindrical* concave mirror with parabolic cross-section. In order to avoid the necessity of moving the boiler tubes, they must be arranged parallel to the earth's axis—that is to say, they must be oriented north-south at an angle of elevation which is equal to the latitude of the place. According to the daily course of the sun, the cylindrical mirrors must be slowly rotated at a rate of 15° per hour, and the seasonally varying declination of the sun can be taken account of by a very small weekly shift of the mirrors parallel to the axis, while the whole system of boiler tubes may remain fixed all the time.

Of course, quite a multitude of boiler tubes and mirrors would be necessary for a power plant of reasonable energy output. The area required depends on the overall efficiency. According to **Trombe** (1953), the efficiency of the nineteenth-century French solar engine was 5 per cent, while an installation built later in Egypt reached 6 per cent. In view of the great progress made as regards efficiency of turbines within the last three decades (see Fig. 16. Chapter III), considerably higher efficiencies might be expected from future installations. A certain difficulty lies in the fact that the

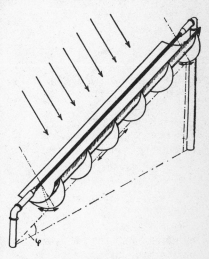

FIG. 79. SOLAR-HEATED BOILER

losses caused by radiation, convection, and conduction of heat from the boiler tubes are the greater the higher their temperature is. Hence the efficiency of collecting solar heat on boilers decreases with increasing steam temperature, while, on the other hand, high steam temperatures are necessary to obtain a good thermal efficiency of the turbine. According to Putnam, it might be necessary to operate the engine with a steam temperature of only 200° C. = 392° F., so that the resulting efficiency, given by the product of turbine efficiency and collector efficiency, would be only 7 per cent.

With appropriate improvements of the collector system, however, fairly good collector efficiencies might be expected even at the high boiler temperatures necessary for efficient modern steam plants. Fig. 80 shows a cross-section perpendicular to the cylinder axis of a design which would meet modern demands of efficiency. The water-tube W is surrounded by a concentric cylinder of hard glass G. The face of the glass cylinder which lies opposite to the focusing mirror M is a reflector which returns to the water-tube its own heat radiation. Losses by convection through air-streams are avoided by the glass tube.

Assuming that with a boiler system as represented in Figs. 79 and 80 a collector efficiency of 50 per cent could be achieved under temperature conditions which lead to an engine efficiency of 20 per cent, the resulting overall efficiency would be

$$0.5 \times 0.2 = 0.1 = 10 \text{ per cent}$$

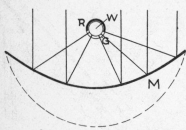

FIG. 80. CROSS-SECTION THROUGH
SOLAR BOILER

M, mirror; W, water-tube G, glass tube
with reflector R.

What would be the size of the collecting system of a solar steam plant which could supply 1000 kW = 1 MW of peak power under the climatic conditions of Mont-Louis? With sunshine power of 1 kW per square metre and 10-per-cent efficiency, the total collector area perpendicular to the sun's rays must be 10,000 m². We may imagine, for instance, that the system consists of 100 boiler tubes each of 20 m length, mounted parallel to the earth's axis and equipped with a cylindrical mirror, according to Figs. 79 and 80, with a projected surface of 20 × 5 m. The ground area occupied by the collector would be about one and a half hectares in medium latitudes, and less in tropical regions.[1]

We may ask further about the dimensions of a larger plant designed for an annual supply of 1 TWh = one thousand million kilowatt-hours situated at a latitude of 30°. According to Fig. 78, the sum of the energies of the 15ths of each month is 5615 gcal/cm², which gives an annual yield of approximately

$$5615 \times 30.4 = 170{,}000 \text{ gcal/cm}^2\text{y} = 1.7 \times 10^{12} \text{ kcal/km}^2\text{y} = 2 \text{ TWh/km}^2\text{y}$$

Assuming again a 10-per-cent efficiency, we find that an annual energy output of 0.2 TWh can be expected per square kilometre horizontal area at 30° latitude if, as in the Sudan, for instance, the percentage of cloudy days can be neglected.

[1] This must not be confused, however, with the area required for a 1000-kW furnace as planned in France. For in this case the figure of 1000 kW refers to the heat input, and therefore little more than one-tenth of the area given above will suffice for a furnace of equal power.

A plant for 1 TWh annual output under the climatic conditions of the Sudan would occupy an area of 5 square kilometres, or approximately 2 square miles. Its installed capacity would be about 430 MW. If iron sheets of 2 millimetres thickness were used as the material of the cylindrical mirrors, about 100,000 tons of sheet iron, and perhaps the same quantity of steel for the frames of the mirrors, would be required. This means that about one-fifth of 1 per cent of the U.S. annual steel production would have to be invested once and for all in a plant which would supply for unlimited periods electric power equal to 1 per cent of the present U.S. hydro-electric output.

We must not overlook, however, another additional investment which is necessary—namely, the storage plant. There are two different possibilities of storing electric energy which *a priori* could be taken into account—chemically in storage batteries or mechanically in high-level water reservoirs. While hundreds of millions of storage batteries are used to store electric power needed for starting and lighting all the motor-cars in the world, no large storage plant for energies up to the order of magnitude of one TWh has ever used batteries. The reason lies partly in the greater losses involved in chemical storing, but mainly in the high investment costs. A battery for 12 volts and 5 ampere-hours capacity is quite heavy and expensive, but seventeen such batteries are necessary to store one kilowatt-hour, and therefore 17 thousand million batteries of that size would be required to store one TWh. With water reservoirs, on the other hand, the storage of vast amounts of energy is no problem. The energy stored in Lake Mead, on the Colorado river, is about 10 TWh. Hydraulic storage seems to be, therefore, the sole practicable way of levelling out the intermittent power supply of a large solar plant. The choice of the site of such a plant is therefore restricted to places where not only sunshine, but water, as well as ground elevations, are available. Even under these restrictions a sufficient number of suitable sites could be found.

With a large bowl, suited to be turned into a water reservoir and situated on a plateau towering some hundred feet over a lake, the primary generating station and the storage plant could be combined in a more economic way. Instead of generating electricity directly from the solar plant, the turbines, fed with solar-heated steam, need only drive the pumps which lift the water from the lower to the higher reservoir. The production of electricity can be done then in a conventional hydro-electric plant fed from the water returning to the lower level.

All this is a technically practicable way of producing power from solar energy requiring neither new physical discoveries nor new inventions. It is quite consoling to know that mankind could take this course before starving when the other energy sources are exhausted. In the present position, with cheaper sources available, it would be difficult to raise the capital necessary for constructing a 1 TWh-a-year plant requiring 5 square kilometres of reflector surface.

Nevertheless, according to a Russian report put before the World Symposium on Applied Solar Energy at Phœnix, Arizona, in November 1955, a 1000-kW solar power station is to be built in a suitable region of the Soviet Union, in which the problem of the heliostatic mounting of mirror surfaces of sufficient sizes will be solved in the following manner.

The boiler is placed on a 150-ft tower, which will be surrounded by a system of twenty-three railway-tracks running in concentric circles. An assembly of many single mirrors is mounted on trucks in such a manner that each reflects the sun-rays just on to the boiler, while the whole group moves slowly along the circle according to the position of the sun. This will be the first solar power station with a capacity reaching the megawatt limit, though, being only an experimental pilot plant, it will be built without restriction from economic considerations.

While large solar power stations do not seem to compete with conventional plants—or even nuclear plants—the case might be quite different with small installations in remote rural areas. A small piston engine made in Italy, with flat-plate collector and sulphur dioxide as the working material, ran a pump suitable for irrigation purposes at the exhibit in Phœnix, and aroused such interest among the farmers of Arizona that at least fifty of them could have been sold on the spot, had marketing channels been established in the United States.

A further improvement was suggested by H. Tabor in a paper read to the World Symposium. It consists of using a flat collector for the production of steam, its surface being covered with a thin layer which is black in the visible spectrum, yet transparent to thermal radiation, so that it makes the collector a ' selective black ' body. By means of electro-deposition of nickel-zinc alloys with added sulphides (' black nickel ') on to polished nickel-plated surfaces it was possible to get a collector which is a good absorber of sunlight and a poor emitter of thermal radiations. When exposed to sunlight such selective black collectors are heated up to higher temperatures than ordinary black bodies, and therefore reasonably good efficiencies might be achieved with the cheaper flat-plate collector system. An overall efficiency of about 10 per cent is seen to be within the range of possibility. Small installations of this kind might be the first to meet the widespread domestic energy demand in rural districts, which remains unsatisfied so far, or is met by the use of fuels with considerable transportation costs.

Direct Conversion of Sun-rays into Electric Power·

The direct method of obtaining electricity from sunshine by means of photo-voltaic cells or thermo-couples appears very attractive to the layman. What an advantage, he argues, to have a conversion plant without any moving parts, and what saving in the installation costs when we can dispense with all the huge machinery of turbines, dynamos, etc.! On sober consideration, however, all these expectations prove to be fallacious. The moving parts enter into the picture again as soon as we remember the necessity for storing the energy and the impracticability of doing it otherwise than in a water-power plant. The expectation of lower installation costs is still more fallacious. True, we should not require big engines, but we need a collector surface densely covered with photocells or thermo-couples, the area of the collector being inversely proportional to the conversion efficiency.

Take as an example the sort of photo-voltaic cells which we have used so far in our exposure meters. According to Putnam, their efficiency is less than 0·5 per cent—that is, one-twentieth of the efficiency we expected from the solar thermal plant. The 1 TWh-a-year photo-electric plant would therefore require a collector area of $20 \times 5 = 100$ square kilometres, or approximately 40 square miles, covered with photo-electric cells. The price of the cells used in exposure meters is of the order of magnitude of one dollar per square inch, and so we find by an easy calculation that the photo-electric surface alone would cost something near 150,000 million dollars, which is about 20,000 times the price of the material for the cylindrical collecting mirrors of the equivalent thermal plant. Unless, therefore, the wholesale price of photo-voltaic cells could be reduced at least to one ten-thousandth of the present retail price all speculations about large-scale photo-voltaic power plants are a waste of time.

Solar Batteries

Rejecting the idea of using photo-voltaic cells as the conversion means for wholesale power supply does not mean to say that they would also be useless for small stations. Between the milliwatt power-level of exposure meters, for which millions of these cells are used to-day, and the gigawatt level of large power stations, for which they can scarcely be taken into consideration, there is still a wide range within which new applications can be expected. Take, for instance, a small observatory cabin on top of a high mountain or on an island far from any electricity supply. If, in addition to illumination (which is supplied most cheaply by kerosene lamps), electric voltage is required for recording instruments, radio receivers, short-wave transmitters, etc., then the use of photo-elements may be taken very seriously into consideration. A new solar battery developed in the Bell Telephone Laboratories delivers enough power per square yard sunlit surface to light an ordinary reading-lamp. A power-converting efficiency of 10 per cent has been attained recently. An arrangement of Bell solar batteries covering the roof of a mountain observatory could do the loading, during daytime, of a system of storage batteries from which the lamps and the electric apparatus might be fed. The high price of the cells, however, prohibits their use in stations of larger power output.

Thermo-couples make use of the so-called thermo-electric effect. If strips or wires of two different metals—for instance, antimony and bismuth—are joined together twice, as in a loop, and if the two junctions are kept at different temperatures, an electric current will flow through the loop. Thus, by heating one junction with concentrated sun-rays and keeping the other shaded and cooled, we can obtain a conversion of solar energy into electric power. Improved thermo-couples have been developed by Maria Telkes which might be applied for miniature power plants in a similar way to the photo-electric solar batteries.

No considerable contribution to the energy system of the world, however, can be expected from direct conversion of solar energy into electric power.

A criterion of the practicability of the different methods of large-scale power production from solar energy may be found by comparing the collector areas required for a plant of given annual production. Table 36 gives the areas which under the assumption made above would suffice for an annual production of 1 TWh. In a solar-heated thermal plant the engine efficiency is supposed to be restricted to 20 per cent on account of the limitations of steam temperature put by the collector efficiency. A thermal plant using Chlorella as a fuel, however, may have an overall efficiency up to 33 per cent, and this value, together with the annual energy yield of Chlorella farms ($1 \cdot 17 \times 10^9$ kcal per hectare according to Table 29), has been assumed in computing the figures of Table 36.

TABLE 36. PLAUSIBLE COLLECTOR REQUIRED FOR A SOLAR POWER PLANT OF 1 TWh ANNUAL PRODUCTION

(Approximate Data)

Method of conversion	km²
Solar-heated thermal plant	5
Thermal plant fuelled from Chlorella	20
Conversion with conventional photo-voltaic cells	100
Conversion with Bell solar batteries	5
Photolysis combined with hydrogen fuel cells	2

As far as present knowledge goes the thermal plant with solar-heated boilers seems to use a given area most economically. In making a choice, however, the fact will be taken into account that the Chlorella-fuelled thermal plant might adapt its power output to the demand of the hour, while the direct methods of conversion can operate only in combination with storage plants or with other stations supplying a fixed amount of power.

Wind Power

According to the data given in Table 34, the fraction of the global energy income from the sun which is converted into wind is 90 Q, or about 26 million TWh, annually. By far the greatest part of this energy is *a priori* unattainable for use by men because it is stored high up at the border of troposphere and stratosphere at altitudes between 30,000 ft and 40,000 ft above sea-level. The so-called jet streams occurring there move with velocities up to 200 m.p.h., and we can congratulate ourselves that they do not blow near the ground, because the destroying force, which is proportional to the square of the velocity, would be about eight times that of the worst tornadoes haunting tropical regions, so that no building on earth except the pyramids would be left standing.

What we can use is the energy of air-streams a few hundred feet above the ground, and, as a matter of fact, extensive use has been made of them for a very long time. The historical importance of wind power in the development of our civilization should never be forgotten. It was through sailing-vessels that the art of deep-sea navigation developed, that the horizon of primitive geography widened, and that Eurasian civilization spread over other continents.

From the times of the ancient Phœnicians to little more than a century ago the overseas trade and traffic was wholly dependent on sailing-craft.

While the practical value of the art of sailing is more or less restricted to recreational purposes to-day, another application of wind power has kept its position better. Wind-wheels, first used by the Chinese, can supply cheap power, and they are still useful to-day in many places outside the range of the electric-supply grid. The main disadvantage of wind power, its irregular and inter-mittent supply, makes itself less badly felt in the kind of work to which it has been mainly applied—namely, milling and pumping. What matters there is the cumulative work rather than continuous operation, and therefore a simply constructed mechanical power transmission from the wind-wheel to the millstones or the pumps serves the purpose without the necessity of providing expensive installations for energy storage. Dutch agriculture owes a good deal of its land to wind-powered pumps, which, doing their work in-defatigably through the ages, pushed the sea back from the earth.

During both World Wars the shortage of fuel and the increased demand for power raised interest in fuelless energy sources, and experiments were made to improve wind-wheels and to extend their use to the production of electric power. Better efficiency and greatly increased power output per wheel were obtained in the experimental plants by applying results of modern aerodynamic research to the ancient technique. The multi-blade wheels familiar to us from Dutch picture postcards were replaced by three-blade and two-blade propeller-shaped wheels, and in view of the fact that the wind velocity increases with the height above ground-level, the wheels were mounted on high steel towers. An experimental wind turbine designed by Smith and Putnam was built in 1941 on a hill in Vermont, U.S.A. The two-blade wheel made of stainless steel had a diameter of 187 ft = 57 m, and was mounted on a tower with its hub 125 ft = 38 m above ground on a pivot, so that it could turn automatically to face the wind. The turbine was coupled directly to a generator, and developed a power up to 1400 kW with favour-able winds. After preliminary tests it delivered power to the local supply grid for some time, but when a structural failure developed the difficulty in obtaining repairs during the War led to its dis-mantling (Putnam, 1948). Cost studies of wind power showed that the economically optimum wind turbine would stand about 165 ft (= 50 m) high, with a diameter of about 225 ft, or approxi-mately 70 m, and a rating of about 2000 kW.

In spite of the high degree of industrialization the U.S.A. is not the country in which great enthusiasm for wind power can be expected, because fuels are relatively cheap there, and abundant water-power resources are still undeveloped. More interest in wind power can be expected from Great Britain, with her poor hydro-electric potentialities. An experimental 100-kW wind turbine with a three-blade wheel on a 75-ft tower is under test on Costa Hill, on the coast of Orkney, where the average wind velocity is estimated at 30 m.p.h, or about 50 km/h (Haldane and Golding, 1950). The western shores of England and Scotland, with fairly constant winds, are particularly suited for electricity production from wind so as to cover the rising power demand in face of declining coal supplies. According to reports of the British Electrical and Allied Industries Research Association, several hundred sites in Western Great Britain are suitable for installation of 2000-kW units, and the total annual production might amount to 4 TWh.

Even greater importance may be attributed to wind power in countries which are either wholly or in parts (on account of transport difficulties) devoid of fuels. The desert land of the Negeb, in Israel, could be turned into fertile land again by irrigation fed from wind-powered pumps. Wide areas of the Soviet Union which are not yet connected by economic means of transport with the centres of fuel production could be provided with rural power from wind turbines long before the network of general electrification is extended to them. A series of seven types of wind turbines, ranging from the smallest units to quite large ones, has been developed in the U.S.S.R. and plans have been announced for the production of 600,000 units of the second largest type, a three-bladed wind turbine with 30-kW rated power. If the production and installation accorded to this plan can be realized the installed wind-power capacity will be 18 GW = 18 million kW, and, with an average of 4000 use hours per year, the contribution to the energy system of the Soviet Union would be about 70 TWh, which is between 30 and 40 per cent more than the present annual production of electricity from all sources in the United Kingdom.

Putnam estimates that the ultimate installed capacity of large wind turbines in a series of windy regions in several countries close to places of growing demands for energy may amount to 100 GW, and that medium units may amount to as much again. Thus, with 200-GW capacity and an assumed number of 4000 annual use hours, the total production would be 800 TWh, which is twice as much as the present global hydro-electric production. It is not quite fair, of course, to compare ultimate future developments of one energy source with the present situation of another. By the time the production figures assumed from wind power have been reached the hydro-electric output will probably have more than doubled. Anyhow, it appears that wind power may rank as a good second to water power among the recurrent sources of energy.

Tidal Power

By catching sea-water at high tide in a basin and releasing it through turbines at low tide an intermittently working hydro-power station can be created. Since the head of such a plant cannot be much greater than the height of the tides, it will be a low-pressure station, which requires a great flow to deliver sufficient power. Applying the formula $P = 8QH$, and assuming a head of $H = 5$ metres (or about 16 ft), we find that a flow of

$$Q = 2500 \ m^3/sec.$$

will be required to produce a power of 100 MW, which corresponds to a medium-sized hydroplant. (The installed capacity of Grand Coulee is 2000 MW.) The magnitude of the required flow will be appreciated better by realizing the fact that there is no river in Europe, except the Danube and the Volga, with an average flow as great as 2500 m^3/sec, and that the average flow of the Thames just outside the range of the tides is 75 m^3/sec. Besides, even with a flow of that magnitude, the full power of 100 MW would be delivered only during a short time of about one hour while the lowest level of low tide is reached. During all other phases of low tide the power is less, and is nil during high tide.

A one-basin tidal plant as described here might therefore be used in connexion with other plants supplying power to a large distributing grid. It could carry the load during periods of productivity—that is, during low tide—and during the rest of the time the supply would have to be undertaken by other plants.

If, on the other hand, a tidal plant were designed to be used as the sole station for supplying a certain region, then provision for storage should be made. This could be done in the same way as in solar thermal plants described above, by pumping water to a higher reservoir and tapping its energy according to demand. Another method is that of the ' two-basin ' system. Two adjacent basins A and B are connected through large gates with the open sea. The gates of A are opened only during the last phases of rising tide, and closed at the time of the peak of high tide, so that the level within A equals the high-tide sea-level. The gates of the other basin B are opened at low tide, and closed at the moment of minimum. The power station which is placed between the reservoirs receives water from A and releases it to the open sea during low tide, and to B during the other time. If the filling capacity of the basins is a multiple of the quantity of water consumed in twelve hours, then the water-levels of both basins will remain almost constant, and the turbines can run permanently with nearly the same head, so that a fixed amount of power can be supplied.

If this power is 100 MW with a head of $H = 5$ m, then a permanent flow of 2500 m³ per second is required. This means that within a tide period of twelve hours about 100 million cubic metres of water flow through the turbines from basin A to B. If the level difference between A and B is not to be decreased appreciably by the loss of this quantity, the capacity of each reservoir must be, say, ten times as great. Therefore a capacity of 1000 million cubic metres, or one cubic kilometre, is required, and the gates must be wide enough to deliver the whole quantity of one cubic kilometre within a time of scarcely more than one hour.

We conclude, therefore, that significant contributions from tidal power can be made only by means of large reservoirs and with dams, gates, and turbines for handling great quantities of water. Although there are tides along all coasts of all continents, only those sites are suitable for the installation of tide power plants where (a) the mean difference between high and low tide is sufficiently great, and (b) the shore-line is favourable for the formation of large basins with dams of reasonable dimensions.

There are a number of places which fulfill these conditions, and detailed projects have been made for the development of the sites.

The best place in the British Isles seems to be the Severn mouth, and the "Severn Barrage Scheme" has been widely discussed. With a dam just above the confluence of the Severn and Wye rivers, a large basin could be built from which the water would be released through thirty-two 25-MW generating units with an aggregate annual production of 2 TWh. The fuel savings achieved by the Severn plant are estimated at about a million tons of coal per annum.

Other favourable places in Europe can be found on the French coast of the Channel, particularly in the part between Brittany and Normandy, and it is probably there that the first big tidal power plant will be built. In 1943 the Société d'Etudes pour l'Utilisation des Marées (SEUM) was founded, which was later incorporated into the nationalized Electricité de France. A report by M. R. Gibrat (1953 and 1956) describes the interesting theoretical work which has been done so far, and will be followed by actual construction of full-scale plants. The first project to be realized within the next few years will be a 342-MW plant yielding 820 million kilowatt-hours annually, on the estuary of the river Rance, near Saint-Malo, on the north coast of Brittany. The dam across the estuary will contain thirty-eight units, each consisting of a Kaplan turbine and a 9-MW alternator fixed on the turbine axis within a water-tight bulb. Every unit is mounted with horizontal axis in tubes across the dam, through which the filling and emptying of the Rance basin will be done. By properly adjusting the pitch of the blades the Kaplan turbines can run in the right direction for producing power, during the periods both of filling and of emptying the basin. On the other hand, by using the alternators as motors, and by consuming power from the grid, the units can work as pumps, and can if desired accelerate the filling and emptying process, and also increase the level difference between the basin and the open sea.

By properly timing the periods of producing and consuming power during the filling and emptying phase some very useful effects, resulting from the specific properties of tidal power, can be achieved. As mentioned repeatedly in this book, the combination of pumps and turbines with basins on different levels is the best means available at present for large-scale energy storage. During the hours of low demand water is pumped to the higher level by dump energy from the run-of-river plants, and is released again through the turbines producing current at peak-load hours. A tidal power basin can under certain conditions perform this task in a much better way, making use of the varying level of the open sea. In ordinary storage basins we can under ideal conditions—that is, neglecting all kinds of losses—recover as much energy as we had stored by the pumping process. In tidal plants, however, the returns may amount to a multiple of the stored energy when a certain tonnage of water is pumped into the basin at a period of small level difference, and released again through the turbines a few hours later when the level difference is near its maximum. Returns with several hundred per cent interest within a few hours can be achieved in this way, and, moreover, the cash value of the returns can be still higher when cheap energy at small load hours is used for pumping while energy is recovered at peak load hours. Of course, this combined effect of increased returns cannot be achieved with a fixed timetable of pumping hours because the electricity consumption has a period of twenty-four hours, while the interval between consecutive high tides is on the average about 12 hours 26 minutes, so that a delay of the tides of nearly an hour daily results. As mentioned, however, the Kaplan units with horizontal axis can be used in both directions of the water-flow for pumping as well as for producing power. If, therefore, the low tide coincides with the small load period (say, between midnight and 5 A.M.) then a small expense of dump energy can be used for lowering the water-level in the basin below that of the open sea, while during the following period of peak load the refilling of the emptied basin from the sea-level—which has risen meanwhile—yields a multiple return. When a few days later the high tide coincides with the small load period the same effect can be achieved by pumping water into the basin and by recovering the energy some hours afterwards.

Thus the two-way action of the horizontally mounted Kaplan units may be used for obtaining more valuable energy at peak load hours than the simple method of one-way power production from the emptying basin.

The experience gained by the Rance plant will be a useful guide for other French projects, and possibly for the British Severn project too. A very large French tidal plant is planned in the bay of Mont Saint-Michel. By means of two dams stretching south and east of the island Chausey, a basin with an area of about 200 square miles miles (500 km²) will be made, and the installed power capacity will be between 10,000 and 15,000 MW, yielding between 24 and 36 TWh annually. On the completion of this and other projects quite a considerable percentage of France's electricity demand could be supplied from tidal power plants.

Similar large projects in other countries concern the Passamaquoddy Bay, on the frontier between Maine and New Brunswick, and also several Russian plants, among them one with an expected annual output of 72 TWh on the Mezen Gulf, opening into the Arctic Ocean near the Arctic Circle at longitude 44°E., another one on the Kola peninsula, and a third on the Okhotsk Sea, between East Siberia and Kamchatka. In Argentina plans have been made to use the estuary of the river Deseado, near Puerto Desiré, on the Atlantic coast, near longitude 66°W., latitude 48°S.

None of these projects has been realized so far, and the competition of nuclear power may cause some delay in the development of tidal power. Still, there is reason to expect that the ultimate success of the Rance project will encourage future enterprises in this field.

Ocean Heat and Earth Heat: the Larderello Plant

According to the Second Law of Thermodynamics, the conversion of heat into mechanical work is possible when two heat reservoirs of different temperature are at our disposal. Actually there are reservoirs on earth with enormous heat capacities and with temperature differences permanently maintained by natural forces. In tropical waters the temperature of the surface layers may be over 30°F. (17°C.) higher than at a depth of several thousand feet. On the Siberian coast the winter temperature of the water below the ice crust is near the freezing-point, while the temperature of the air may be $-35°C$. to $-40°C$. The interior of the earth is hotter than the surface, and the energy transported by heat conduction to the surface and radiated away is estimated at about 180,000 TWh per annum, or seven times the present world's consumption.

All these reservoirs offer the possibility of using the temperature differences for operating steam-engines by placing a boiler into the medium of higher temperature and a condenser into the colder one. A suitably chosen working fluid with sufficient vapour pressure can drive a steam turbine, and after condensation be returned to the boiler again. Butane, for instance, with a boiling-point at $-10°C$. under atmospheric pressure, will develop enough gauge pressure when heated to freezing-point by sea-water to drive a turbine, and will, after expansion in the turbine, readily condense in an air-cooled condenser at $-35°C$.

The French inventor Georges Claude made the first successful large-scale experiments in 1929 and 1930 with an ocean-heated steam-engine off Havana. The boiler was heated by surface water circulated through its ' flame '-tubes, while the condenser was cooled by water sucked up from the depths. The experiments in the West Indies were broken up after a gale destroyed the metal tube which was used for sucking up the cold water. New experiments with improved outfit have been started recently off the Gold Coast of West Africa, near Abidjan. The boiler is heated by a circulation of 9000 gallons per minute of warm surface water in a lagoon where the mean temperature is 30°C. = 86°F., while the condenser is cooled by a circulation of cold water sucked up from a depth of 500 metres. Instead of the rigid sucking-tube used off Havana, a flexible rubber pipe of 2 m diameter reinforced with steel hoops draws up the cold water. The expected gross power will be 11 MW, giving a net output of 7 MW after deducting auxiliary and pumping requirements. If disturbances from heavy storms can be avoided a continuous twenty-four hours daily operation would be possible, supplying steady power to a grid. The plant can be considered as an indirectly heated solar steam-engine with energy storage (in the lagoon) before conversion into mechanical work. A disadvantage of the method lies in the low efficiency of the process caused by the small temperature difference. Large installations are therefore necessary, and the plant belongs to the same category of power stations as hydro-electric plants requiring considerable capital investment per TWh annual output, but operating without fuel costs.

Better efficiencies could be achieved by using the earth's heat, because the difference of the temperatures in the interior and on the surface is much higher. The difficulty lies in extracting the heat from the depths. The heat is there, billions and trillions of B.Th.U., and at temperatures high enough for efficient conversion into power—but how to collect it and extract it from innumerable cubic kilometres of earth and rock and to concentrate it upon a boiler or an air-heater? Thousands of widely ramified underground tubes like the blood-vessels of a living tissue would be necessary to do the job.

A practical application of tapping heat from the earth has been made, therefore, only in places where Nature was kind enough to do the collecting by itself, and to present the extracted heat in a nice and handy form. In the steam wells of Larderello, in Tuscany, near Florence, a continuous natural steam-flow comes from holes in the ground, and it turned out that it is not only hot, but under pressure as well, so that after being properly harnessed it can be fed directly to turbines. (A certain complication arose from the fact that the volcanic steam contains chemicals which are corrosive to normal turbine steel. It was therefore only when blades from special non-corrosive alloys were fitted into the turbines that the heat exchangers formerly used could be dispensed with.) The plant is working now with a capacity of 261 MW, and from 1904, when the development began, to 1950 a total of about eight million tons of valuable chemicals like borax and ammonia has been recovered, in addition to the large-scale power production.

Unfortunately such gifts of Nature are very rare. Larderello and another yet undeveloped site in New Zealand seem to be the only places where power can be produced directly from volcanic steam. Still, there are numerous places where the Claude process might be performed with far better efficiency than with lukewarm lagoon water as the heat source. Iceland not only possesses her famous geysers, but also numerous ' heat fields' where steam or hot water emerges in a regular flow from fissures in the rocks, and can be delivered to consumers through pipe-lines. The total output is sufficient to provide space heating for about 30,000 of the inhabitants of the capital city of Reykjavik. Although abundant amounts of hot water are left over, no attempt has been made yet to utilize them for power production.

Surveying all possibilities for power supply from recurrent sources, we can say: The very rich energy income of the earth and its offshoots are spread in various different forms all over the globe, but mainly in a very diluted form, so that economic recovery is not easy. In some places, as, for instance, at the sites used for water-power plants or in Larderello, a natural concentration has occurred, so that exploitation can compete with power production from fossil fuels. Local energy demands can be met by water-power, fuel from vegetation, volcanic heat, etc., but no attempts at a world-wide supply from recurrent sources will be made as long as the use of energy stored in coal, oil, or natural gas is so much cheaper.

Most of the recent attempts to find new methods for energy production from inexhaustible sources, like solar heat, wind, and tides, have been overshadowed by the advent of the atomic age. The expectation of obtaining cheap energy from nuclear fuels has a paralysing effect on the pursuance of other projects.

As regards efforts to live off income, mankind is in the situation of a carefree young man who for some time has been living well on a rich inheritance. Feeling that his wealth will come to an end, he is just about to find a job with a regular small income, but is stopped from doing so because of news telling him that a rich uncle is going to die. So we are stopping plans with solar and other power in the expectation that the uranium reserves on the earth will prove to be a new, rich heritage.

CHAPTER XIV

Atomic Energy

Units and Constants used in Atomic Physics

ALONG with the usual units of physics, special atomic units of mass and of energy will be used in the following pages. In comparing the masses of atoms and molecules of different substances an atomic unit of weight has been introduced. It was originally intended to be equal to the weight of the hydrogen atom, and since early measurements had shown the oxygen atom to be sixteen times heavier, the atomic weight of oxygen was taken as 16. Although later measurements proved that the ratio of the two weights is not exactly 16, but a little less, it was found convenient to define the unit of atomic weight as 1/16 of the weight of the oxygen atom, so that the atomic weight of oxygen is exactly 16·0000. The reason for this was the fact that many fundamental determinations of atomic weights of other elements had been made by using oxides—that is, oxygen compounds—of these elements, so that all international tables of atomic weights contained data based on this unit. The atomic unit of mass is called, shortly, *mass unit*, abbreviated to MU.

An important fundamental constant of physics is the number L, giving the ratio between the microscopic unit gramme and the mass unit MU:

$$1 \text{ gramme} = L \text{ MU}$$
$$L = 6 \cdot 025 \times 10^{23} \quad \dots\dots\dots\dots\dots\dots\dots\dots\dots(41)$$

This number L is called the *Loschmidt number*, or the *Avogadro number*. We obtain the weight of the atom of any element expressed in grammes by dividing its atomic weight, which is always given in MU, by the Loschmidt number $L = 6 \cdot 025 \times 10^{23}$.

Since the atomic weight of hydrogen is very nearly equal to 1, the reciprocal of L gives approximately the mass m_H of the hydrogen atom, expressed in grammes:

$$m_H = 1 \cdot 67 \times 10^{-24} \text{ grammes} \dots\dots\dots\dots\dots\dots(42)$$

The hydrogen atom is the lightest among the atoms of all elements, but the electron, which can be considered to be the atom of negative electricity, is much lighter still. Its mass m_e is given by:

$$m_e = 0.0055 \quad MU = 0.91 \times 10^{-27} \text{ grammes } \dots\dots\dots(43)$$

The energies released or consumed in elementary processes between single particles like atoms or electrons will not be expressed in B.Th.U. or calories, but in an appropriate unit called an *electron-volt* (abbreviated, ev). The electron-volt is the energy acquired by a particle with an electric charge equal to the charge e of the electron when it is accelerated by falling through an electrical potential difference of 1 volt. The conversion into absolute C.G.S. units is

$$1 \text{ ev} = 1.6 \times 10^{-12} \text{ erg} = 1.6 \times 10^{-19} \text{ joule} \dots\dots\dots(44)$$

In nuclear physics the million-fold unit *Megaelectron-volt* (abbreviated, Mev) is used. The conversion formulæ are:

$$1 \text{ Mev} = 10^6 \text{ ev} = 1.6 \times 10^{-6} \text{ erg} = 1.6 \times 10^{-13} \text{ joule}$$
$$= 1.52 \times 10^{-16} \text{ B.Th.U.} = 3.82 \times 10^{-17} \text{ kcal}$$
$$= 4.45 \times 10^{-20} \text{ kWh} \dots\dots\dots\dots\dots\dots\dots\dots\dots\dots\dots(45)$$

Another physical constant which plays an important rôle in nuclear physics is the velocity of light. It is given by

$$c = 3 \times 10^{10} \text{ cm per second.}$$

The Atom is not 'Atomos'

We owe the possibility of using atomic energy for power production to the fact that the atoms are not what their name meant originally. The Greek word *atomos* means 'indivisible,' and the ancient conception of the atom was the smallest particle into which matter can be split.

Experience since has shown that all elements are actually composed of very small individual particles the properties of which determine the physical and chemical behaviour of the element. These particles have been identified with the atoms of the ancients, but towards the end of the nineteenth century it was found out that they are not indivisible but can be split into fragments with opposed electrical charges. Thus, on the one hand, a number of atoms of one or more elements can unite to form a molecule, and, on the other hand, a single atom can be decomposed into its constituents, which are the electrically positive nucleus (one for each atom) and a number of electrically negative electrons. All these processes of building up larger complexes or of decomposition into smaller units involve either release or absorption of energy in the form of radiation or thermal agitation (heat).

The discovery that atoms are not indivisible was not the only surprise to upset ideas of the atom conceived in the first half of the nineteenth century. The credo of practically all serious physicists and chemists up to about 1880 ran as follows:

(1) The atoms are indivisible.
(2) The atoms of different elements are basically different from one another, have different masses, and cannot be transmuted into one another.
(3) All the atoms of any single element are exactly equal to one another, and in particular have equal masses.

Doubts regarding the truth of dogma No. 1 arose when the phenomena of electrolysis and of electrical discharges in gases proved that neutral molecules and atoms can be decomposed into positive and negative electrically charged particles.

The discovery of radioactivity made by the French scientists **Pierre** and **Marie Curie** in 1898, and the subsequent investigations made by **Rutherford** and his school in Cambridge first reduced dogmas Nos. 2 and 3 to mere rules to which certain exceptions existed, but soon after World War I it was recognized that dogma No. 3 was altogether wrong, and that the rule about the impossibility of transmuting elements into one another is valid only as long as the atoms are not exposed to bombardment with electrically charged particles accelerated by sufficiently high voltages.

Prout's Hypothesis

The idea of the atom as a simple, compact, indivisible, and unchangeable particle had therefore to be abandoned in favour of the conception of a more complex structure, which we shall describe later in this chapter. An early pioneer of the modern conception of the atom was the British scientist and physician **William Prout** (1785–1850), who a century and a half ago drew the attention of contemporary chemists to the striking fact that by choosing the weight of the hydrogen atom as the unit we obtain for the atomic weights of most of the light elements very nearly whole numbers. He rightly concluded that this fact could not be mere chance, and proved, therefore, that the atoms of all the other elements are aggregates of hydrogen atoms. We know to-day that Prout was very near the truth, but during his lifetime his hypothesis seemed to be definitely disproved by the fact that small deviations of the atomic weights from whole numbers were established beyond doubt by very exact measurements. Thus, if—as had been internationally agreed upon—the unit is chosen so that the atomic weight of oxygen is exactly 16·0000 MU, then the atomic weight of hydrogen, for instance, is not 1, as was originally believed, but 1·008 MU, and the ratio between the two numbers is not the integer 16, but 15·873. Still larger deviations from whole numbers were found with other elements; thus chlorine has an atomic weight of 35·46.

A full explanation by which the observed deviations from whole numbers were reconciled with Prout's basic idea could be found from the knowledge of the structure of the atoms gained at the beginning of this century. Details about the gradual development of that knowledge are given in a good many popular books on the atom, among which *Modern Ideas of the Atom,* by S. Lucas,[1] may be recommended to the reader. Instead of telling the whole story once more, we shall here summarize the results.

The Nuclear Theory of the Atom

According to ideas conceived by Rutherford in Cambridge, and later successfully extended by **Niels Bohr** in Copenhagen, the atoms are something like planetary systems, about 10^{22} times smaller than our solar system. The rôle of the sun as the central body of the system is played by the so-called nucleus of the atom, which is a minute particle positively charged and containing virtually all the mass of the atom. The planets of the system are represented by the electrons describing circular or elliptical orbits round the nucleus. In the normal state of the atom, which is electrically neutral, the positive charge of the nucleus is fully compensated by the negatively charged orbital electrons. Since their charges are all alike, it follows that the charge of the nucleus is a whole number if we take the electronic charge as the unit. Experience shows that this number is at the same time the so-called *atomic number* Z—that is, the ordinal number attributed to the element when we arrange all the elements in the sequence of their atomic weights, taking the lightest as No. 1, the next lightest as No. 2, and so on. Thus, for instance, the hydrogen atom, which is the lightest of all, consists of a nucleus (called *proton*) with the charge + 1 surrounded by one orbital electron with charge − 1. The second lightest atom is that of helium, consisting of a nucleus (also called *alpha particle*) with a charge + 2 surrounded by two electrons, and so on, as

[1] Harrap, 1948.

indicated in Table 37, which gives a few examples of other elements.

TABLE 37. THE ATOMIC NUMBERS OF SOME OF THE ELEMENTS
(Atomic number = number of electrons in the neutral atom.)

SYMBOL	ELEMENT	ATOMIC NUMBER (Z)
Li	Lithium	3
C	Carbon	6
N	Nitrogen	7
O	Oxygen	8
Na	Sodium	11
Cl	Chlorine	17
Fe	Iron	26
Cu	Copper	29
Ag	Silver	47
Au	Gold	79
Ra	Radium	88
U	Uranium	92

Apart from its enormously vaster size, our solar system differs from the atomic planetary systems by the orientation of the orbits. All the nine planets of our sun move along orbits which lie nearly in a plane. In the heavier atoms like gold, however, the numerous electrons are whirling round in orbits of quite different orientations.

In electrical discharges or in chemical processes it can happen that a neutral atom loses one or two of its electrons, and thus turns into a positive *ion*. Electrically charged atoms or molecules are called ions. The atoms of the alkali metals lithium, sodium, potassium, rubidium, and cæsium are particularly inclined to lose the outermost electron of their planetary electron shell, while others, like the atoms of the halogen elements fluorine, chlorine, bromine, and iodine, are, on the contrary, inclined to attach an odd electron to their shell, thus becoming negative ions.

Energies involved in Chemical Processes

Mutual interchange of electrons between atoms plays a fundamental rôle in building and decomposing chemical compounds, and the heat released in chemical processes, like those of burning coal or gas, stems ultimately from the electric forces between the electrons and the atomic nuclei. When, for instance, carbon dioxide molecules are built by burning coal, the electric energy of the three nuclei and the total of 22 electrons in the molecule CO_2 is less than the sum of the energies of the same particles assembled in a single atom C and an oxygen molecule O_2. The energy difference released during the formation of the compound CO_2 is used for enhancing the thermal agitation of the carbon dioxide molecules, or, in other words, for raising the temperature of the gas. The total energy turnover achieved in chemical processes is of the order of magnitude of a few electron-volts (see p. 290) per molecule of the resulting compound. Thus, for instance, the process of burning coal given by the formula

$$C + O_2 \rightarrow CO_2 \quad \dots\dots\dots\dots\dots\dots\dots\dots\dots(46)$$

releases 4·17 ev per carbon atom, which corresponds to a heat value of

8050 kcal per kg, or 14,500 B.Th.U. per lb,

of pure carbon. Similarly, the process of compounding hydrogen and oxygen to water

$$2H_2 + O_2 \rightarrow 2H_2O$$

releases 2·5 ev per molecule of H_2O, corresponding to a heat value of 29,000 kcal/kg = 52,200 B.Th.U. per lb of hydrogen.

Strictly speaking, all the energy obtained by burning fuels like coal, oil, wood, etc., is a kind of atomic energy, because it is derived from the electrical forces binding the electrons in the outer layers of the atom. The energy output of all such reactions is restricted to a few electron-volts per elementary process, and no radical increase can be expected from any new chemical reactions, which consist of the building up or decomposing of molecules without affecting the nuclei themselves.

What has been called ' atomic energy ' since 1945 is, however, the energy released by *nuclear processes* in which not only the electronic shells of the atom are affected, but also the internal structure of their nuclei. The energy turnover of nuclear processes is millions of times greater than that of the very ' tame ' chemical reactions involving only the shells. The following account will help us to understand this better.

The Teachings of Radioactivity

Three fundamental dogmas of the nineteenth-century chemistry were shaken by discoveries made in the field of radioactivity during the first two decades of this century, and at the same time Prout's idea of a common origin of all elements from simple primary particles had an unforeseen revival. What we learnt from studies connected with radioactivity was:

(1) Atoms are not indivisible. They can be 'mildly' decomposed by peeling off one or more electrons from their outer layers, or even radically split by disintegrating their nuclei.

(2) The rule that elements cannot be transmuted into one another holds only for the ordinary chemical processes. By the more radical operation of nuclear disintegration we can transmute one element into another.

(3) It is not true that all atoms of a given element are exactly equal to one another. Most elements are mixtures of atoms of slightly different masses.

The process of nuclear disintegration connected with transmutation of elements was first observed in radioactive elements. It was found that the so-called alpha radiation of radium consisted of the emission of an alpha particle from the nucleus of the radium atom. The alpha particle could be identified with the nucleus of the helium atom, and what is left over of the original radium after splitting off an alpha particle from the radium nucleus is an atom of another element called radium emanation, or radon, which, according to its chemical properties, belongs to the same group as helium—namely, to the rare gases. Thus the element radium, which is a metal of the earth alkaline group, like calcium, strontium, and barium, transmutes itself by the process of alpha radiation into two sorts of rare gases. The process of transmutation is not finished at this stage, however, because the radium emanation is not a stable element either, but transmutes itself, by emitting again an alpha particle, into another radioactive element, radium A, which in its turn becomes radium B. Thus after a number of further changes the series ends up with an element radium G, which is stable at last and no longer changes. Chemical analysis showed that radium G is identical with lead as regards all its physical and chemical properties, except the fact that its atomic weight is 206, while that of ordinary lead is 207·2.

Isotopes. The Revival of Prout's Hypothesis

The case of radium G and lead was the first known instance of atoms of different masses belonging to the same element. Other examples were thorium C, radium C, and bismuth. It was soon found that the characteristic property of an atom determining the chemical and physical behaviour of an element is not, as had been assumed, its mass, but its *nuclear charge,* which is equal to the number of orbital electrons in the neutral state of the atom, and also equal to its *atomic number* Z. The term *isotopes* has been adopted for substances with equal atomic number, but different mass of the atom. Thus radium G is an isotope of lead, radium C an isotope of bismuth. Isotopes are something like twins with great similarity and the same family name (' lead,' for instance), but slightly different masses.

In 1920 **Aston** in Cambridge made the discovery that the phenomenon of isotopy is not restricted to substances at the top end of the table of elements with high atomic weights, but occurs also in lighter elements. The rare gas neon was the first which could be proved to consist of two isotopes with the masses 20 and 22, and subsequently it was found that nearly all elements are mixtures of isotopes. Curiously enough, hydrogen was among the last elements found to be a mixture of two isotopes. Besides the ordinary, or light, hydrogen, with a nucleus of mass 1, there is also a heavy isotope with mass 2. The nuclear charge is, of course, 1 in both cases, or else the atom would not belong to the element hydrogen. The heavy isotope of hydrogen has been given a separate name, *deuterium*, and a separate symbol, D. The nucleus of deuterium, with mass 2 and charge 1, is called *deuteron*. The oxide of deuterium, D_2O, is *heavy water*; it is mixed in small concentration with all water existing on earth. In pure form heavy water has a slightly greater density than ordinary chemically pure water.

It was of particular interest to see that chlorine, for instance, consists of two isotopes of masses 35 and 37 in such proportions, with a higher percentage of the lighter one, that the resulting average weight of the mixture is 35·46. It will be remembered that chlorine, with its atomic weight lying just between the two whole numbers, was a kind of Crown witness in the case against Prout's hypothesis. Clearing up this large deviation from the whole-number rule made the path free for a revival of Prout's idea in terms of modern conceptions of the structure of the atom, although isotopy alone could not account for certain smaller deviations. The explanation of the latter, given by Einstein's law and revealing very important properties as regards the energy content of the atoms, will be discussed later.

The Radioactive Families of Elements; Alpha, Beta, and Gamma Rays

Radium, which is one of the most prominent members of the generation series of radioactive elements, is not the aboriginal ancestor, but is in itself the descendant of other radioactive elements.

The ultimate source is *uranium*, which among all elements occurring in nature has the highest atomic weight and atomic number. (The latter is 92.) The whole process of radioactive decay of a series of elements results, therefore, in a slow transmutation of the radioactive element uranium into the stable—that is, non-radioactive—element lead. Two more series of radioactive elements exist also in nature; they are called the *actinium series* and the *thorium series*. Lead is the common end product of all three series, uranium is the ancestor of the radium and actinium series, thorium the starting-point of the thorium series. Hence all the uranium and thorium in the world will be ultimately turned into lead. But we shall not live to see it, because it takes 4·5 thousand million years to disintegrate half the existing quantity of uranium, and about 14 thousand million years to do the same with thorium.

Among the members of the three radioactive families there are some like radium, radon, and radium A which are alpha emitters. This means that their nuclear disintegration consists of the emission of an alpha particle from the nucleus. Other members of the series undergo a different kind of decay; their disintegration consists of the emission of an electron from the interior of their nuclei. The kind of radiation which consists of fast-moving electrons ejected from atomic nuclei had been called *beta radiation* before the identity of its particles with electrons was recognized. We know to-day that beta particles are nothing else than electrons emitted with great speed from nuclei of those among the radioactive elements which are beta emitters.

Both the alpha and beta radiation belong to the class of corpuscular radiations consisting of fast-moving atomic missiles. In some of the radioactive elements the corpuscular radiation is accompanied by an electromagnetic wave radiation of extremely short wave-length of the order of magnitude of 10^{-10} cm. These additional rays, which belong to the same class as X-rays, are called *gamma rays*; they are considerably more penetrating than the alpha and beta rays. While the corpuscular rays of all radioactive elements can be screened off by a few sheets of paper, some of the more penetrating sorts of gamma rays will go through a wooden board of several inches thickness and will be only partly absorbed by a single-brick wall—that is, one of about ten inches thickness. One of the problems of nuclear engineering is the protection of personnel against radiation hazards arising from gamma rays, which may cause lethal burns if applied in overdoses.

It should be noted that any radioactive element is either an alpha

emitter or a beta emitter, while gamma radiation is an additional
effect occurring in certain elements of either type.

Spontaneous and Artificial Nuclear Disintegration

The emission of alpha and beta particles by radioactive elements
are processes which occur quite spontaneously without being in-
fluenced by any action from the surroundings. One atom out of
842,000 atoms of radium, and two out of eleven atoms of radium
emanation, decay daily by emitting an alpha particle, and no
chemical or physical process short of bombardment with very fast
particles could accelerate or retard the rate of disintegration. Radioactivity is therefore a kind of congenital disease of the nucleus,
a certain instability of its internal structure which makes it prone
to disintegrate by emitting either an alpha or a beta particle. Among
all the naturally occurring elements, with one exception, only those
with high nuclear charges—that is, with atomic numbers between
81 and 92—suffer from this peculiar instability. The one exception
is the element potassium, which shows a very faint beta radiation.
' Faint ' means here that the number of atoms among which a
single one decays daily is some million times larger than in the case
of radium. Although, therefore, the potassium atoms are not
absolutely stable, they are much less inclined to disintegrate than
those of the heavy and highly charged atoms of the radioactive
series. Some of the latter are so unstable that within a fraction of
a second half of their quantity is disintegrated. They would not
exist in nature unless they were permanently new-born from their
radioactive parents, which in their turn are descendants of the long-
living ancestors uranium or thorium.

Until the time of World War I it seemed as if the phenomenon
of nuclear transmutation was restricted to a series of very rare
elements with atomic numbers in the neighbourhood of $Z = 90$,
while atoms of the vast majority of elements were stable enough
to withstand any attempts at transmutation. However, this opinion
proved to be erroneous; we know to-day that any element from
$Z = 1$ (hydrogen) to the heaviest, $Z = 92$ (uranium), can be transmuted by appropriate measures, and that, moreover, a number of
still-heavier radioactive elements with nuclear charges from 93 to
100, non-existent in nature, have been made artificially.

Rutherford was the first to achieve a successful transmutation of
a light element by bombarding nitrogen with alpha particles of
radium C. What happens in a collision of this kind can be described
best by the formulæ of nuclear chemistry in which the number A,

giving the mass of the atom, is placed at the top left of the chemical symbol, and the nuclear charge Z (atomic number) at the bottom left. Thus, for instance, the symbol for the nucleus of ordinary helium, with mass 4 and charge 2, is $^4_2$He; it is sometimes also indicated for simplicity by the Greek letter α. The main isotope of nitrogen, with mass 14 and charge 7, is symbolized by $^{14}_7$N. The lower right-hand side is left for the ordinary numerals indicating chemical combination. Thus, for instance, the nuclear formula for ordinary water would be written $^1_1$H$_2$. $^{16}_8$O, and for heavy water $^2_1$H$_2$. $^{16}_8$O, or, simpler, D$_2$O.

What Rutherford observed in his 1919 experiments led him to the conclusion that the nuclear reaction occurred thus:

$$^4_2\text{He} + {}^{14}_7\text{N} = {}^{17}_8\text{O} + {}^1_1\text{H} \dots\dots\dots(47)$$

In words: the fast alpha particle $^4_2$He, impinging vehemently on a nitrogen nucleus, entered its interior and stuck there while a hydrogen nucleus (which is called a *proton*) was emitted. The remaining nucleus, enriched by an alpha particle and bereft of a proton, has a charge 7 + 2 − 1 = 8, and is therefore the nucleus of an oxygen atom (compare Table 36), but not of the main isotope with mass 16, but of the rarer and heavier isotope $^{17}_8$O. In this way a reaction has taken place which, starting from helium and nitrogen, produced hydrogen and oxygen. The old dream of the alchemists of transmuting elements which during the whole of the nineteenth century was believed to be absolutely impossible has been realized in Rutherford experiments—although with an output of only hundreds, or just a few thousands, of atoms of transmuted elements, which is about a million times less than the smallest weighable quantity.

Rutherford's discovery of artificial nuclear disintegration, made in 1919, inaugurated the new science of nuclear physics, and opened the era of ' atom-splitting,' as nuclear processes were commonly called. Ever more nuclear reactions were studied, and less than two decades later many hundreds of reactions were known by which all elements could be transmuted into elements with neighbouring atomic numbers.

Progress in ' Atom-smashing ' Machines

The rate of progress was increased in 1932, when two of Rutherford's disciples, **Cockcroft** and **Walton,** succeeded in transmuting lithium into helium by bombarding a thin layer of lithium with

Note. In American publications the mass number is placed on the top right.

hydrogen ions (protons) which were accelerated in an electrical field. The advantage of this method lies first of all in the fact that the number of bombarding particles is far greater in an electrical discharge of ionized particles than in the relatively faint beam of alpha rays emerging from a radioactive source. Another advantage of using electrically accelerated particles for the bombardment of atomic nuclei was achieved by the development of modern accelerating machines in which the energy of the single missiles could be increased far beyond the energy of any particles emitted by radioactive substances. The energy of the alpha particles used by Rutherford in his pioneer experiment was a few Megaelectron-volts. Cockroft and Walton could not yet achieve this energy. The accelerating potential in their machine was 800,000 volts, and since the absolute value of the charge of the proton is equal to that of the electron (the former being positive, the latter negative), the energy per particle available in the 1932 Cambridge experiments was 800,000 electron-volts, or 0.8 MeV. Therefore the bombarding device used by Cockcroft and Walton was in comparison to Rutherford's radioactive bombarder superior only as regards the number, but not as regards the energy, of projectiles. Soon, however, the ratio of available particle energy was reversed by the astonishing progress made in building 'atom-smashing' machines of ever-increasing power.

The common feature of all these machines is that they are devices for accelerating particles like protons, deuterons, or alpha particles by means of electrical fields with the aim of attaining energies as high as possible. The higher the energy per particle the higher is the chance of hitting and disintegrating a nucleus. A first step in this development was the construction of electrostatic accelerators. **Van de Graaff** in the U.S.A. succeeded in building an electrostatic generator for 5 million volts D.C., which connected to a big discharge tube, accelerated protons to energies up to 5 MeV.

Although electrostatic generators are used with satisfactory success in a number of laboratories, the modern development leading to far greater energies went in other directions. Voltages up to a few million volts can still be handled within buildings of reasonable size, but a radical increase up to hundreds or thousands of million volts is out of the question. Still, even with relatively modest voltages of 100,000 or so, enormous energies can be achieved by applying the same accelerating field to one and the same particle repeatedly, over and over again. Imagine a merry-go-round on an absolutely frictionless pivot and a device which gives it twice in each revolution an accelerating kick. After a sufficient number of kicks it will gain quite a considerable momentum. It is exactly this method which has been adopted in the system of reiterated acceleration. Among these types of atom-smashing machines the *cyclotron,* constructed by **Ernest Lawrence** in Berkeley, California, and its further developments, named synchrotron, bevatron, and cosmotron, achieved the greatest progress towards endowing atomic particles with great energies. In these accelerators ions of hydrogen or of deuterium or of helium moving in an evacuated ring-shaped tube are forced by means of a deflecting magnetic field to run along a circular path, and a periodic electric field applied in a tangential direction gives the particles an accelerating kick each time they pass through the field. The sky-rocketing increase of particle energies achieved within the last two decades, and the still further increase expected in the near future, is represented in Table 38, which gives the particle energy records for a number of years.

TABLE 38. RECORD OF PARTICLE ENERGIES ACHIEVED AND EXPECTED
IN MODERN ACCELERATORS

YEAR	ACCELERATOR	MeV
1932	Cascade generator	0·8
1932	Cyclotron	1·2
1936	Electrostatic generator	5·0
1936	Cyclotron	6
1939	,, ...	10
1942	,, ...	60
1948	Synchrotron	380
1952	,, ...	1200
1955	,, ...	6000
1960	,, ...	26,000

It is interesting to compare these accomplishments with the progress made in other fields of technology. During the same period the speed and carrying capacity of aeroplanes, the power of engines, the range of radio transmitters, the efficiency of electric lamps, have all made definite progress, but the relative increase during that period is given by a factor lying between 1 and 2. Even reviewing the progress made in two millennia, we find, for instance, that the velocity of man-carrying vehicles shows an increase of about 50 : 1 if we compare the speed of a modern jet plane with the horse-driven Roman *quadriga*. However, the energies of electrically accelerated atom-smashing particles have increased to-day more than 7000 times over the 1932 performance, and will show an increase of 30,000 : 1 as soon as the giant cosmotron now under construction at the European Centre for Nuclear Research (CERN) in Geneva is in operation. A good deal of inventive genius combined with mathematical and technological skill was necessary to achieve this progress. It should be borne in mind, however, that these giant machines—the largest projected now having a diameter of 660 ft—do not directly contribute to power production, but are, on the contrary, consumers of power. They are mere tools for probing into the structure of atomic nuclei—that is, for making experiments from which greater detail about nuclear forces can be gathered. The results of these investigations will throw some light on the fundamental problem of the structure of matter, and may in addition help to solve the biggest problem of future power production—namely, to arrange controlled thermo-nuclear chain reactions. This will be discussed in the final chapter of the book.

The Composition of Atomic Nuclei. Protons and Neutrons

Modern conceptions as to the internal make-up of atomic nuclei are based on a very important discovery made by **James Chadwick** at Cambridge. By critically analysing the results of earlier pioneer experiments of **Bothe** and **Becker** in Germany and of **Joliot** and **Curie** in Paris he found that these experiments could be explained only by assuming (as Rutherford had suggested tentatively long ago) the existence of an elementary particle with a mass equal, or very nearly equal, to that of the proton (the nucleus of the hydrogen atom), but with zero charge. This electrically neutral particle was called a *neutron,* and we know to-day that it plays a fundamental rôle in the nuclear processes which are used for releasing atomic energy. In addition, it is one of the primary building-stones of all matter. According to our present knowledge, the nuclei of all elements are combinations of only two primary particles—the *proton* and the *neutron*. In this way Prout's idea is revived in a slightly modified form. While he supposed that all atoms were aggregates of hydrogen atoms, it seems to be established now that they are aggregates of protons and neutrons. Therefore Prout's basic idea that the whole numbers giving the atomic weights of so many elements could be explained by their composition of primary particles of equal mass was verified at last—although more than a century after its publication.

Indicating the nuclear charge ('atomic number') by Z and the mass number—which is the atomic weight made a whole number —by A, the very simple rule about the composition of the nuclei runs thus:

Each nucleus of an isotope of atomic number Z and mass number A is an aggregate of Z protons and A-Z neutrons.

The collective name of the two particles proton and neutron which constitute the nucleus is *nucleon*. The dimensions of the nucleons, as well as of the electrons, are extremely small, their diameters being of the order of magnitude of 10^{-13}cm. The atom of the main isotope of hydrogen ('light hydrogen') consists simply of a proton as the nucleus and an electron circling round the proton in an orbit of about 10^{-8}cm diameter. (It should be noted that this diameter is between ten thousand and a hundred thousand times larger than the diameter of the nucleus itself. Viewed from the planet electron, the 'sun' proton would therefore not look like the size of our real sun, but like a somewhat enlarged star.) The atoms of the heavy hydrogen (deuterium) consist of a deuteron and one orbital electron. The deuteron itself is the combination of one proton and one neutron. The nucleus of ordinary helium (the alpha particle) with mass number 4 is composed of two protons and two neutrons; the complete neutral helium atom consists of this nucleus and two orbital electrons. Table 39 gives the composition of the nuclei of some elements which will play a rôle in the discussions of the next two chapters.

The table contains just a few examples of isotopes to which we shall refer later on in this book; it should be clearly understood, however, that elements like carbon, barium, radium, etc., possess many more isotopes than listed here. The total number of all isotopes known so far exceeds one thousand.

According to a terminology proposed in the United States in 1947, the term 'isotope' should be restricted to denote the members of a family of different atoms with equal atomic numbers, and therefore belonging to the same element. Thus, for instance, ordinary hydrogen and deuterium are isotopes, and also radium G and lead. The term for *all* sorts of atoms with different weights and different atomic numbers, however, should be *nuclides*. The existence of 274 stable nuclides has been established, and in addition there are 53 naturally occurring radioactive nuclides, like the isotopes of uranium, thorium, and their descendants. The largest families among the

TABLE 39. COMPOSITION OF THE NUCLEI OF SOME ISOTOPES OF A FEW ELEMENTS

Symbol	Element	Number of Protons	Number of Neutrons
$^3_2$He	Helium	2	1
$^4_2$He		2	2
$^6_3$Li	Lithium	3	3
$^7_3$Li		3	4
$^{12}_6$C	Carbon	6	6
$^{13}_6$C		6	7
$^{136}_{56}$Ba	Barium	56	80
$^{138}_{56}$Ba		56	82
$^{226}_{88}$Ra	Radium	88	138
$^{230}_{90}$Th	Thorium	90	140
$^{232}_{90}$Th		90	142
$^{235}_{92}$U	Uranium	92	143
$^{238}_{92}$U		92	146
$^{239}_{92}$U		92	147
$^{239}_{93}$Np	Neptunium	93	146
$^{239}_{94}$Pu	Plutonium	94	145

stable elements are tin, with ten isotopes, and neon, with nine. Gold, however, is lonely, with only one stable isotope, $^{197}_{79}$Au. The artificially made radio-nuclides will be dealt with in the next section.

In view of the fact that the term ' nuclides '—though preferable in principle—is not yet generally accepted, the expression ' isotope,' as more commonly used in the European literature, will be retained in this book, both in the restricted sense, and also for the more general notion of the nuclides.

Radioactive Isotopes

The majority of all isotopes investigated within the last two decades have no natural occurrence, but are made artificially by nuclear processes changing the composition of the nuclei. Such artificially made isotopes are not stable elements, but disappear sooner or later by radioactive disintegration, which transmutes them into other isotopes. These are either radioactive again or finally stable. The important discovery of artificial radioactivity was made by **Irene and Frederick Joliot-Curie** in 1934, when they found that an aluminium nucleus $^{27}_{13}$ Al bombarded with alpha rays absorbed the alpha particle and emitted a neutron, turning thereby into a nucleus which was a radioactive isotope of phosphorus not existing before in nature. The formula of this nuclear reaction is

$$^{27}_{13}\text{Al} + ^{4}_{2}\text{He} \rightarrow ^{30}_{15}\text{P} + ^{1}_{0}\text{n} \dots\dots\dots(48)$$

(n is the symbol for neutron.) In contrast to the ordinary phosphorus $^{31}_{15}$P, the newly built isotope is not stable, but a so-called positive beta emitter, or positron emitter, which emits an elementary particle of equal mass to that of the electron, but of opposite— that is, positive—charge, called a *positron*. Emitting a positron from the nucleus reduces its charge by one unit, while the mass number is not changed. With e^+ as the symbol of the positron, the formula indicating the spontaneous decay of radioactive phosphorus is

$$^{30}_{15}\text{P} \rightarrow ^{30}_{14}\text{Si} + e^+ \dots\dots\dots(49)$$

The final result of bombarding aluminium with alpha rays is therefore the production of an isotope of silicon with the mass number 30. Joliot's discovery was the first step into a new land of nuclear science of ever-growing importance and fertility. Subsequent research work done within the last two decades has shown that not only one but four radioactive isotopes of phosphorus with different masses can be made artificially, and that, moreover, suitable nuclear processes can be devised which not only create radioactive isotopes of all the 92 elements existing in nature, but also of further elements No. 93 to No. 100 which so far had not existed anywhere. Among the latter, which are called *transuraniums,* the element No. 94, plutonium, has won the sad fame of being the terrible explosive used in the A-bomb which destroyed Nagasaki. But, quite apart from the so-called fissionable, or fissile, elements, which we shall discuss later, the artificially made radioactive isotopes of light- and middle-weight elements—shortly called radio-isotopes—proved to be very valuable materials with an astonishing versatility of their application. This is what Gordon Dean, former chairman of the U.S. Atomic Energy Commission, says in his book *Report on the Atom* about the radio-isotopes: "Actually radio-isotopes constitute perhaps the happiest chapter in the story of the atom. They are used to treat the sick, to learn more about disease, to improve manufacturing processes, to increase the productivity of crops and livestock, and to help man to understand the basic processes of his body, the living things around him, and the physical world in which he exists."

Some of the radio-isotopes are negative beta emitters, like those among the families of the natural radioactive elements decaying by the emission of electrons from the nucleus. Others are positron emitters, as, for instance, radiophosphorus $^{30}_{15}P$, which was the first radio-isotope ever observed. Attentive readers may ask how an electron can be emitted from the nucleus, which, as had been stated before, is composed only of protons and neutrons. The answer is: We have reasons to assume that even the primary particles proton and neutron are not eternally unchangeable, but can undergo transmutations as well. One of these processes is the conversion of a neutron into a proton and an electron; the other is the transformation of a proton into a neutron and positron. Any such process taking place in a nucleus will alter its charge, and therefore transmute it into the atom of a different element.

The best survey of the existing stable isotopes and of the radio-isotopes which have been produced so far can be given in the two-dimensional isotope chart in which the atomic number Z is plotted along the ordinate and the number of neutrons $(A - Z)$ along the abscissa. Since the chart is too large to be printed here in full, only the beginning and the upper end containing the elements usable as nuclear fuels are shown in Figs. 81 and 82. All nuclei

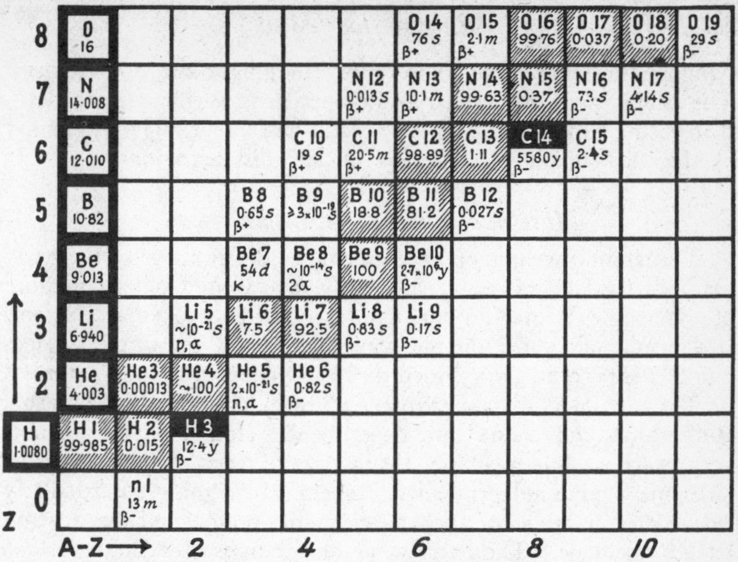

FIG. 81. ISOTOPE CHART OF ELEMENTS NOS. 1 TO 8

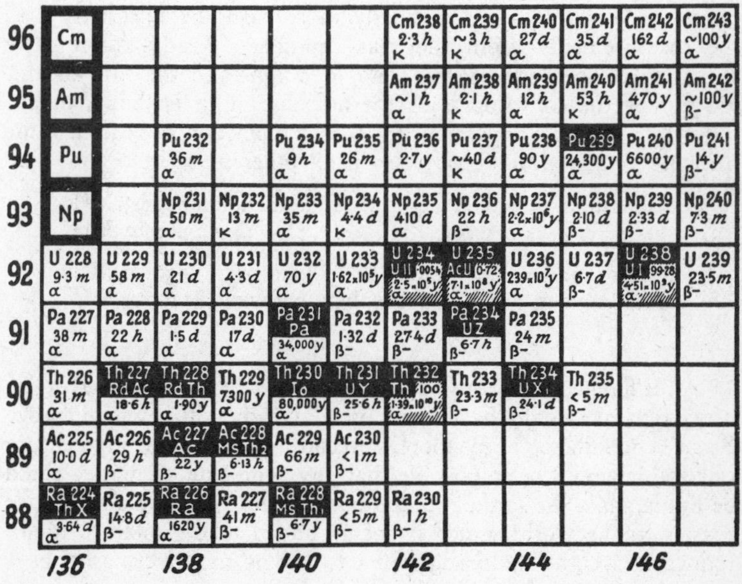

Z	Element	A−Z=0	1	2	3	4	5	6	7	8	9	10	11
8	O 16							O 14 76 s β+	O 15 2·1 m β+	O 16 99·76	O 17 0·037	O 18 0·20	O 19 29 s β−
7	N 14·008						N 12 0·013 s β+	N 13 10·1 m β+	N 14 99·63	N 15 0·37	N 16 7·3 s β−	N 17 4·14 s β−	
6	C 12·010					C 10 19 s β+	C 11 20·5 m β+	C 12 98·89	C 13 1·11	C 14 5580 y β−	C 15 2·4 s β−		
5	B 10·82				B 8 0·65 s β+	B 9 ⩾3·10⁻¹⁸	B 10 18·8	B 11 81·2	B 12 0·027 s β−				
4	Be 9·013				Be 7 54 d K	Be 8 ~10⁻¹⁴ s 2α	Be 9 100	Be 10 2·7×10⁶ y β−					
3	Li 6·940			Li 5 ~10⁻²¹ s p,α	Li 6 7·5	Li 7 92·5	Li 8 0·83 s β−	Li 9 0·17 s β−					
2	He 4·003		He 3 0·00013	He 4 ~100	He 5 2×10⁻²¹ s n,α	He 6 0·82 s β−							
1	H 1·0080	H 1 99·985	H 2 0·015	H 3 12·4 y β−									
0			n 1 13 m β−										

A−Z →

FIG. 82. ISOTOPE CHART OF ELEMENTS NOS. 88 TO 96

This is part of a complete chart designed by Friedlander and Pirlman.

Z	Element	136	137	138	139	140	141	142	143	144	145	146	147
96	Cm			Cm 238 2·3 h K	Cm 239 ~3 h K	Cm 240 27 d α	Cm 241 35 d α	Cm 242 162 d α	Cm 243 ~100 y α				
95	Am		Am 237 ~1 h α	Am 238 2·1 h K	Am 239 12 h K	Am 240 53 h K	Am 241 470 y α	Am 242 ~100 y α					
94	Pu			Pu 232 36 m α		Pu 234 9 h α	Pu 235 26 m α	Pu 236 2·7 y α	Pu 237 ~40 d K	Pu 238 90 y α	Pu 239 24,300 y α	Pu 240 6600 y α	Pu 241 14 y β−
93	Np			Np 231 50 m α	Np 232 13 m K	Np 233 35 m α	Np 234 4·4 d K	Np 235 410 d K	Np 236 22 h β−	Np 237 2·2×10⁶ y α	Np 238 210 d β−	Np 239 2·33 d β−	Np 240 7·3 m β−
92	U	U 228 9·3 m α	U 229 58 m α	U 230 21 d α	U 231 4·3 d K	U 232 70 y α	U 233 1·62×10⁵ y α	U 234 UII 0·0054 2·5×10⁵ y α	U 235 AcU 0·72 7·1×10⁸ y α	U 236 2·39×10⁷ y α	U 237 6·7 d β−	U 238 UI 99·28 4·51×10⁹ y α	U 239 23·5 m β−
91	Pa	Pa 227 38 m α	Pa 228 22 h α	Pa 229 1·5 d α	Pa 230 17 d α	Pa 231 Pa 34,000 y α	Pa 232 1·32 d β−	Pa 233 27·4 d β−	Pa 234 UZ 6·7 h β−	Pa 235 24 m β−			
90	Th	Th 226 31 m α	Th 227 RdAc 18·6 h α	Th 228 RdTh 1·90 y α	Th 229 7300 y α	Th 230 Io 80,000 y α	Th 231 UY 25·6 h β−	Th 232 100 1·39×10¹⁰ y α	Th 233 23·3 m β−	Th 234 UX₁ 24·1 d β−	Th 235 <5 m β−		
89	Ac	Ac 225 10·0 d α	Ac 226 29 h β−	Ac 227 Ac 22 y α	Ac 228 MsTh₂ 6·13 h β−	Ac 229 66 m β−	Ac 230 <1 m β−						
88	Ra	Ra 224 ThX 3·64 d α	Ra 225 14·8 d β−	Ra 226 Ra 1620 y α	Ra 227 41 m β−	Ra 228 MsTh₁ 6·7 y β−	Ra 229 <5 m β−	Ra 230 1 h β−					

lying in a horizontal line are isotopes belonging to the same element. The times given for the radio-isotopes are their half-life times— that is, the time during which exactly half of any given quantity of the substance has decayed by nuclear disintegration.

Mass and Energy. Einstein's Law

The atomic weight of an element, as determined by the exact methods used in chemistry, is the average weight of the mixture of isotopes contained in the natural occurrence of that element. It depends, of course, on the mass numbers of the single isotopes and of the percentage in which they are contained in the mixture.

The so-called " mass spectroscopy " introduced by **J. J. Thomson, Rutherford,** and **Aston** into experimental atomic physics permits very exact measurements of the masses of each single isotope of an element quite independently of chemical weighings. Particularly exact determinations of the masses of the proton and the neutron have been made. The exact value of the mass of a single isotope is called its *isotopic weight*, while the nearest whole number is its *mass number*. The measurements show that the isotopic weight differs very slightly—but distinctly more than the uncertainty of the measurements—from the mass number. Besides, there is a slight difference between the isotopic weight and the sum of the weights of the constituents of the nucleus. This is shown in the following example referring to the composition of the helium nucleus; similar deviations occur in all other nuclei, too.

Exact mass of proton	1·007593 MU
„ „ „ neutron	1·008982 MU
Sum of masses of 2 protons and 2 neutrons	4·033150 MU
Exact mass of $^4_2$ He nucleus (alpha particle)	4·002776 MU
Difference	0·030374 MU

A difference like this, far surpassing the possible error of measurement, would have been used in former times to disprove Prout's hypothesis once more. According to nineteenth-century physics, it would be impossible that any aggregate of masses could be lighter than the sum of its components, for there are no negative masses in the world which by being added to the mixture might reduce its weight. Although even to-day no negative masses have been discovered, the possibility of an aggregate of masses being lighter than the sum of its components can be reconciled with the basic concepts of modern physics. Half a century ago **Albert**

Einstein, the greatest physicist of our age, derived a law by mathematical deductions from his theory of relativity which upset two more dogmas firmly believed by all serious nineteenth-century scientists. The law of the conservation of matter stated that matter can be neither created nor annihilated, and the law of the conservation of energy stated the same for energy. All apparent losses of energy are only conversions into other forms of energy—for instance, dissipation into heat—and all apparent losses of matter are only conversions into other states of matter, occurring, for example, in the process of evaporation, oxidation, and so on. According to former concepts of physics, the sum of all masses in the universe should be eternally constant, and the sum of all energies should be constant as well.

Einstein proved that these laws are only approximately true, and only so far as ordinary chemical processes are concerned. In nuclear processes, with their enormously greater turnover of energy, deviations from both laws can be observed, and in accordance with Einstein's prediction of 1905 *mutual conversions of matter into energy and of energy into matter* occur. The reason why deviations from the law of conservation of matter were not observed before lies in the fantastically high exchange rate by which matter is converted into energy. Einstein's law is the simple formula

$$E = mc^2 \quad\dots\dots\dots\dots\dots\dots\dots\dots\dots\dots\dots\dots\dots\dots\text{(50)}$$

where m is the increase or decrease of the mass of a body the energy of which is increased or decreased by the amount E. The mass m is expressed in grammes, the energy E in ergs, and c is the velocity of light expressed in centimetres per second. Since $c = 3 \times 10^{10}$ cm/s and therefore $c^2 = 9 \times 10^{20}$, or

$$c^2 = 900,000,000,000,000,000,000 \dots\dots\dots\dots\dots\dots\text{(51)}$$

the conversion factor is so large that all the energy we can supply to a body by heating it or by other means will not suffice to obtain a measurable increase of its mass. A body gains kinetic energy when it moves, and will therefore become heavier than in its state of rest. But even a jet plane breaking through the sound barrier will, according to equation (50), increase its mass by not more than a ten-thousand millionth of 1 per cent. A greater gain in mass can be achieved by heating a body. But even when we heat a piece of iron from freezing-point to 1000 degrees centigrade its mass increase will be only one half of a millionth per cent of its mass in the cold state.

The reverse side of the excessive exchange rate in the conversion between matter and energy is the result that any weighable change of the mass of a body which is not simply caused by adding or taking away a number of atoms or molecules must involve enormous quantities of energy. This is what happens in building up atomic nuclei from primary particles.

The fact that the mass of the helium nucleus (or alpha particle) is distinctly less than the sum of the masses of its constituents appeared to be the last obstacle against accepting Prout's hypothesis in its amended form. The difference, which is called the *mass defect*, was found to be 0·03037 MU per atom of helium, or 30·37 grammes per 4 kilogrammes of helium gas. We have good reasons for assuming that the process of fusing two pairs of protons and neutrons to form a larger cluster (the alpha particle) will just as well release energy as is done, for instance, in the chemical binding processes forming water molecules from hydrogen and oxygen or carbon dioxide molecules from carbon and oxygen. The released energy is radiated away, and causes a loss of mass according to Einstein's law. By identifying this mass loss with the observed defect of 30·37 grammes per 4 kilogrammes of synthesized helium we can compute the energy output of the process. According to equation (51) the energy expressed in ergs is 9×10^{20} times the mass difference. Performing this multiplication sum, and reducing to 1 kg, we find that the production of each kilogramme of helium by direct synthesis from protons and neutrons would be accompanied by an energy release of $6·83 \times 10^{21}$ ergs = 683 thousand million kilojoules = 647 thousand million B.Th.U. = 163 thousand million kilocalories = 190 million kilowatt hours. The amount of hydrogen consumed in this process is 500 grammes—that is, slightly more than a pound. Burning the same quantity of hydrogen would release an energy of 14,500 kcal, and so we must conclude that the nuclear reaction releases about ten million times more energy than the chemical process of binding the same quantity of atoms to form molecules.

This is the result of theoretical considerations based on Einstein's law of 1905. Its reality has been demonstrated very ostensibly by the vast energies released in the hydrogen-bomb tests. Long before this, however, the validity of equation (50) had been quantitatively checked by numerous experiments in which differences of mass and energies released in nuclear processes had been compared.

In view of these striking facts it seems quite natural to ask naïvely: Why not use the helium synthesis for power production?

The reserves of hydrogen present in all the oceans, multiplied by six hundred-odd thousand million B.Th.U. per pound, are a capital of energy many billion times larger than the heat content of all the coal reserves of the world.

The answer is: We can obtain energy in the usual way by igniting a mixture of hydrogen and oxygen, but for two reasons we cannot perform an analogous procedure with a mixture of protons and neutrons. First of all, free neutrons do not exist in weighable quantities, and, secondly, ignition of nuclear processes requires temperatures of hundreds of millions of degrees Centigrade. Still, although a direct synthesis according to a formula

$$2n + 2p = \alpha \dots\dots\dots(52)$$

(n, p, and α denoting respectively neutron, proton, and alpha particle) cannot be performed, the process of building up helium atoms from hydrogen atoms goes on continuously in the interior of the fixed stars, including the sun. It supplies the vast amount of energy which is radiated from the sun's surface, and is, therefore, the ultimate source of all life on the earth. We know also that since 1952 man-made reactions converting hydrogen into helium in a somewhat different way have been performed in exploding hydrogen bombs. Both processes, however—the natural one occurring in the stars and the artificial one occurring in the hydrogen bomb—are more complicated than the reaction $2n + 2p = \alpha$, which cannot be realized in its simple form.

Nuclear reactions by which heavier nuclei are built up from their constituents are called *fusion* reactions. The creation of matter in the form of elements as they now exist has been done by fusion reactions. On account of the large energy release of nuclear reactions, the problem of fuel resources will be solved once and for all as soon as we succeed in handling well-controlled—that is, not explosive—fusion reactions. Details about these reactions can be obtained from knowledge about the forces acting in the interior of the nucleus.

Nuclear Forces and Stability of Nuclei

Comparing chemical reactions building up molecules from atoms with fusion reactions building up composite nuclei from nucleons, we find some common features and some characteristic differences. The most important common feature is that in both cases energy is released by processes of building up certain kinds of molecules and of nuclei, and, on the other hand, also by processes of disintegrating other kinds of molecules and other kinds of nuclei. A difference lies in the fact that the number of building-stones in synthesizing chemical compounds from the elements is about 90, and the number of possible combinations to form molecules is more than a million, while the number of nucleons is only 2, and the number of possible combinations to form nuclei is only about a thousand. Denoting the two nucleons proton and neutron by the symbols p and n, and taking again Z as the atomic number and A as the mass number, the composition of any nucleus can be described in a kind of chemical formula

$$p_Z \, n_{A-Z} \dots\dots\dots\dots\dots\dots\dots\dots\dots\dots\dots\dots (53)$$

where Z ranges between 1 and 100 and A – Z between 1 and about 150.

Certainly the combination of all possible values of Z with all possible values of A – Z would result in a number of $100 \times 150 =$ 15,000 different nuclei. Experience teaches, however, that A and Z cannot be chosen quite independently of each other for building up a nucleus. It can be seen from Table 39 and from Figs. 81 and 82, representing the beginning and the end of a complete isotope chart, that only those nuclei exist in which a certain balance between the numbers of protons and neutrons is upheld. Among the lighter elements those isotopes are most stable in which the numbers of protons and neutrons are equal. Thus, for instance, the main isotope of oxygen, with 8 protons and 8 neutrons, is by far the most abundant in the natural mixture, while the two heavier isotopes, with 9 and 10 neutrons in addition to the 8 protons, together make less than one quarter per cent of the natural occurrence of oxygen. Other isotopes of oxygen with more than 10 or less than 8 neutrons are not stable, and do not, therefore, occur naturally. When they are made artificially they decay after a short time by radioactive disintegration. If there are too many neutrons in the nucleus the disintegration will be a negative beta decay, consisting of the conversion of a neutron into a proton and an electron whereby the latter is emitted from the nucleus, which in its turn becomes the nucleus of the element with the next higher atomic number. If, on the contrary, there are too few neutrons in the nucleus a positive beta decay will occur in which a proton is converted into a neutron and a positron. The latter is emitted, and the rest is a nucleus of the element with the next lower atomic number, as shown, for instance, in the process represented by equation (49), in which radio-phosphorus is transmuted into a stable silicon isotope.[1]

---

[1] It may be mentioned incidentally that both kinds of positive and negative beta decay are accompanied by the emission of a still much lighter particle, the *neutrino*, with a charge zero and a mass which is less than a millionth of the electron's.

We find, therefore, that there is a tendency within the nuclei to arrange the proportions of their two constituent sorts of nucleons so that a proper balance between protons and neutrons is established. While, however, in the light elements the balance is reached with equal numbers of both sorts, the case is different with elements of medium and high atomic numbers. The further we proceed to heavier elements the more a certain excess of neutrons is required to build up stable nuclei. Thus, for instance, the element gold has only one stable isotope $^{197}_{79}$ Au with 79 protons and $197 - 79 = 118$ neutrons. The difference between the number of neutrons and protons, given by

$$(A - Z) - Z = A - 2Z \dots\dots\dots\dots\dots\dots(54)$$

is called the *neutron excess*. It is zero in the case of the main isotopes of helium, carbon, nitrogen, and oxygen, and increases with higher atomic numbers. Table 40 gives the data for some elements in which a certain isotope excels in abundance.

TABLE 40. NEUTRON EXCESS OF THE MAIN ISOTOPES OF SOME ELEMENTS

Symbol	Element	Number of		Neutron Excess $(A-2Z)$	Percentage Excess $(100(A-2Z)/Z)$
		Protons (Z)	Neutrons $(A-Z)$		
O	Oxygen	8	8	0	0
Co	Cobalt	27	32	5	18·5
Rh	Rhodium	45	58	13	28·9
Cs	Cæsium	55	78	23	41·8
Au	Gold	79	118	39	49·4
Bi	Bismuth	83	126	43	51·8
U	Uranium	92	146	54	58·7

The necessity for a growing number of excess neutrons with increasing atomic number can be understood by assuming that the rôle of the neutrons is that of cementing together the protons, which repel each other violently on account of their positive electric charge when they are compressed within the extremely small volume of the nucleus. Although our knowledge about the forces within the nuclei is still very incomplete—the quest for more knowledge necessitates large expenditures for giant cosmotrons—we have reason to believe that the electrostatic repulsion between the protons is compensated by attracting forces of a hitherto unknown nature which do not decrease with the inverse square of the distance, like gravity and electricity, but with a much higher inverse power. They do not make themselves felt, therefore, at distances larger than 10^{-11} cm, but within a certain much smaller range they grow very rapidly with decreasing distance, so that they hold the nucleons together with an iron grip. Strong attracting nuclear forces with a very small range seem to provide an interaction between protons and neutrons, as well as between protons and protons and between neutrons and neutrons. It is to these forces that we owe the existence of all elements other than hydrogen.

With the exception of $^1_1$H and $^3_2$He, no stable nucleus exists with an excess of protons. This is quite understandable because with protons in the majority the repelling forces between them would be stronger than the cementing action of the neutrons, and would therefore disrupt the nucleus. Why, on the other hand, nuclei with a too great excess of neutrons are not stable cannot be explained within the frame of this elementary exposition.

A good deal of our general knowledge about stability of nuclei and of possible energy production from nuclear transmutation processes has been gained from Einstein's law, together with precise measurements of isotopic weights. Experience shows that composed nuclei are lighter than the sum of their constituent nucleons; an example is the *mass defect* of helium which we used for calculating the energy released in the formation of the nucleus. In quite the same way as we computed what may be called the ' heat of formation ' of helium from its mass defect we may proceed to ask whether and how much energy could be further released by building up ever larger clusters of protons and neutrons. By packing together, for instance, four alpha particles into a single nucleus we should obtain an assembly of 8 protons and 8 neutrons, which is obviously the nucleus $^{16}_8$O of the main isotope of oxygen. The energy released by the process is 9×10^{20} times the mass difference. To compute the latter we must bear in mind that 16·0000 MU is equal to the mass of the *neutral oxygen atom*. In order to find the exact mass of its nucleus we have to subtract the mass of the 8 electrons contained in the neutral atom. The mass of the electron is $m = 0·00054876$ MU, and the mass of the alpha particle as given above is 4·002776 MU. Therefore:

Mass of 4 alpha particles	16·0111 MU
Mass of nucleus $^{16}_8$O	15·9956 MU
Difference	0·0155 MU

This means that in synthesizing 16 kilogrammes of oxygen from helium a mass defect of 15·5 grammes would result. A calculation analogous to that done before shows that the heat of formation of oxygen from helium is 24·2 million kWh per kilogramme of synthesized oxygen. The process of packing four alpha particles into an oxygen nucleus is, therefore, what the chemists call an *exothermic* reaction—that is, one which does not consume but releases energy, as is done, for instance, in the processes of burning coal or gas. Although such reactions can proceed by themselves after being ignited, they do not start spontaneously because a certain threshold of energy must be passed first. Thus, to ignite a gas flame a minimum temperature is required; the corresponding temperature for igniting a fusion reaction for synthesizing oxygen from helium would be many million times higher, and is practically unattainable so far.

Let us imagine, however, for the moment that the technique of nuclear engineering was advanced far enough to master fusion reactions at discretion. Then the result just obtained concerning the oxygen synthesis would lead to a practical consequence. Suppose that after consuming all reserves of fossil fuels, as well as of uranium and thorium, mankind had to depend on fusion reactions for power production. The helium synthesis from hydrogen would supply vast amounts of energy, but after some billion years it might be considered unwise for climatic reasons to deplete the oceans still more, so that the further transmutation of hydrogen had to be stopped. In that case a prolonged energy supply could be secured by using the vast accumulated helium reserves for synthesizing oxygen and deriving energy from the process of building larger nuclei.

The question is now: How far does the tendency go for nucleons to fuse together in ever larger clusters and to pay for it by converting a part of their mass into energy? We know that much heavier nuclei do actually exist, but their abundance on earth and in the universe seems to be much smaller than that of the medium-weight and light-weight class. Besides, the urge to form larger aggregates decreases as we proceed towards greater atomic numbers. This can be inferred from the binding energies computed from the mass defects by Einstein's law. The combination of an oxygen nucleus $^{16}_{8}O$ with an alpha particle forming a nucleus of neon $^{20}_{10}Ne$ would be one more exothermic reaction, and the formation heat can be computed from the exactly determined isotopic weight of $^{20}_{10}Ne$. The result is an energy output which is still immensely greater than that of chemical reactions, but distinctly smaller than that of the fusion reactions producing helium and oxygen. This can be seen from the following synopsis:

Imaginary fusion reaction forming:	kWh per kg
Helium from protons and neutrons	190 million
Oxygen from helium	24·2 million
Neon from helium and oxygen	6·25 million

Considering the energy release of a reaction as a yardstick of the readiness of the participant nuclei to perform the reaction, we may conclude from the data given here that the urge to aggregate inherent in protons and neutrons leads very impetuously to the creation of the nuclei of the lighter elements, but relaxes gradually with increasing size of the clusters. The existing precision measurements of isotopic weights allows us to continue the list of heats of formation farther along the table of elements, and the result is that up to the elements in the neighbourhood of $Z = 26$ (iron) the synthesis of nuclei from smaller aggregates is still an exothermic reaction, although with ever smaller energy release. Between the atomic numbers 30 and 70 there is scarcely any further tendency for lighter nuclei to fuse into heavier ones, and, finally, the tendency is reversed among the heaviest elements. Energy is released, not by fusion, but by disintegration of nuclei! With more than about 80 protons in the nucleus, the mutual electric repulsion of the protons becomes so strong that the cementing action of the neutrons no longer suffices to ensure permanent stability. A comparison of the isotopic weights of the heaviest elements with the sum of the masses of any two smaller nuclei from which they could be composed shows that the mass difference has the reverse sign. The composed nucleus is no longer lighter, but somewhat heavier, than the sum of its two components, and therefore, according to Einstein's law, the reaction fusing the two components together is no longer exothermic, but endothermic. Work must be done to perform the fusion while the inverse process of disintegration releases energy, and can therefore proceed by itself as soon as there is a possibility of passing a certain threshold of energy.

In addition to these theoretical predictions, experience shows that the heaviest nucleus which exists in a stable form without spontaneous radioactive decay is that of the bismuth isotope $^{209}_{83}Bi$. None of all the naturally existing or artificially produced nuclei with atomic numbers above 83 (bismuth) is stable any longer.

They are all radioactive, disintegrating by either alpha or beta decay, and some of them—the so-called fissionable isotopes—show in addition to their radioactivity the remarkable phenomenon of *fission*, which consists in disintegration into two nearly equal halves under the action of impinging neutrons. This will be discussed more fully in the next chapter. What we have learnt so far from applying Einstein's law to the results of exact isotopic mass measurements are the following conclusions, which are of vital importance for the problem of future power production.

The matter of the world, as it is at present composed from elements No. 1 to No. 92, is far from being in a state of saturation or definite stability. The greatest part of all matter in the universe seems to consist of lighter nuclei with an unsaturated urge for further aggregation which can release vast quantities of energy. Another reserve of nuclear energy, much smaller than the former, but, taken absolutely, also very great, is contained in the heavy nuclei which, as if by a mistake of the Creator, have been overcharged, and are tending now to disintegrate and to discharge the surplus energy accumulated in the process of their formation.

The first category of nuclear energy is the source of all the heat and radiation of the sun and the stars. The other one seems to play a minor rôle in cosmic processes, but it is that sort of atomic energy which mankind is just beginning to tap for technical applications.

CHAPTER XV

Nuclear Reactors

Why was Atomic Energy the Great Unknown till 1945?

ALL that has been said in the preceding sections about atomic energy, stability of nuclei, and nuclear-energy reserves in the universe has been common knowledge to most physicists for more than twenty years. But, with the exception, perhaps, of comic papers illustrating Einstein's curious idea of an ocean liner crossing the Atlantic with just a pound of matter turned into energy, little publicity was given to the fantastic prospects opened by the law $E = mc^2$. The reason for this did not lie, perhaps, in a tendency to keep these matters secret, but rather in the fact that it was not until the end of 1938 that the possibility of a technical use for nuclear energy became visible. Until that time nuclear physics was certainly of stimulating interest for basic science, but no way could be seen to apply it to power production.

It has been known for about half a century that the heat produced by the radiation of a given quantity of radium during its lifetime is about 60,000 times greater than the heat obtained by burning the same quantity of coal. But, quite apart from the high price and insufficient availability of radium, its large heat content is of no use because the rate of delivery is far too small and cannot be accelerated by any means.

It has also been known since Rutherford's pioneer experiments of 1919 that we can produce artificial disintegration of nuclei accompanied by an energy release which is a million times greater per atom than the energy obtained from any chemical reaction. But all that seemed practically useless as long as nuclear reactions were restricted to small-scale laboratory experiments. Even with the specifically high energy output of about a million electron-volts or more per atom in nuclear reactions no measurable amount of energy could be produced in experiments in which the number of atoms taking part in the reaction was of the order of magnitude of some hundred, or perhaps a few thousand, per second. In spite of its poor specific energy output of only a little over 4 electron-volts per atom of carbon, the process of burning coal can be used for large-scale heat production because the quantity of reacting material is nearly unlimited. The number of carbon atoms oxidized to CO_2 in a medium-sized thermal power station is about 10^{27} per second, and this is so immensely more than the quantities involved in nuclear reactions performed prior to 1942 that a competition of nuclear with chemical processes in the field of power production appeared to be out of the question.

Chain Reactions

The reason why nuclear reactions were restricted to small-scale experiments with an immeasurably small energy output did not, perhaps, lie in the lack of available material. Many of the substances which can undergo nuclear transmutations with great specific energy release per atom exist in sufficient quantities. The restriction preventing the extension from the laboratory scale to the industrial process lay in the fact that, instead of being self-maintaining, the reaction could proceed only with those few atoms which were hit by very fast bombarding particles from a radioactive source or from an accelerating machine. Compared with the conditions of chemical processes, the situation was somewhat similar to that of a fuel which would burn only as long as we held a flaming match to it. The possibility of using our common fuels like wood, coal, or oil for heat production is based on the—sometimes unwelcome— fact that fire tends to spread by itself. Expressed in chemical terms, fire is a self-maintaining chemical *chain reaction*.

Generally speaking, a chain reaction is a process in which a primary action A and a secondary effect B are interrelated in such a way that they assist each other in growing. In the case of fire the primary action is the heat igniting a part of the fuel, and the secondary effect is the oxidation process which again produces more heat and raises the temperature of neighbouring parts of the fuel over the point of ignition, so that the flame spreads farther. More heat is developed, and thus the reaction becomes self-maintaining, instead of being restricted to the originally ignited spot. Some other sorts of self-maintaining chain reactions occur in nature, and are used also in technical appliances. The production of oscillations in an electric circuit by feed-back is an example from quite a different field.

The obstacle against the use of nuclear energy for the production of heat and power was the apparent impossibility of inducing nuclear processes to become self-maintaining chain reactions. The decisive discovery which, against all expectations, opened the way for nuclear chain reactions was made in 1938 by **Otto Hahn** in Berlin.

The Fission of Uranium

In 1934 **Enrico Fermi**, who was then working in Rome, but who emigrated to the U.S.A. after 1938, had observed that by bombarding uranium with neutrons new kinds of radioactive isotopes were created the chemical nature of which he could not identify. Up to that time all nuclear transmutations—both the spontaneous radioactive disintegrations and the artificially produced transmutations— consisted of splitting off from the nucleus, or inserting into it, a light particle like an electron, proton, neutron, or alpha particle. Therefore the end product of a nuclear transmutation was in all cases the nucleus of an element with an atomic number differing from that of the parent element by not more than two units. For adding or splitting off a neutron leaves the nuclear charge unchanged; an electron or a proton changes it by one unit, an alpha particle by two units. Since uranium, with atomic number 92, is, among all elements in nature, that with the highest nuclear charge, it was natural to expect that the new radioactive element produced from uranium was an isotope of uranium itself, or of its immediate neighbours, No. 91, protactinium, or No. 90, thorium. Chemical analysis showed, however, that the unknown element was certainly not an isotope of either of these elements, and therefore it was assumed for some time that it was a *transuranium*—that is, an artificial element with an atomic number higher than 92.

This assumption failed, too, however. The final solution of the riddle was found in the autumn of 1938, when Hahn and **Strassmann** working in the then Kaiser Wilhelm Institut für Chemie (now the Max Planck Institut) in Berlin-Dahlem, established beyond doubt that among the mixture of new radioactive substances produced by bombarding uranium with neutrons there was one which turned out to be an isotope of the element *barium*. This result was so surprising that Hahn hesitated some time before he published it. For, as mentioned just now, all the nuclear disintegrations known up to that time consisted of splitting off quite a tiny fraction of the bombarded nucleus, an alpha particle at the utmost. But here in the case of the disintegration of the uranium nucleus the end product was an isotope of barium, whose atomic number is $Z = 56$, which differs by 36 units from that of uranium ($Z = 92$).

This was quite a different kind of disintegration—much more radical than any known before. It could be explained either by assuming that neutron bombardment of uranium starts quite a chain of alpha decay processes in which 18 alpha particles were emitted successively—the charge of the alpha particle being 2, this would account for the difference of 36—or by assuming a one-step disintegration by which the nucleus would be radically split into two halves, with charges of 56 and 36 respectively. The element $Z = 36$ is the rare gas krypton, and later investigations showed that one of the products of the uranium disintegration was actually a krypton isotope. The second assumption proved, therefore, to be true. The nucleus is split nearly in the middle; we are made acquainted here with an entirely new type of nuclear disintegration, for which the term *fission* was applied. It was soon found out that fission can also yield end products with a different break-down of the nuclear charge. It is only a part of the nuclei hit by neutrons which is split into the fractions with atomic numbers 56 and 36. Various other combinations, as, for instance, 50 + 42 (tin and molybdenum), 52 + 40 (tellurium and zirconium), 54 + 38 (xenon and strontium), 57 + 35 (lanthanum and bromine), have been observed as well, so that a whole gamut of isotopes—all strongly radioactive—of elements of medium atomic weights is the product of uranium fission.

The fission process differs from the formerly known nuclear disintegrations not only by the greater size of the particles split from the parent nucleus, but also by its greater specific energy production. The kinetic energy of an alpha particle emitted from a radium nucleus in its decay is 4·79 Mev—which in itself is more than a million times the energy of 4·17 ev per carbon atom in burning coal—but the sum of the kinetic energies of the fission fragments flying apart is about 160 Mev. To this must be added the energy of the accompanying gamma radiation and the energy released by the subsequent radioactive decay of the fission products, so that the total energy release per fissioned uranium atom is approximately 200 Mev.

But the true significance of Hahn's discovery was neither the radical splitting of the nucleus nor the large energy output, but still another feature of the fission process. It was explained in the preceding chapter that in order to ensure stability of atomic nuclei a certain balance between protons and neutrons must be established, and it can be seen from Table 40 that this balance is reached with an appropriate percentage of neutron excess which from zero among the group of light elements rises to about 30—40 per cent among the medium-weight stable nuclei, and ends with nearly 60 per cent in the case of uranium. If now a uranium nucleus is split into two halves—say, for instance, a barium and a krypton nucleus —the proportion between protons and neutrons can be kept either equal in both fractions or unequal. In the first case the high neutron excess of nearly 60 per cent is retained in both fission products, which, being much lighter, belong to a class of nuclei which can exist in a stable form only with a 30-to-40-per-cent neutron excess. In the second case, with unequal allotment of neutrons, one of the fractions may have the proper proportion of neutrons, while the other is still worse off than in the first case by having a neutron excess of more than 60 per cent.

Thus, at any rate, at least one, but probably both, of the fission products will be loaded with more neutrons than adequately adapted for a well-stabilized nucleus of their size. Therefore it was to be expected that the newly built and badly adjusted nuclei would try to get rid of their overload by jettisoning a number of the surplus neutrons. The process of fission would thus result not only in two lighter nuclei flying apart with high speed, but also in the emission of a number of free neutrons. Experiments which were made soon after the publication of Hahn's discovery showed that this expectation was right, and that the fission process actually released between two and three neutrons per fissioned uranium nucleus.

This fact was the cardinal point for the possibility of technical applications of nuclear energy. For the fission process, unlike the radioactive decay or the artificial nuclear disintegration by bombardment with protons or with alpha particles, turns out to be a reaction which creates the tools for continuing the process with other nuclei of the same substance. One first neutron causes a fission which, besides producing energy, releases two or three other free neutrons, which may cause fissions again, releasing thereby still more neutrons, causing in their turn more fissions—and thus a real chain reaction is set in motion, which can finally spread out over the whole amount of material which is assembled. While, therefore, all artificially made nuclear reactions were formerly restricted to the relatively small number of nuclei hit by bombardment, a way could be seen now to extend the process in the form of a *nuclear chain reaction*. We remember that the energy output is estimated to be nearly 200 Mev per fissioned uranium atom, as against 4·17 ev per carbon atom in burning coal. That means nearly fifty million times more energy per atom, or, considering the different atomic weights, about 2·4 million times more energy per pound of fuel.

As early as in the spring of 1939 it was therefore clear to most of the nuclear physicists in the world that, contrary to former expectations, a theoretical possibility existed for industrial and military use of nuclear energy. The problem was how to carry over theory into practice.

Uranium 235

If the expectation is right that successive emission of neutrons from the fission products of uranium will release a self-maintained chain reaction, then a single neutron might suffice to start a process which on account of the high energy production would have explosive character. We know, however, that neutrons are continuously set free by the cosmic rays which in penetrating the atmosphere and the uppermost layer of the earth's crust cause nuclear disintegrations. Although free neutrons are not stable particles, but either are absorbed in an atomic nucleus or decay after a short lifetime by emitting a beta particle and becoming thereby a stable proton, the equilibrium between neutron production by cosmic rays and their absorption or conversion is yet such that there is a sufficient number of free neutrons always present in every body on the surface of the earth to start a chain reaction if the conditions for such a reaction are given.

Why, now, can pure uranium metal exist on earth without being exploded by a self-maintaining chain reaction of fission processes kindled by one of the omnipresent free neutrons? The reason will soon be understood by realizing clearly the chances of a newly born neutron continuing the chain by starting another fission process in a neighbouring nucleus. A piece of metal looks very compact, seen with our macroscopic eyes. But we know that the dimensions of the atomic nuclei and the electrons are of the order of magnitude 10^{-13} cm, while the distances between neighbouring nuclei is some 10^{-8} cm in solid bodies and fluids, and about ten times more in gases under normal pressure. Seen from the standpoint of a neutron, any solid body, including, of course, a piece of uranium, is mostly empty space occupied only by tiny nuclei with diameters which are about one hundred thousandth of the distance to the next neighbour. Free neutrons, unbiased by any electrical charge, are neither attracted nor repelled by the atomic particles, and will therefore freely move over relatively long distances within any physical body before they happen to collide directly with one of the widely scattered tiny objects. It is only in case of a direct hit that the short-range nuclear forces will catch it in a firm grip. The chance of hitting a nucleus with an uncharged particle traversing the space of an atom is less than that of hitting a single fly in a large theatre with a pistol-shot without aiming. Many millions of atoms must therefore be traversed by a neutron before it happens to hit a nucleus and to cause a fission.

We see, therefore, that one reason for interrupting the chain of a snowball-like multiplication of neutrons may lie in the probability that any fission-born neutron will traverse the whole block of uranium without encounter, and escape before generating descendants by another fission process. A condition for producing a chain reaction is, therefore, that a sufficiently large volume is filled with fissionable matter. The minimum mass required for maintaining a chain reaction of fissions is called the *critical mass*.

But even a whole mountain of pure natural uranium would not explode. The reason is that uranium in its natural occurrence is an isotope mixture composed of 99·274 per cent U 238, 0·720 per cent U 235, and 0·006 per cent U 234. (It has been said so often in Chapter XIV that the atomic number of uranium is 92 that we can omit the lower left index in the nuclear symbol, and add for the sake of simplicity the mass number behind the chemical symbol U.) It was soon found out after 1938 that it is not the whole isotope mixture which takes part in the uranium fission, but only the nuclei of the rare isotope U 235, which constitutes only $\frac{1}{140}$ of the natural uranium. The vast majority of all the nuclei contained in a block of pure uranium metal consists of U 238 nuclei which, when hit by a neutron, do not suffer a fission, but capture the neutron and undergo a subsequent transmutation which, as was found later, proves also to be a process of high technical importance. This will be further discussed below. What matters here is the fact that the presence of U 238 nuclei in the natural uranium acts as a kind of fire-extinguisher by capturing the fission-born neutrons and thus suppressing the propagation of the chain reaction. If the composition of natural uranium from the isotopes U 238 and U 235 were reversed, with the latter 140 times more abundant than U 238, then the first metallurgist welding together a lump of pure uranium metal the size of a coconut would have started a terrific explosion.

If, on the contrary, it is intended to arrange an explosive nuclear chain reaction, then it is necessary to assemble a sufficient quantity of fissionable matter in a pure form—that is, without any enclosure of nuclei which can interrupt the chain by absorbing the new-born neutrons before they succeed in causing another fission. In the case of uranium this means the removal of the U 238 nuclei, or, in other words, the separation of the isotope U 235. As a matter of fact, U 235 is a suitable substance as an explosive for atom bombs, and it has been separated from natural uranium. But in order to do this job it was necessary to build one of the largest single industrial plants of the world, and pure U 235 is not only an extremely terrible substance, but also one of the most expensive ones, incomparably more costly than gold or platinum.

The reason for the vast expenditure necessary for large-scale isotope separation lies in the perfect chemical identity of different isotopes of the same element, which makes all conventional chemical methods for analysis and separation useless. It is very easy for chemists to separate elements which differ strongly from each other; it is just a little bit less easy to separate elements with similar chemical properties, like strontium and barium, for instance, but it is hopeless to try to separate isotopes of the same element from each other by chemical methods. A number of physical methods exist, however, by which isotope separation can be accomplished; they work best with pairs of isotopes with a relatively great difference in mass numbers. The separation of deuterium (heavy hydrogen) with the nucleus ${}_1^2 H$ from ordinary hydrogen is less difficult than that of other isotopes, because the mass ratio is $2 : 1$, and it gives, therefore, among all separation plants, the greatest yield with the least expenditure—although even in that case the price is still very high. On the other hand, the mass of U 235 differs from that of U 238 by little more than 1 per cent, and therefore the task of performing the separation in weighable quantities is an enormous one. Among all the methods tried during the War diffusion of a gaseous compound, uranium hexafluoride, through porous membranes, proved to be the most efficient one, and the large separation plant of the Clinton Engineering Works in Oak Ridge, Tennessee, using this method is supplying both U 235 for A-bombs and uranium isotope mixtures enriched in U 235—called shortly, *enriched uranium*—for certain types of nuclear reactors, as will be explained in the next sections of this chapter.

With the successful production of pure, or strongly enriched, U 235, the path was clear for explosive nuclear chain reactions. By quickly bringing together two lumps of U 235 which have separately less, and jointly more, than the critical mass a state is reached in which the first of the omnipresent cosmic-ray-born free neutrons causes the fission creating two or three more neutrons, which continue the process, so that the chain reaction spreads like an avalanche. Within a time of far less than a millisecond the huge energy of many million kilowatt-hours is released. The first secret test in Alamogordo, in the desert highlands of New Mexico, on July 16, 1945, confirmed the correctness of the theoretical predictions.

The vast energies released by nuclear chain reactions would, however, have been of little avail for the problem of power production unless it had been possible to harness the forces of the atom and to implement well-controlled nuclear reactions with arbitrarily adjustable energy output. This decisive accomplishment had been achieved in all secrecy about two and a half years before the first A-bomb brought terrible disaster over Hiroshima.

Nuclear Cross-sections. Fast and Slow Neutrons

If an obstacle prevents us from reaching an aim we can try either to remove it or to by-pass it. In the case of the fission chain reactions in natural uranium the obstacle was the presence of the many U 238 atoms which prevented the neutrons born from a U 235 nucleus reaching other U 235 nuclei and generating more neutrons. Removing these obstacles meant separation of the isotopes, a very expensive process, as we have seen. If it were possible to by-pass the obstacles, then a chain reaction might be accomplished in natural uranium too. But how to do that?

Systematic measurements of the so-called nuclear *cross-sections* enabled the physicists to devise a clever trick by which the hindering U 238 nuclei could be by-passed. The trick is based on the fact that the apparent size of an atomic particle which determines the probability of its colliding with other particles is not a fixed magnitude, but varies according to circumstances. Macroscopic solid bodies have a given magnitude; thus, for instance, a billiard-ball has a diameter of 6 cm, and a collision between two billiard-balls will occur in all circumstances when the distance between their centres becomes as small as six centimetres. In the case of atomic particles, which are more or less centres of forces, or aggregates of such centres, no smooth surface of a rigid body exists. What we call a collision, between a nucleus and a neutron, for instance, is not a contact between two surfaces, but a constellation in which the mutual repelling, or attracting, or other forces begin to act. Thus when we say that the collision of a neutron with a U 238 nucleus results in the capture of the neutron into the nucleus it means that the neutron will be caught in the grip of the nuclear forces and drawn into the interior of the nucleus as soon as a certain small distance between the centres of the two bodies is reached. In the same way, a fission will occur as soon as the distance between the centre of a neutron and the centre of a U 235 nucleus gets as small as a certain characteristic distance. If we denote this distance by r, then the number of collisions occurring in a given physical system will be the same as in an imaginary system in which all the neutrons are replaced by mathematical points, and the nuclei by balls with a radius r (diameter $2\,r$). It is obvious that the probability of collisions will depend on the magnitude of the balls; big balls are easier to hit than small ones. The mathematical probability of collisions will be proportional to the cross-section of the ball, which is πr^2.

It has been possible to determine quite a series of cross-sections of different nuclear processes by suitable observations of the numbers of processes occurring in given circumstances. If we put the observed cross-sections equal to πr^2, the length r means the distance between the reacting nuclei at which the process in question will start.

While now microscopic rigid bodies have fixed dimensions, and therefore fixed cross-sections, the case is quite different in atomic particles. The latter seem to have quite peculiar predilections for certain processes, and certain partners of collisions, and peculiar aversions against others. And therefore the cross-section of a nucleus, which is a kind of yardstick for its readiness to react, is not simply a fixed given property of this particular sort of nucleus, but depends also on the kind of reaction following the encounter, on the partner, and even on the way the partner is approaching. In the particular case of the uranium nuclei the cross-section for capture of neutrons by U 238 is different from the so-called fission cross-section of the U 235 nuclei, and both of them depend also very essentially on the velocity of the colliding neutron!

Now we come back to the point of our story. The task was to produce nuclear chain reactions in natural uranium, and the problem was how to by-pass the far more numerous U 238 nuclei, which by absorption swallow the new-born neutrons before they can generate others by hitting a U 235 nucleus. With a numerical superiority of 140 : 1 of the capturing nuclei, a solution would have seemed hopeless unless the experiments had shown how strongly the cross-sections of capture by U 238 and of the fission of U 235 depend on the neutron velocity. Imagine a large room in which black and white balls are floating freely in the air. If all the balls are of equal size and the black balls 140 times more numerous than the white ones, and if we shoot at random into the space filled by the balls, then, of course, the probability of hitting a black one will be 140 times greater than that of hitting a white one. If, however, with the same number of balls, the black ones are of the size of a pea and the white ones are balloons of one foot diameter, then, in spite of their smaller number, the chance of hitting a white ball will be greater. We can therefore actually by-pass the hindering U 238 nuclei by arranging the reacting assembly so that the fission cross-section of the U 235 becomes very much greater than the capture cross-section of the U 238. This can be done by *slowing down* the neutrons, because the experiments proved that the fission cross-section increases strongly with decreasing neutron velocity. Fig. 83 (not drawn to scale) shows how the ratio between the U 238 and U 235 cross-sections varies with the velocity of the neutrons.

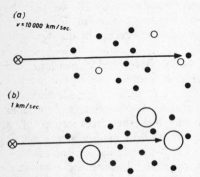

(a)
v = 10 000 km/sec.

(b)
1 km/sec.

FIG. 83. DIFFERENT CROSS-SECTIONS OF THE URANIUM ISOTOPES FOR SLOW AND FAST NEUTRONS

(a) Fast neutrons, (b) slow neutrons. Empty circles, U 235; black circles, U 238.

Before proceeding further it should be explained what fast and slow means in connexion with neutrons. Slow neutrons are those which have a velocity of just a few thousand miles per hour, while fast neutrons move with a speed more than a thousand times faster. It is customary to characterize moving atomic particles not by their speed, but by their energy. Expressed in these terms, fast neutrons are those with energies in the order of magnitude of 1 Mev, while neutrons with energies of 1 ev and less are slow[1].

The energies with which the neutrons are emitted from the fission products of U 235 are more than 1 Mev; they are therefore fast neutrons. In traversing matter they are partly absorbed by the U 238 nuclei, partly slowed down by elastic collisions in which they are not absorbed by the partner of the collision, but just rebound, whereby a fraction of their kinetic energy is transferred to the kicked nucleus. Thus any neutron will be slowed down after a while, and will finally end up with the thermal velocity correspond-

[1] The corresponding speeds are about 14,000 kilometres per second for 1 Mev neutrons and 14 kilometres, or 8.7 miles, per second for 1 ev neutrons. (The energy of particles moving with velocities small compared to the velocity of light c is proportional to the square of the velocity, according to the formula of classical physics $E = \frac{m}{2} v^2$. Deviations occur at speeds comparable to $c = 300,000$ km/sec.

ing to the temperature of the uranium block. All atoms of a body are permanently in the state of thermal agitation moving irregularly with an energy which is proportional to the absolute temperature. As soon as the neutrons have been slowed down to thermal velocities—we are speaking then of ' thermal neutrons '—the state is reached illustrated in Fig. 83 (b), where the cross-section of the U 235 nuclei is so much larger than that of the capturing U 238 nuclei. If, therefore, the neutrons would slow down to thermal velocities their chances of reproduction by fission of U 235 would be good. But unfortunately on their way down from the high energies they are exposed to the risk of being captured by U 238. This risk is particularly great when they pass a certain critical energy range near 7 ev, for with neutrons of that energy the capture cross-section of the U 238 reaches a sharp peak.

In order to secure by-passing in spite of these traps, a method was devised to remove the new-born neutrons as soon as possible from the danger-zone, put them into a medium where they can safely slow down in non-absorbing surroundings, and bring them back into the uranium for spawning as soon as they have passed the critical energy range of 7 ev and have slowed down to thermal velocities.

The Uranium Pile

In order to implement this scheme a group of eminent physicists, including **Enrico Fermi** and **Leo Szilard**, working in the laboratory of A. H. Compton, in Chicago, set to work to build up a structure, which they called a ' pile,' consisting of alternate layers of pure metallic uranium and chemically pure graphite. The idea was that the neutrons produced by fission should be given the opportunity of leaving the uranium by diffusion through the metal, then to diffuse through the graphite, where they lose velocity by elastic collisions with the carbon nuclei, and to enter finally the next layer of uranium bricks after being sufficiently slowed down. By a proper choice of the thickness of the layers, based on calculations about the diffusion of neutrons in uranium and graphite, one could expect that a sufficient number of neutrons would return to the uranium with the proper velocity to find the U 235 nuclei ready for fission with the large cross-section shown in Fig. 83.

The assembly of layers in the pile used for slowing the neutrons is called the *moderator*. The reasons for choosing graphite as the moderator material were the following: (1) The material must not contain any nuclei which, like U 238, absorb neutrons by capture.

(2) The nuclei of the material should have a mass nearly equal to the mass of the neutron, because both theory and experience teach us that the aim of losing as much energy as possible by as few as possible collisions can be reached best by elastic collisions with partners of equal, or nearly equal, mass. Since the masses of protons and neutrons are virtually equal, hydrogen would be the best moderator material, but, unfortunately, protons capture neutrons, too, by marrying them, converting thereby into deuterons. The next light nuclei after protons are deuterons, which, being saturated, are no longer anxious to attach further neutrons. Therefore they are best suited as collision partners for slowing neutrons, and heavy hydrogen, or pure deuterium, would be an ideal moderator were it not for the fact that deuterium, as well as ordinary hydrogen, is a gas with very low density, while a proper moderator should be a substance with appropriate density. Failing heavy hydrogen itself, heavy water, D_2O, is a really good moderator, and it was only the difficulty of providing sufficient quantities of heavy water during war-time which prevented heavy water from being the moderator of the pile in which the first chain reaction has been realized. Lacking heavy water, graphite was taken in the Chicago experiments because the carbon molecules of mass number 12 are still reasonably light, and have the advantage of not absorbing neutrons. Among the elements lighter than carbon helium is not suitable because it exists only in gaseous form at ordinary temperatures, lithium and boron are neutron absorbers, and beryllium was not available in sufficient quantities.

The pile was constructed, therefore, with graphite as a moderator by alternately piling up bricks of graphite and lumps of metallic uranium, and also of uranium oxide, to form a regularly spaced lattice. As long as the pile was still small during the assembly of its parts, no start of a chain reaction was to be expected, because the percentage of escaping neutrons was too large to permit an increase of neutron numbers by production. But at a certain critical size which could be approximately foretold from theoretical computations the moment would be reached when the turnover from single fission processes (caused by a neutron source placed in the centre) to a chain reaction would occur. In order to avoid a surprise by an explosion, instruments were placed at various points in the pile and near it to indicate the neutron intensity. In addition, the so-called *control rods* were inserted in properly spaced slots in the pile. These control rods consist of a material which is a particularly effective absorber of neutrons. Boron steel and cadmium metal are two of such substances. By inserting the control rods more or less deeply into their slots the neutron flux within the pile can be regulated.

On December 2, 1942, in assembling the pile, shortly before its scheduled size was reached, a strong increase of the neutron radiation indicated that the historical moment had come when the first man-made nuclear chain reaction was initiated. But no publicity was given then to this great achievement of human genius because all the work was kept strictly secret until August 6, 1945. The existence of the chain reaction was at first revealed only by the neutron records, while the heat production—which is the ultimate aim in power reactors—was immeasurably small at that time. It is estimated that the power-level on December 2 was about half a watt. Ten days later, with the pile completed and the control rods cautiously drawn out a little more, a power-level of 200 watts was reached; eleven months later a larger pile, built in Clinton, operated at a power-level of 500 kW (heat); in May, 1944, 1800 kW was reached, and in 1945 three reactors were operating at Hanford, Washington, with a total power of several hundred Megawatts. At the time when the existence of the atom bomb was revealed to the public in August, 1945, the possibility of large-scale energy production from nuclear chain reactions was established beyond doubt, although the enormous heat produced could not yet be utilized, but was wasted at that time.

Construction and Function of Reactors

The uranium pile built in Compton's Metallurgical Laboratory by E. Fermi is still the prototype of many of the present devices for effecting nuclear chain reactions. Only the name has been changed meanwhile into *reactors*. A reactor is a suitably built assembly of fissionable material in which a controlled nuclear chain reaction can be maintained. A cross-section through a graphite reactor of a type similar to the first Chicago pile is given in Fig. 84.

The essential parts of all reactors of this type are:

(1) Rods or slugs of uranium as the fissionable material (fuel).

(2) A moderator (graphite or water).

(3) Control rods for regulating the power-level.

(4) A circulating fluid which acts either just as a coolant or as a means for transporting the produced heat to the boilers of a power plant.

(5) A reflector shield surrounding the active part of the reactor in order to reflect back some of the neutrons that normally would leak away.

(6) A thick shield of concrete or other strongly absorbing material for protecting the operating personnel from radiation damage.

Fig. 84 is a diagrammatic sketch of a graphite-moderated reactor of the type used in the British atomic factories and nuclear-power plants. In addition to the neutron-absorbing cadmium steel rods (*c*), by which the power-level of the reactor is controlled, the safety rods (*d*) of the same material are arranged vertically, and kept in their lifted position by electromagnets. In case of emergency the current feeding the electromagnets is interrupted, and the rods (*d*) fall into the holes, under the influence of gravity, and stop the chain reaction by absorbing a great deal of the neutron flux.

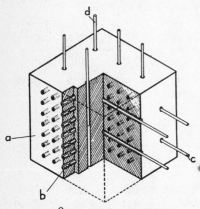

FIG. 84. CORE OF GRAPHITE-
MODERATED REACTOR

(*a*) Graphite pile, (*b*) uranium slugs, (*c*) control rods, (*d*) safety rods. Reflector shield and concrete wall are not shown.

Reactors have three functions of quite different nature, and therefore many reactors serving quite different purposes have been built since 1945. The three main functions are:

(1) *Heat Production*. The primary function of each reactor—which it cannot help fulfilling—is heat production. As we have seen in a preceding section of this chapter (p. 321), the energy release per fissioned U 235 atom is about 200 Mev. The energy consists mainly of the kinetic energy of the fission fragments flying apart, and of the neutrons, and partly also of the radioactive radiation emitted from the fission products. All that energy is ultimately converted into heat—and thus *atomic energy viewed from the standpoint of power economy is nothing else than heat produced by nuclear chain reactions*. The yield of 200 Mev per split atom corresponds to about 29,000 million kilocalories, or 23 million kWh, per kilogramme fissioned U 235. An enormous amount of heat will

therefore be released before the whole mass of several tons of uranium assembled in a reactor is used up. The problem is not so much that of getting more power out of a reactor as of holding its power-level down to a reasonable level at which the removal of the heat from the reactor can keep pace with its production. In contrast to the explosive chain reaction in the A-bomb, with its limitless multiplication of reproducing neutrons, the fission reaction in the reactor is multiplying its neutron generation only during the starting period. Later on a stationary state is reached in which the number of fissions occurring per second is kept constant. By shifting the control rods different power-levels can be attained.

(2) *Plutonium Production.* Although heat production will be the main purpose of future reactors, it was not the object of the first large and powerful reactors which were constructed during 1944–45 in the Hanford works. The military necessities of the War and the possible danger that the enemy might possess and use the atom bomb first made it imperative to obtain sufficient quanti-ties of nuclear explosives at the earliest date possible. Since the output of U 235 from the isotope separation plant in Oak Ridge was limited, the attempt was made to obtain still another material which had the peculiar property of being fissionable and of emitting neutrons along with the fission process. Setting aside the regret-table fact—due to the situation of the world at war—that all the efforts tended to produce means of mass destruction, one can say that it was a great accomplishment of human genius that nuclear physicists were able to predict from theoretical considerations that a hitherto unknown element, which does not exist in nature at all, could be made artificially in such and such a way, and that it would after its creation turn out to be a terrible explosive which could be detonated under certain yet untried conditions.

The new fissionable substance is the nuclide, with charge 94 and mass number 239. It is an isotope of the element No. 94, which was called plutonium (symbol Pu), and which is generated as a by-product of the nuclear process in a reactor. It has been mentioned in the section on uranium 235 that the more abundant nuclei U 238 after capturing a neutron undergo a radioactive trans-mutation which is of particular practical importance. Denoting by the chemical symbol Np the artificially created element with its atomic number 93 lying between uranium and plutonium, we can describe the process following a neutron capture in ordinary uranium by a reaction formula:

$$^{238}_{92}U + ^{1}_{0}n = ^{239}_{92}U \xrightarrow[23\ m]{\beta} {}^{239}_{93}Np + e^{-} \quad \dots(55)$$
$$\xrightarrow[2.3\ d]{\beta} {}^{239}_{94}Pu + e^{-}$$

The meaning of this formula is: After swallowing a neutron the U 238 nucleus becomes a U 239 nucleus, which is far more unstable than the alpha-ray emitter U 238. It decays by negative beta emission with a half-life of 23 minutes, and is converted thereby into the isotope $^{239}_{93}Np$ of the artificial element neptunium, which is still short-lived, and decays by beta emission with a half-life of 2·3 days into the plutonium isotope $^{239}_{94}Pu$. This is a long-living radioactive alpha emitter with a half-life of 24,300 years. As had been suspected from theoretical considerations, it turned out to be fissionable, too, and chain reactions can develop in a sufficiently large lump of pure plutonium. Thus two different substances, U 235 and Pu 239, are available now both for nuclear power production and as atomic ammunition. Remember the way by which plutonium is produced in the reactor: A part of the neutrons created by fission hits, after being slowed, a U 235 nucleus and, by fissioning it, provides for reproduction, and thus for the continuation, of the chain reaction. Others of the neutrons hit U 238 nuclei, stick there, and turn them by way of neptunium into plutonium. Production of heat and of plutonium are therefore coupled with one another and with consumption of U 235. The Smyth report (1945), in which for the first time official information on atomic energy was released after the War, states that in order to produce a kilogramme a day of plutonium a reactor must be releasing energy at the rate of 500,000 to 1,500,000 kilowatts. Taking into consideration that one kilogramme of U 235 produces about 23,000,000 kilowatt-hours, and that a kilowatt-day is as much as 24 kWh, we arrive by an easy calculation at the result: If the figure of 1,500,000 kWh gives the true value of the energy output accompanying the production of 1 kg of plutonium per day, then a greater quantity of U 235 is consumed than Pu 239 produced. However, assuming that the first figure (500,000 kW) given by Smyth is right, then more plutonium is produced than uranium 235 consumed. In the latter case we speak of a *breeding reactor*, or, shortly, *breeder*. This will be discussed later.

(3) *Isotope Production*. Besides being a producer of heat and of plutonium, a reactor is also a very strong and large neutron source. Neutrons are, on the other hand, efficient tools for effecting nuclear reactions, and particularly for producing radioactive isotopes,

and therefore a large-scale production of isotopes on a commercial scale has been made possible by using the neutron radiation of the reactors. By placing samples of suitably chosen elements like sodium, phosphorus, or others into holes in the shield of the reactor and subsequently exposing them for some time to the strong radiation, a wide assortment of radio-isotopes for all possible purposes can be produced. In view of the widespread and beneficial use which is made of radio-isotopes to-day the significance of this function of the reactors, along with their importance for power production, must not be underrated.

In addition, however, to this well-planned intentional isotope production, an unintentional, and with powerful reactors even most unwelcome, production of radioactive isotopes occurs because the fission fragments of the split U 235 atoms are nuclei of isotopes of medium-weight elements ranging from No. 30 (tin) up to No. 63 (europium). Many of the fission products are neutron absorbers, and therefore they would stop the chain reaction after some time if they were left in the reactor. For this reason they have to be removed from the reactor at regular time intervals, in much the same way as in an ordinary steam plant the ash has to be removed. But with this important difference: Whereas ash is, apart from its dirt, quite a harmless material, and may be deposited anywhere, radioactive isotopes in any large quantity are very dangerous. Hence provision has to be made to bury them in very deep pits dug especially for the purpose.

In calculating the amount of this unintentional isotope production one has to consider the fact that, when producing and storing over an unlimited period, the stock of radioactive material does not increase proportionally to production time, because the radioactive material begins to decay the moment it is formed. When, by switching off the reactor, production is stopped, the stock of radioactive material will decrease by nuclear disintegration and accompanying transmutation into non-active isotopes. The relevant quantity is therefore not the amount of material produced *monthly* but the *equilibrium* which will be reached when further production is balanced by the decay of the material. The whole set of radioactive elements produced in a reactor is a mixture of about 300 isotopes. These isotopes have quite different lifetimes, ranging from a fraction of a second to several years. Neglecting the quickly decaying isotopes, and taking into account only those with a half-life of one week at least, the calculation shows that the radioactivity of the equilibrium quantity which is proportional to the

thermal power corresponds to about one pound of radium for a
1-kW reactor.

This is a strikingly high figure in view of the fact that the
quantity of all the radium in the world used so far for scientific
or medical purposes scarcely exceeds a few pounds, and that the
power of a medium-sized municipal station is not 1 kW but about
100,000 kW electric power, or nearly 500,000 kW thermal power.
The activity of the by-products of an atomic plant of that size
will, therefore, correspond to that of 250 tons of radium, or *250
million grammes.* Each single gramme of radium is a strong source
of radiation which must be handled very carefully, and is nowadays
stored, when not in use, in the strong-rooms of hospitals in order
to avoid harm to personnel. In the case of war the equilibrium
amount of the fission products (the ' atom ash ') of a single middle-
sized atomic power plant could be used for delivering radioactive
dust by air raids on enemy target areas in sufficient quantities to
make all cities of a whole continent uninhabitable.

One of the consequences of our entering the atomic age is,
therefore, the fact that even after destruction of all A-bombs and
H-bombs every state which uses uranium for power production will
automatically still possess terrible means for mass annihilation.
Therefore not just the prohibition of certain weapons, but the
radical abolition of wars, will be the necessary condition of the
further existence of mankind.

Types of Reactors

In the reactors built similar to the Chicago prototype the fissionable material is localized in lumps or rods forming a structured
unit imbedded in the moderator, which is used for slowing the
neutrons. This class of reactors is called *heterogeneous,* while
homogeneous reactors are those in which the fissionable material is
uniformly distributed throughout the moderator. An example of
this type is the so-called ' water boiler ' built in Los Alamos, which
is a much simplified model of a nuclear reactor. It consists of a
soluble salt (uranium nitrate) of enriched uranium (containing
about 90 per cent U 235) dissolved in ordinary light water. The
whole active part of the water boiler is contained in a stainless-steel
sphere $\frac{1}{16}$ in. thick, 12 in. in diameter. With enriched uranium as
a fuel, ordinary distilled water can be used as a moderator because
the loss of neutrons from absorption by the protons of light hydrogen is compensated by the greater percentage of fissionable U 235
nuclei present in the mixture.

Reactors, both heterogeneous and homogeneous, are also classified according to the fuel they use (natural uranium, uranium enriched in U 235, plutonium, or uranium 233), further according to the moderator material (graphite, heavy water, or light water), and finally according to the coolant (air, water, liquid-metal cooling). In addition, three different types are distinguished according to the average velocity of neutrons that propagate the chain reaction. *Thermal reactors* are those in which most of the neutrons, before causing new fissions, are slowed down to thermal velocities (about a mile a second). *Fast reactors* are those which, containing sufficiently enriched uranium, do not use any moderator but maintain the chain reaction with fast neutrons (energy 1 Mev or more, speed over ten thousand miles a second). In some reactors the neutrons are slowed down to energies between 1 and 10^4 electronvolts; these are called *epithermal* or *intermediate reactors*.

Another method of classification is based on the purpose to which the reactor is to be put. Some reactors serve for fundamental and applied research, as, for instance, the study of neutron radiation, production of isotopes, testing of radiation damage of materials, etc. Another class of reactors serves to produce plutonium for A-bombs. A third class is power production, and the fourth is power production combined with breeding of fissionable material.

Most of the first large reactors—among them the powerful ones of Hanford—were heterogeneous, graphite-moderated, water-cooled thermal reactors using natural uranium. Even those of them which operate at low power-level, as, for instance, the first British reactor GLEEP (" Graphite Low Energy Experimental Pile "), are rather large structures of the size of a small house, and are very expensive.

Since, however, enriched uranium is available and can be loaned by the U.S. Atomic Energy Commission to scientific institutions, an increasing number of simpler and smaller reactors have been built. The Oak Ridge National Laboratory developed nuclear-fuel elements of enriched uranium which, properly assembled in groups of 16 in a water-tank, constitute a reactor with ample neutron flux for research purposes. This so-called " Swimming Pool " reactor is illustrated in Fig. 85. It consists of an assembly of fuel elements, together with control rods and reflector, submerged in a pool of water that acts simultaneously as moderator, coolant, and protective shield. The fuel elements are contained in the ' active lattice ' shown in Fig. 85. Each fuel element is made up of five fuel plates

24 in. long that consist of uranium-aluminium alloy ‘ sandwiches ’ clad in aluminium. A fuel plate contains between 30 and 40 grammes of U 235; the total fuel load is about 3 kg of U 235. After production of about 6 million kWh heat, which corresponds to a 10-percent burn-up, the fuel elements must be reprocessed for removing the fission products.

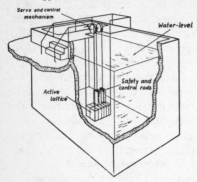

FIG. 85. "SWIMMING POOL," HETEROGENEOUS RESEARCH REACTOR, COOLED AND MODERATED BY LIGHT WATER WITH AN ALUMINIUM-URANIUM 235 ALLOY AS THE FUEL

Fig. 86 shows a homogeneous fluid-fuel reactor in which, as in the Los Alamos " Water Boiler," the fuel (enriched uranium) and the moderator are uniformly mixed. It can consist either of an aqueous solution of the fuel or of alloys of U 235 or Pu 239 with metals of low melting-points. In the construction shown in Fig. 86 the cooling is done by circulating the liquid fuel mixture through a heat exchanger in which the heat is transferred to a boiler system or to the working medium of a closed-cycle gas turbine. The chain reaction is maintained only in the spherical part of the reactor, which is properly dimensioned for reaching the critical size.

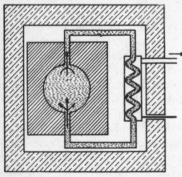

FIG. 86. HOMOGENEOUS REACTOR WITH CIRCULATING FUEL MIXTURE

Other types of homogeneous reactors use a solid homogeneous mixture of fissionable material with a suitable moderator like graphite. Thus, for instance, a sintered solid cylinder of a mixture of powdered enriched uranium and graphite, fitted with holes parallel to its axis for circulating the coolant, could serve as the core of a small and compact reactor. Certainly a lower limit to the external dimensions of any reactor is given by the necessity for surrounding it with a radiation barrier. Whatever progress could be achieved as regards reducing the size of the active zone of the reactor, there is no way of diminishing the thickness of the shield if any appreciable amount of power is produced. Even if we could succeed in constructing a 1000-kW reactor core of the size of a coconut the necessity for surrounding it with a concrete shield of about 6 ft thickness on all sides gives a definite lower limit to the overall dimensions and its total weight. That is why the atom-driven motor-car cannot be realized.

Among the great variety of reactors built since 1945 a distinct tendency towards use of enriched fuel can be observed in the U.S.A. The reason for it is simpler construction, smaller size, and considerably smaller costs. While the Oak Ridge reactor X-10, built in 1943, and after successive improvements developing 3800 kW heat power, cost 5·2 million dollars, a swimming-pool reactor fitted with the new ORNL fuel elements using U 235-aluminium alloys could be obtained for little more than 100,000 dollars. Although a reactor of that size is on account of the limited capacity of its cooling system not built for large power output, it can be very efficient as regards that particular property which is most relevant for research purposes—that is, its *neutron flux*. The number of neutrons passing per second a unit area within the active zone of a reactor, or at the inner surface of the reflector, is called the *neutron flux*. A flux of $10^{12} n/cm^2/sec$ is an average value obtained in any good modern reactor, while the record figure is $4 \times 10^{14} n/cm^2/sec$, measured within the beryllium reflecting shield of MTR (Materials Testing Reactor, in Arco, Idaho).

A good deal of the progress made within the last decade is due to the availability of uranium enriched in U 235. It is one of the beneficial consequences of the regrettable international tension that the apparent necessity for stockpiling atomic ammunition has led to a very substantial expansion programme for U 235 separation plants. In addition to the one thousand million dollars separation plant built in Oak Ridge during the War, recent allocations made for expansions and new installations of gaseous diffusion plants are as follows (given in the usual American abbreviation 1 Mb = one Megabuck = one million dollars): 464 Mb expansion of Oak Ridge, 459 Mb at Paducah, Kentucky, and 1300 Mb at a new Ohio plant. Thus with a large stockpile of U 235, which it is hoped will be eventually transferred from the ammunition depots to the power industry, ever more efficient and relatively handy reactors will be built in future.

Breeder Reactors

The decisive step for making nuclear fuel an energy source far more ample than the world's fossil-fuel reserves was done when a group of scientists working at the National Reactor Testing Station, Arco, Idaho, succeeded in breeding more Pu 239 than U 235 consumed. This means a multiplication of usable heat content of all the world's uranium reserves by a factor 140. The reason can be easily understood. The fissionable component of natural uranium is the rare isotope U 235, which is contained in the natural mixture with a percentage of only 0·72 per cent, while the remainder of 99·28 is U 238, which is practically useless. (A very small percentage of fast neutrons succeeds in fissioning U 238 as well, but the 'fission cross-section' of the U 238 nuclei is so very small compared to its 'capture cross-section' that the contribution of these processes to the total energy balance is negligible.) Although, therefore, the heat value of U 235 is about 23 million kWh per kilogramme, we cannot obtain that much directly from one kg of natural uranium, but only 0·72 per cent, or about $\frac{1}{140}$ of this amount.

On the other hand, we have seen in the section on the construction and function of the reactor that one of its functions is plutonium production. Plutonium is the second fissionable material, and can be used not only as an explosive in A-bombs, but as a nuclear fuel in a reactor. If, therefore, the fuel load of a reactor using natural uranium is depleted after the burn-up of its U 235 content, it might be reprocessed again, and after removal of the neutron-absorbing fission products be loaded with Pu 239, instead of U 235, and then be re-fed to the reactor. In the new charge the plutonium will be burned up by fission, but at the same time a new supply of Pu 239 will be produced from U 238 by neutron capture according to equation (55). As a matter of fact, the Hanford reactors do produce plutonium on a large scale. Suppose the plutonium were not withdrawn for A-bomb production, but re-fed as fissionable material for replacing the burned-up U 235, could the process (which permanently produces heat energy) then be continued until all the U 238 is converted into Pu 239?

This is not the case in the Hanford reactors, because the number of Pu 239 nuclei produced within a given time is less than the number of nuclei consumed by fission for maintaining the chain reaction. Defining the ratio between produced and consumed quantity as the *regeneration ratio*[1] of the reactor, we may consider

[1] The expression *conversion ratio* or *breeding ratio* is also used.

an example where the regeneration ratio is 0·8. Let us assume that we could start a chain reaction in an imaginary producing reactor using natural uranium with an initial charge of only 1,000,000 U 235 nuclei. When they are burned up we have obtained 800,000 Pu 239 nuclei, with which we refuel the reactor. These produce 0·8 × 800,000 = 640,000 Pu 239 nuclei, so that the next charge will be 640,000 nuclei producing 512,000 new ones, which in their turn produce 409,600, and so on, until the charge is too small for further maintenance of the reaction. The result can be easily calculated from the well-known formula for the sum of a geometrical progression. Instead of obtaining energy from only one million fissions without plutonium we have increased the energy output by the factor

$$1 + 0.8 + 0.8^2 + 0.8^3 + 0.8^4 + \ldots = \frac{1}{1-0.8} = 5 \ldots\ldots(56)$$

Thus with a regeneration ratio of 0·8 we obtain five times more energy than without using the new-bred plutonium at all.[1] Similarly, a regeneration ratio of 0·9 would increase the total energy yield derived from a given quantity of natural uranium by a factor of 10.

If the regeneration ratio can be made greater than 1 more fuel will be produced in the process than consumed. But the total gain is limited to the multiplication factor 140, because the cycle of re-fuelling can be continued only as long as any of the original material U 238 is left, which was 140 times more abundant than the initial charge of fissionable material.

A reactor which along with power production generates plutonium with a regeneration ratio of 1 or more is called a *breeder reactor*. The news that the breeding problem has been solved was broadcast in June 1953 by an announcement stating that the experimental breeder reactor EBR at the Reactor Testing Station at Arco, Idaho, had reached a regeneration ratio surpassing 1. The reason why this figure had not been obtained earlier, although plutonium production on a scale as large as possible was the sole task of the giant Hanford plants, lies in the small margin of the neutron economy of the fission reaction. The average number of neutrons released per fissioned nucleus is 2·5 with U 235 and 2·9 with Pu 239. Consider a reactor in a stationary state with constant power output; then the number of neutrons produced per unit time must remain

[1] It has been assumed here for simplicity that both the energy release per fissioned nucleus and the regeneration ratio is equal for U 235 and Pu 239. Any difference would, however, make little alteration to the result, because it is only *the initial charge* which consists of U 235.

constant. Assume that 1000 fissions occur within a certain short time interval. Then 2500 neutrons are produced at the same time. One thousand of these are used to cause new fissions for maintaining the chain reaction at constant power-level. Of the remaining 1500 neutrons a part is captured by U 238 and produces plutonium, but another part is absorbed in impurities[1] of the metal and the moderator, while still another part escapes through the surface. These two latter groups are lost for generating plutonium. Let us assume that in the example just given 700 of the 1500 neutrons not needed for maintaining the reaction are lost by absorption and leakage. Then 800 remain for plutonium production, and therefore, having started with 1000 neutrons, we obtain a regeneration ratio of only 0·8. Something of this kind seems to have been the situation at Hanford, and apparently it has not been possible yet to reduce the losses in heterogeneous graphite-moderated reactors so far that a regeneration ratio of 1·0 can be reached.

With plutonium as the fuel, the conditions are more favourable. First of all, the average number of neutrons released by a Pu 239 fission is 2·9, instead of 2·5 with U 235. In addition a plutonium-enriched reactor can be operated without moderator as a fast reactor. It can be considered as the kind of throttled atom bomb in which the chain reaction is slowed by mixing the fissionable material with a suitable diluent that absorbs just the right percentage of neutrons to facilitate the maintenance of the reaction at a constant power-level. Instead of a neutron-reflecting shield, a mantle of natural or of depleted[2] uranium is provided in which all the neutrons escaping from the reactor core are captured by U 238 nuclei, which are subsequently converted first into $^{239}_{93}Np$ and finally into $^{239}_{94}Pu$. By a suitable choice of the composition of the plutonium-containing core and by properly designing the dimensions of core and mantle the important step of realizing a *breeder reactor* has been performed. The general arrangement of the Experimental Breeder Reactor No. 2 which is to be built under the five years' plan of the U.S. Atomic Energy Commission is shown in Fig. 91 (p. 354).

[1] How strongly the operation of the reactor can be affected by the presence of impurities can be recognized from data about the neutron capture cross-section of some nuclei. Expressed in barns (1 barn = 10^{-24} cm²), the neutron capture cross-section of the boron nucleus $^{10}_{5}B$ is 4000 barns, which is 890,000 times more than that of carbon (0·0045 barns). Considering that the isotope $^{10}_{5}B$ comprises 18·8 per cent of the isotope mixture constituting natural boron, we find that an impurity of only 0·1 per cent boron in the graphite would increase its neutron absorption by a factor of 167.

[2] Depleted means a smaller percentage of U 235 than the natural content of 0·72 per cent.

Although a very small fraction of the U 238 nuclei also under-goes fission with fast neutrons, it contributes too little to the heat output to be considered as a fuel. U 238 is therefore called the *fertile material* of the reactor from which the proper fissionable fuel is bred.

Breeding can also be performed by using the element thorium as the fertile material in a blanket surrounding the reactor core. Natural thorium consists of a single isotope Th 232 with a half-life of $1\cdot39 \times 10^{10}$ years. By the capture of a slow neutron a Th 232 nucleus is converted into Th 233, which is a negative beta emitter with a half-life of 23 min, the product being protactinium 233; this is also beta active with a half-life of $27\cdot4$ days, and dis-integrates into U 233. The reaction and disintegration scheme is thus

$$^{232}_{90}\text{Th} + ^{1}_{0}\text{n} \longrightarrow ^{233}_{90}\text{Th}\frac{\beta^-}{23\text{m}} \rightarrow ^{233}_{91}\text{Pa}\frac{\beta^-}{27\cdot4\text{d}} \rightarrow ^{233}_{92}\text{U}. \quad (57)$$

The uranium isotope U 233 is the third known substance which is fissionable with slow neutrons. It is also a radioactive alpha emitter decaying into Th 229 with a half-life of $1\cdot6 \times 10^5$ years, which is very short compared with the age of the earth. Hence it does not occur in natural uranium, and is, like plutonium, an artificially produced nuclear fuel. With breeder reactors using natural thorium as fertile material being included in the programme of the American five years' plan, we may take it safely for granted that the conversion of Th 232 into U 233 will be made eventually with a regeneration ratio of 1.

These results are of great importance for the power economy of the world, because they lead to the conclusion that the total energy content of nuclear-fuel resources is represented by the product of the number of atoms in the recoverable tonnage of uranium plus thorium and the energy released per fission of Pu 239 and U 233, which is of the order of magnitude of 200 Mev. An estimate of the total energy made available in this way is given at the end of this chapter.

Atomic Power Plants

It should be emphasized that atomic energy, although opening numerous new possibilities by the application of radio-isotopes in biology, medicine, and industry, is by no means a new, mysterious force from the standpoint of power production, but simply another source of *heat*. Atomic power plants differ, therefore, from con-

ventional thermal power stations only in this one point. The boiler[1]
is not heated by a coal- or oil-fired furnace, but by means of a
heat exchanger fed from a reactor. Using atomic power means
utilizing the heat (which any reactor cannot help producing) in
a conventional thermal power plant. It may appear surprising,
therefore, that since 1944 not fewer than about a hundred thousand
million kilowatt-hours of heat energy generated in the Hanford
Works have been wasted by uselessly heating the Columbia river,
instead of producing power. The reason lies partly in the armaments
race, which left little time to experiment while also making the
urgently needed atomic weapons, partly in the inherent difficulties
of building reactors for economic power production, and last but
not least in the fact that the U.S.A. is a rich country with ample
coal and oil resources, where fuel-waste makes itself less felt than
anywhere else.

The difficulties are caused partly by the high boiler temperature
necessary to operate a steam plant economically and partly by the
strong radioactivity which is induced by nuclear transmutation in
all materials which are in direct contact with the reactor core. The
construction of a nuclear power plant could be simplified by using
the interior of the reactor directly as the boiler, so that the coolant
fluid is evaporated in the core and directly fed to the turbine. This
was at first believed to be impossible, because the steam which is
exposed to the dense neutron flux contains a not too small percentage
of atoms that have been transmuted into radioactive nuclei. It
was feared that their strong beta and gamma radiation would soon
destroy the turbine-blades.

The transfer of heat from the reactor to the steam is therefore
usually done by means of an intermediate fluid or a gas which circulates between the reactor and the heater windings in the boiler.
The temperature of this medium must be higher than that of the
steam entering the turbine, which according to present standards
of efficiency should be at least 700° F. or higher. If water is used
for the heat transfer, then its high temperature involves also high
pressure, so that the whole reactor with its neutron-reflecting shield
should be encased in a vessel which must resist the high temperature and pressure as well as the strong neutron radiation, causing
quite new kinds of radiative corrosion. Too little experience was

[1] For the sake of simplicity we shall assume in the following that the conventional part of the plant is a steam-driven turbo-generator. It may, however, turn
out to be advantageous to use a closed-cycle AK gas turbine instead of a steam
turbine (see Chapter V). In that case the boiler will be replaced by an air-heater.
This would make scarcely any difference in the construction of the reactor itself.

available on the behaviour of the conventional sorts of steel alloys under such extraordinary conditions.

Another difficulty is connected with the 'canning' problem. The uranium slugs used as the fuel in a heterogeneous reactor are surrounded by the cooling water, and would corrode unless they were covered by a protecting layer. As Smyth (1945) states in his report, it was not a simple matter to find a sheath that would protect uranium from water corrosion, would keep the fission products out of the water, would transmit heat from the uranium to the water, and would not absorb too many neutrons. After a good many headaches a solution of the canning problem was found which worked satisfactorily under the conditions obtaining in Hanford—that is, with a moderate temperature of the cooling water. But what would be the performance of the protective cover under the much more severe conditions of high pressure and high temperature?

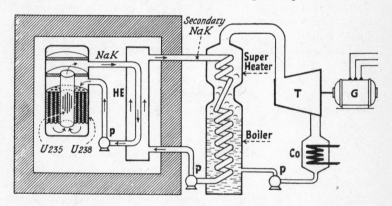

FIG. 87. COOLING SYSTEM OF EXPERIMENTAL BREEDER REACTOR

Still another difficulty lies in the need to remove and reprocess the fuel from time to time. The usual way to do it is to push the contaminated uranium slugs out of the horizontal channels running through the graphite block. This can be done by remote control in the reactors of the Hanford type. But how to do it when the whole lattice is enclosed in a high-pressure boiler vessel? All these technological difficulties retarded the development of the long-expected first atomic power plant.

A workable solution for the heat transfer from the reactor to the boiler was found in using a liquid metal as the coolant and heat-transmitting medium. Fig. 87 shows the cooling system of the experimental breeder reactor EBR of Arco, Idaho, which is described by W. H. Zinn in *Nucleonics* (1952). An alloy of sodium and potassium which is liquid at room temperatures, but boils at 1500° F., was chosen as the coolant because it does not absorb neutrons appreciably, and has good characteristics as a heat-transfer medium on account of its great heat conductivity and high boiling-point.

The primary NaK coolant flows first through the U 238 blanket surrounding the core, and then through the core itself. It gets hot there, in a double sense, both thermally and radioactively. On account of the latter it is not fed directly to the boiler, but to a heat exchanger where the heat is transferred to another circuit of sodium-potassium alloy which is no longer radioactive, since the two streams of alloys are separated by the walls of the tubes in the heat exchanger, through which no transfer of radioactivity occurs. The secondary system of hot liquid flows to a steam boiler and superheater, from where steam is fed at 550° F., 400 psi to a 250-kW turbo-generator.

Liquid-metal coolant is also reported to be used in the first Russian atomic power plant, which has been operating since June 27, 1954. With a power of 5 MW, it was the first land-based nuclear power plant surpassing the 1000-kW mark. Taking into account ship-propelling engines as well, it was superseded by the engine of the U.S. submarine *Nautilus*.

Heat Rating and Irradiation Level

None of the hitherto existing nuclear plants generates electric power at a cost competitive with electricity from hydroplants or from conventional fuels. The thermal efficiency is still below 20 per cent, and the installation costs are very high. A characteristic feature of nuclear power plants is the fact that capital must be invested not only in the construction of the buildings, reactors, turbines, and engines, but also in the *fuel inventory* of natural or enriched uranium, which must be assembled in the reactor to maintain the chain reaction. The capital invested in the fuel inventory is a not negligible percentage of the total installation costs. Thus, for instance, Britain's first nuclear power station was built at Calder Hall, Cumberland, at a cost of sixteen and a half million pounds and contains an initial charge of fabricated uranium costing about five million pounds, which makes quite a considerable addition to fixed charges of production costs due to capital interests. One of the factors determining the economy of a nuclear power plant is therefore the *heat rating*, which is the rate of production of heat from each ton of fuel charge in the reactor. The heat rating of the experimental power reactor BEPO at Harwell is 150 kW per ton of the charge consisting of a natural uranium. With improved designs of gas-cooled, graphite-moderated thermal re-actors, like that under construction at Calder Hall, the heat rating will be increased to 1 MW per ton of fuel. With an assumed over-all efficiency of 20 per cent, a plant with an installed capacity of 50 MW electric power would require 250 MW of heat, and there-fore a fuel inventory of 250 tons of natural uranium. Better heat ratings can be expected from heavy-water-moderated reactors, and still more from those using enriched fuel (up to 10 MW per ton). With fast reactors heat rates of 100 MW per ton can be achieved, but there are considerable difficulties in securing the necessary quick and reliable cooling and heat-transfer system.

Another relevant factor of the operation of a reactor is its *irradi-ation level*, which means the *total heat energy* obtained per ton of the fuel charge before it must be replaced by a new one. While, therefore, the heat rating is the *rate of heat production* (to be ex-pressed in kilowatts or megawatts) per ton of the fuel charge, the irradiation level is the sum of all the heat energy (to be expressed in kilowatt-hours, or megawatt-hours, or megawatt-days) which can be extracted from the charge before discarding it. The fixed charges due to capital interests depend on the heat rating (decreas-ing with increasing heat rating), while the running fuel costs are inversely proportional to the irradiation level. Although the energy content of each ton of U 235 is a fixed quantity, the irradiation level actually achieved in a reactor may be considerably lower, because for several reasons not all the fissile U 235 contained in the fuel charge can be used. It is quite easy to compute the 'ideal irradiation level' of natural uranium. Since each metric ton of natural uranium contains 7·2 kilogrammes of U 235, and each kilo-gramme of U 235 releases by fissioning an amount 23 million kWh = 23,000 MWh, the irradiation level of a reactor with a fuel charge of natural uranium would be 7·2 × 23,000 MWh = 165,600 MWh per ton if all the U 235 could be burned up in the charge. This is not possible, however, because the fission process is accompanied by a strong beta, gamma, and neutron radiation, which causes damage to the fuel elements and to their corrosion-protective layers. Besides, as mentioned earlier in this chapter, some of the fission products are neutron absorbers, so that their accumulation would cause a self-poisoning of the reaction. Unless, therefore, a continuous replacement of fuel is provided (as may be practised in homogeneous liquid-fuel reactors) the fuel charge has to be removed and replaced by a new one before complete burn-up of the fissile U 235 content. For these reasons the actual irradiation level is lower than the above given figure; the estimate for Calder Hall is 3000 Megawatt-days of heat per ton. This is 72,000 MWh per ton, or about 44 per cent of the ideal irradiation level.

Detailed programmes for further work on power reactors have been published in Great Britain and in the U.S.A. The new American installations are not designed primarily to increase the power-generating capacity of the country, but to obtain information about the relative merits of the different types of reactors. According to the British programme, however, about 5 per cent of the total electricity demand of this country will be covered from nuclear power stations by 1965, and about 25 per cent by 1975.

Britain's Ten-year Plan

A Programme of Nuclear Power was presented to Parliament by the Lord President of the Council and the Minister of Fuel and Power in February 1955 and was published in a White Paper. The following section gives a short summary of this report.

The programme is based on the consideration that:

Improved living standards, both in advanced industrial countries like our own and in the vast under-developed countries overseas, can only come about through the increased use of power. The rate of increase required is so great that it will tax the existing resources of energy to the utmost.

Without nuclear power the rate of consumption of coal (or its equivalent as oil) by the power stations alone would increase . . . by perhaps 2½ times over the next 20 years, reaching about 65 million tons a year by 1965 and 100 million tons a year in the 1970's, and would at that time be rising by 4 or 5 million tons a year.

The provision of enough men for the mines is one of our most intractable problems, and is likely to remain so. In order to meet the present demand for coal recourse has been had to voluntary Saturday working as well as to opencast production and to imports: but the demand continues to increase. Any relief that can come from other sources of energy, such as nuclear power, will do no more than ease the problem of finding and maintaining an adequate labour force

A forecast of domestic power consumption is given in the table opposite (Table 41):

TABLE 41. ELECTRICITY CONSUMPTION IN GREAT BRITAIN

	TWh (thousand million kilowatt hours)				
	1925	1950 (Estimated)	1954	1965 (Forecast)	1975
Industrial	3·7	23·4	32·0	61	107
Domestic and agricultural	0·6	14·9	19·6	37	63
Commercial	0·9	6·1	9·5	16	27
Traction	0·5	1·5	1·4	2	4
Total Sales	5·7	45·9	62·5	116	201
Total units sent out (including transmission losses, etc.)	6·4	51·9	69·0	130	223

The report adds:

This forecast is unlikely to be a serious overestimate, and there are good reasons that it might be conservative. The implied annual rise in industrial production is modest compared with the annual increases that have proved possible since the war. No allowance is made for any significant narrowing of the gap between United States productivity—and electricity consumption—and that of the United Kingdom, and it is possible that full employment will lead to a more rapid rise in consumption than in the period before the war.

To meet the growth in demand from about 64 TWh a year to-day to over 200 TWh annually in 1975 an adequate rise of the installed generating capacity will be necessary. The present figure is 20,000 MW as a total of all plants in the United Kingdom, and it is estimated that the capacity will have to be increased to 35–40,000 MW by 1965, and to perhaps 55–60,000 MW by 1975. This would involve the high figures of coal consumption mentioned above, with all the inherent difficulties of providing an adequate labour force. Therefore it is planned to obtain an ever-increasing part of the additional power requirements from nuclear plants. The figures given in the Government's programme are 1500 to 2000 MW of nuclear power by 1965, and somewhere between 10,000 and 15,000 MW by 1975. A provisional time-table of this development is given on page 350. The periods during which the single items of the programme will be constructed are indicated by brackets.

PROVISIONAL TIME-TABLE OF THE BRITISH PROGRAMME OF NUCLEAR POWER

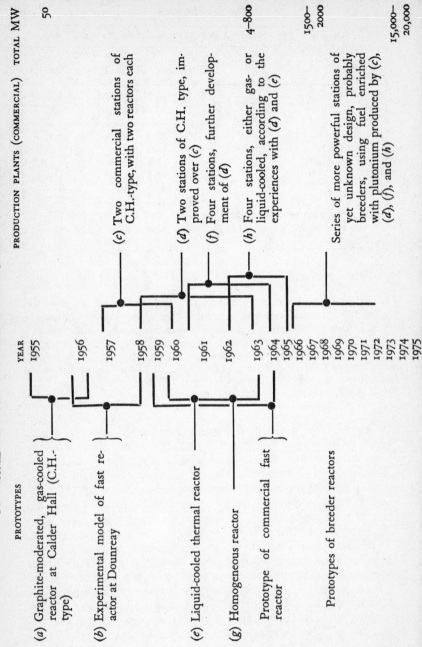

The design of the Calder Hall plant, which when completed in 1956 was the world's most powerful nuclear electric station up to that time, is based on the experiences made with BEPO at Harwell, a graphite-moderated thermal reactor, rated at about 6 MW heat, fuelled with natural uranium and cooled with CO_2 gas (carbon dioxide) under pressure. Although heat transfer by a fluid medium like heavy or light water is more efficient from the purely thermal standpoint, the gas-cooling has been chosen for the Calder Hall station as well as at Harwell for safety reasons. This will be understood by considering that liquid water is a stronger neutron absorber than gaseous CO_2, and decreases, therefore, the reactivity—that is, the rate of reproduction of neutrons in the chain reaction. When the reactor operates at a certain power-level—say, for instance, 250 MW heat—an equilibrium is reached in which the number of neutrons generated by fission per second is kept constant. If now for some reason a failure in the cooling mechanism occurs the absence of the neutron-absorbing water might—according to the fears of cautious constructors—upset the reproduction equilibrium so quickly that the mechanical shut-off mechanisms would not act promptly enough to prevent dangerous overheating, or even exploding, of the reactor. This cannot happen with gaseous CO_2 as a coolant, because carbon dioxide is a very modest neutron absorber, so that its absence would alter the reactivity of the system in such a small degree that the control mechanism could easily operate in time.

The disadvantage of gas-cooling lies in the fact that in order to transfer heat from the reactor to the boilers of the plant at a rate of more than 200 MW very large quantities of gas must be set in motion, or, in other words, quite a tornado must be kept flowing through the cooling ducts of the reactor core. Two reactors have been operating for some time in the Windscale Works near Sellafield, Cumberland, of a similar design to those of Calder Hall—namely, gas-cooled, graphite-moderated, but serving a different purpose by producing plutonium without generating power. Each of the reactors is provided with two blower-houses containing a number of radial-flow centrifugal blowers, each driven by an electric motor of 2400-3000 horse-power. Air is used as the coolant, and its draught is so strong that a horizontal force of about 600,000 pounds tends to blow the graphite block towards the chimney.

The Calder Hall plant is also equipped with two large reactors, each flanked on its four corners by cylindrical steam-raising towers about 80 ft high, which contain the boiler and superheater tubes.

The circulating carbon dioxide, coming hot from the reactor, enters at the top of the towers, gives up its heat to the banks of boiler tubes, and is returned via powerful blowers to the reactor. From Fig. 88, which shows a sketch of Calder Hall Power Station, it can be seen that the steam-raising plant is immensely more bulky than the turbines and generator units.

This is what **Sir Christopher Hinton**, who was chosen to direct the design, construction, and operation of Britain's atomic factories, says about Calder Hall:

> What we are doing to-day will look as clumsy and costly in a hundred years' time as one of Watt's early steam-engines looks to us, but we may well be opening the door to similarly important advances in power plant engineering. . . . However, the graphite-moderated, gas-cooled reactor is the one for which the greatest amount of knowledge and design experience are available, and the cost of producing power from it appears to compare favourably with the estimates for the cost of power from alternative types of thermal reactor.

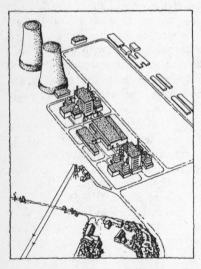

FIG. 88. SKETCH OF CALDER HALL
NUCLEAR POWER STATION

The first two or four commercial plants to be built according to the British programme during the next eight years are, therefore, likely to use reactor designs of the well-tried type of BEPO at Harwell and of the Windscale works or the Calder Hall plant. At the same time, however, experiments will proceed with pilot plants of different construction. Among the experimental reactors to be built in the next ten years are:

Liquid-cooled thermal reactors.
Thermal reactors using enriched fuel.
Homogeneous reactors.
Fast breeders.

It is hoped that better heat ratings and irradiation levels will be obtained in these reactors, and that, therefore, the commercial plants, to be built according to the experiences gained with pilot plants, will be able to generate electricity at competitive costs. The reason for the delay in the construction of these more advanced types of nuclear power plants lies both in the lack of experience and of material. It will take some time until the isotope separation plant at Capenhurst provides sufficient amounts of U 235, and the Windscale and Calder Hall plant the necessary amount of plutonium, to cover the demand of fissile material for the military authorities, as well as for the inventory of full-scale power plants using enriched fuel. When the process of electricity production from natural uranium has gone on for some time the plutonium gained from the discarded fuel charges will serve to prepare enriched fuel for the more advanced types of reactors.

Since those latter plants are also producers of plutonium—the output being proportional to the electricity units sent out—there will possibly be enough plutonium available within a few decades to enable the clumsy big graphite reactors to be replaced step by step by smaller models possessing enriched fuel and higher heat rating.

The first British fast breeder is under construction now at Dounreay, near Caithness, in Scotland.

The expected costs of electric power from nuclear plants will be between 0·6 and 1 *d*./kWh at the bus-bars of the work. The ratio between this price and the consumer's price will be the same as is the case with electricity from conventional plants.

If the development of new power stations proceeds as scheduled in the present programme about 20 to 25 per cent of Britain's demand for electricity will be produced from nuclear power stations by 1975. (For later developments see p. 367.)

The American Five-year Plan

The programme of the U.S. Atomic Energy Commission for 1954–59 provides for development of five distinct technological approaches with the following types of experimental reactor power plants:

(1) pressurized water reactor; (2) boiling-water reactor; (3) sodium-graphite reactor; (4) homogeneous reactor; (5) fast breeder reactor.

Figs. 89–91 give diagrammatic sketches of the first two and the last of these five projects, together with technical data as far as they are available.

The technical data of the other reactors are:

SRE (Sodium Reactor Experiment). Heterogeneous, thermal,

graphite-moderated, sodium-cooled. Slightly enriched uranium fuel (regeneration ratio ~ 0·9), or U 235 as fuel and Th 232 as fertile material, with a regeneration ratio slightly better than 1.   Only experimental, no turbo-generator, 20 MW heat exhausted to atmosphere.

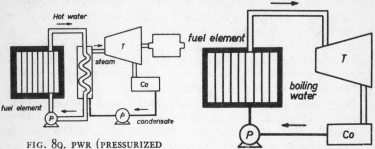

FIG. 89. PWR (PRESSURIZED WATER REACTOR)

Moderated and cooled by water, slightly enriched uranium fuel (1·5 — 2·0 per cent U 235). Temperature: fuel > 600°F; coolant 500–600°F. Pressure: reactor core 2000 psia; boiler 600 psia. Pressure vessel surrounding reactor core 9 ft diameter, 28 ft height. Heat output 300 MW; electrical output 60 MW.

FIG. 90. EBWR (EXPERIMENTAL BOILING WATER REACTOR)

Heterogeneous, thermal, moderated, and cooled by ordinary water. Enriched uranium fuel. No heat exchanger; coolant evaporated and directly fed to turbine. Heat output 20 MW; electrical output 5 MW. To be completed in the winter of 1956.

HRE–2 (Homogeneous Reactor Experiment No. 2). Fuel and moderator in aqueous uranyl sulphate solution. About 500°F. and 1000 psia in the reactor core. Heat output 5 MW (Arrangement analogous to that in Fig. 86.)

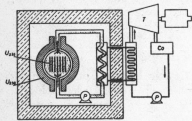

FIG. 91. EBR–2 (EXPERIMENTAL BREEDER REACTOR NO. 2)

Fast reactor, no moderator. NaK coolant. Loaded first with U 235, later with Pu 239 bred in U 238 blanket. 62·5 MW heat; 15 MW electrical power.

HTR (Homogeneous Thorium Reactor). Planned to be built on the basis of the experiences with HRE-2. Uranyl sulphate fuel enriched with U 233. Breeder blanket Th 232. 65 MW heat; 16 MW electrical power.

A total sum of about 200 million dollars will be invested into this experimental programme, which comprises only one full-scale plant (PWR), being built at Shippingport, Pennsylvania, while the others are medium-sized pilot plants for testing the feasibility of different designs. An obvious advantage of the boiling-water reactor is the simplicity of its construction, which might reduce the installation costs. On the other hand, fuel economy can be achieved (*a*) by using high steam temperatures and pressure for obtaining good thermal efficiency and (*b*) by breeding. The condition (*a*) will be best fulfilled by using fluid metals as the coolant and heat-transfer medium. The SRE (Sodium Reactor Experiment), which produces only steam, and no electricity, is intended to discover the maximum permissible fuel element and structure temperatures, and to determine performance limitations.

A serious disadvantage of all heterogeneous reactors with solid fuel elements lies in the fact that long before full burn-up the fuel must be removed from the reactor for reprocessing in order to separate the strongly radioactive and neutron-capturing fission products. To perform this process with solid fuel involves stopping the operation of the plant and opening the high-pressure shell, which is full of radioactive matter. In this respect the homogeneous reactors are distinctly superior. Instead of batchwise processing, continuous operation will be possible in the experiments HRE and HTR, in which chemical plants for reprocessing the fuel are included.

It is tempting, of course, to combine the advantages of the homogeneous reactor, permitting cheaper fuel processing, with the liquid-metal heat transfer, yielding higher thermal efficiencies. Preliminary research work has been done, therefore, in the Nuclear Engineering Department of the Brookhaven National Laboratory, Upton, N.Y., with the aim of constructing a Liquid Metal Fuel Reactor (LMFR). The most important feature of this type, described by C. Williams and F. T. Miles (1954), is integration of chemical processing with the reactor. This can be accomplished by providing both the fuel and the fertile material in liquid form, so that both substances can be withdrawn, processed, and refed to the reactor during full operation. Liquid-metal reactor fuel and breeding systems having reasonable melting-points can be obtained by adding low-melting-point diluents to high-melting thorium and uranium. A necessary condition is that the diluents should not absorb an appreciable amount of neutrons. Because bismuth, lead, and tin have both low melting-points and low absorption cross-sections, alloys and dispersions of these metals were investigated as possible LMFR fuel and reactor materials. One of the combinations taken into consideration is a solution of about 0·06 – 0·1 per cent U 233 in bismuth as the fuel, and a slurry of thorium bismuthide (Th_3Bi_5) dispersed in liquid bismuth as the fertile material. Fig. 92 gives a somewhat simplified illustration of one of the possible LMFR power systems described by Williams and Miles. The heterogeneous, thermal, graphite-moderated reactor differs basically from the Sodium Graphite Reactor SRE in that the fuel is not contained in

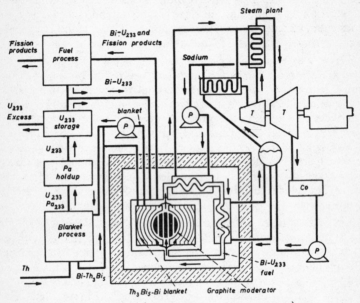

FIG. 92. LMFR (LIQUID METAL FUEL REACTOR)

solid metal slugs, but is dissolved in the coolant. Both the spherical reactor core and the surrounding breeder blanket consist of graphite blocks fitted with channels for circulating the fluid. The uranium solution is circulated by a powerful pump (not shown in the figure) through channels in the core and the blanket, and then through heat exchangers, while the thorium bismuthide slurry flows only through separate channels in the blanket. The uranium fuel coming hot from the core delivers a part of its heat at high temperature in the first heat exchanger (shown in Fig. 92 on the top of the core) to liquid sodium, and another part at lower temperature in a second heat exchanger to the water and steam of the boiler. The high-temperature heat carried away by the sodium is used for

superheating the steam before entering the high-pressure turbine
T_1 and for reheating it before it enters the low-pressure turbine T_2.

In addition to the quick circulation for heat transfer, there is
a far slower circulation of the fuel through a reprocessing plant for
separation of the fission products. In the same way, the fertile
thorium solution is slowly pumped through another plant for
separating the newly bred U 233, which is then added to the fuel
refed to the reactor.

The advantages of this system are: (a) high-temperature operation, and thus high thermal efficiency, of over 35 per cent, but
without high pressures in the reactor; (b) good enough neutron
economy to give some breeding, with resulting low fuel costs; (c)
relatively simple construction; (d) easy fuel handling by pumping;
(e) cheap chemical processing to remove fission products, integrated
in the reactor design.

The fuel costs are composed of mining and processing from
ore to metal, of separating the newly bred fuel from the fertile
material, and of separating the fission products from the partly
burned-up fuel. These latter processes are considerably cheaper
with liquid fuel and breeder substance than with solid uranium
and thorium, and therefore the construction shown in Fig. 92 might
be expected to compete economically during the next decade with
fast breeder reactors. (More is given about U.S. projects on p. 363.)

Radiation Hazards

If proper precautions for safe disposal of the fission products
are taken the nuclear power station will have the decisive advantage
over conventional thermal plants that it does not contaminate the
atmosphere with smoke and soot. A fully nuclear powered and
heated city would enjoy a much clearer atmosphere than present
cities and industrial places. On the other hand, however, even
small leakages of the stored fission products would have immeasurably more serious consequences than the worst pollution of the
atmosphere by the combustion products of conventional fuel. For
the equilibrium activity of the fission products (see p. 335) of a
municipal power plant of a big city would be more than a hundred
times larger than the amount necessary to kill the whole population
of that city. An elaborate system of precautions against hazards
caused by negligence or criminal misuse will be necessary to protect
the people living in nuclear-powered countries. Geiger counters or
similar monitoring devices may ultimately become as familiar
instruments in households as thermometers are to-day.

How efficient the protective measures taken by the British authorities are can be seen from a report, "Atomic Energy in 1954," published in *The Engineer*, vol. 199, p. 86, making the following statements about Harwell:

> It is indeed reassuring to know that so slight is the degree of radiation injury caused to the staff employed at Harwell that risk to life and health arises more from falling down holes, getting splinters into eyes, and, most of all, from accidents on the main road outside the establishment.

Economy of Nuclear Power Plants

It would be a mistake to believe that the use of nuclear fuels would make power production cheaper to the extent that energy stored in a pound of uranium 235 exceeds that in a pound of coal. First of all, uranium is vastly more expensive than coal. While the coal prices vary in different parts of the world between 5 and 35 dollars per ton, according to the transportation costs, U 235 costs about 20 million dollars per ton. And even if the fuel costs could be neglected, the kWh rate of electricity bills would by no means drop to almost zero, because the fixed charges due to the capital invested in the plants, as well as the operation and maintenance costs of the plant and of the transmission and distribution system, are in most cases higher than the fuel costs.

The following considerations refer to the generation costs of electricity—that is, the cost of the kilowatt-hour at the bus-bar of the plant. Transmission and distribution costs are virtually the same in nuclear plants as in conventional thermal plants. In contrast to some hydro-power stations, the site of a nuclear plant can be chosen so that long-distance transmission lines over several hundreds of miles can be avoided. The generating costs are composed of three items: fixed charges, fuel, operation and maintenance.

The fuel costs depend on the overall efficiency of the plant and on the fuel price. The connexion between these variables is given in Fig. 93, representing the variation of fuel cost per kWh in relation to plant efficiency and coal price in a conventional power station.

In nuclear plants the fuel costs not only depend on the plant efficiency in converting heat into power, but also on the percentage of burn-up of fissionable material in a given charge of fuel which is feasible before the chain reaction is stopped by the depletion of U 235. In a discussion of nuclear-power costs given by W. H. Zinn, Director of Argonne National Laboratory (1952), a non-regenerative reactor using enriched uranium at a cost of 20 dollars per gramme U 235 is considered as a first example. In this type of reactor, which will propel atomic submarines, no use is made of the plutonium generated in the U 238, and no reprocessing plant is included in the reactor system. The enrichment of the fuel is assumed to suffice to uphold the chain reaction until half

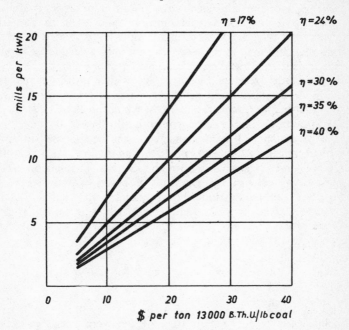

FIG. 93. VARIATION OF FUEL COST PER kWh ELECTRICITY WITH PLANT
EFFICIENCY AND COAL PRICE

of the U 235 is burned up, after which recharging with new fuel is necessary. With an energy content of 24 million kWh per kilo-gramme of U 235 (one-half of which is actually used), and assuming a conversion rate of heat to power (plant efficiency) of 24 per cent, we obtain a fuel cost of 7 mills[1] per kWh. This is considerably higher than that of modern conventional steam plants using 7 dollars a ton, 13,000 B.Th.U./lb coal, which is respectively 3·5 and 2·8 mills/kWh at 24 per cent and 30 per cent plant efficiency.

The fuel economy is better in regenerative reactors in which

[1] A mill is a thousandth of a dollar. Conversion into British currency gives approximately: 1 mill = 0·08 pence; 1 penny = 12·5 mills.

some of the plutonium bred in the U 238 is used along with the initial content of U 235. No proper breeding with a regeneration ratio 1 is supposed in this type of reactor. Zinn assumes that in a heavy-water reactor 1 per cent of the total charge of natural uranium—that is, a part of the 0·7 per cent U 235 content and a part of the new-bred plutonium—might be used without chemical processing involving additional costs. With a price of 35 dollars per lb of pure metallic natural uranium, and again 24 per cent efficiency, the nuclear-fuel cost is 1·3 mills/kWh, which is below that of coal, and makes, therefore, extra money available for capital for the construction of the nuclear portion of the plant.

In a breeder reactor, moreover, in which ultimately all the uranium or thorium could be converted into fissionable material and burned in the reactor the fuel cost would be 0·013 mills/kWh, which is quite negligible compared with all other costs.

The operation and maintenance costs can be supposed to differ so little between coal-fuelled, nuclear, and hydro plants that the difference can be neglected in a rough estimate. The data on which the generating costs of various kinds of power stations, shown in Fig. 94, are computed include the figure of the operation and maintenance cost given by J. A. Lane (1954), which is 1·0 mills/kWh at 80 per cent load factor (7000 annual use hours). It varies inversely with the load factor, as shown below:

Annual load factor per cent	40	50	60	70	80	90	100	
Operation and maintenance	2·0	1·6	1·33	1·14	1·0	0·9	0·8	mills/kWh
	0·16	0·128	0·106	0·091	0·08	0·072	0·064	pence/kWh

The fixed charges are taken by Lane with 15 per cent of the investment, by others with 12 per cent. The data underlying Fig. 94 are computed on fixed charges of 12·5 per cent. If I is the investment expressed in dollars per kW capacity of the plant, and n is the number of annual use hours, then the fixed charges of 12·5 per cent can be found by the equation

$$\text{Fixed charges} = \frac{0·125}{n}I \ \$/\text{kWh} = \frac{125}{n}I \ \text{mills/kWh} = \frac{10}{n} \ I \ \text{pence/kWh}$$

$$\dots\dots\dots\dots\dots\dots\dots\dots\dots\dots\dots(58)$$

Fig. 94 shows how the generating costs of different kinds of plants depends on efficiency, fuel price, and load factor. To each curve is added: the capital investment in dollars/kW, the type of the plant, and in thermal plants also the price and kind of fuel and overall efficiency of conversion from heat to power. Among the examples of the hydro plants one is added in which the

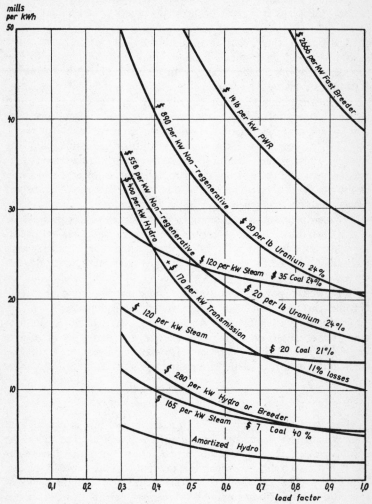

FIG. 94. ELECTRIC POWER COSTS OF NUCLEAR POWER PLANTS COMPARED WITH CONVENTIONAL STEAM AND HYDRO PLANTS

Abscissæ, load factor; ordinates, mills/kWh. First four from top, nuclear plants; others, conventional steam and hydro plants.

generating costs are given on the consumer's end of a 400-miles 220-kv transmission line. Another curve inscribed " amortized hydro " refers to a hydro plant with fixed charges corresponding to a capital investment of $70/kW for renovation and modernization of the plant in excess of normal operation and maintenance costs.

Apart from the trivial truth that the variation of generating costs with the load factor is stronger in expensive plants with low fuel costs than in cheaper plants with high fuel costs, the following facts can be learned by looking at Fig. 94:

(1) Power plants with non-regenerative reactors costing as much as the Shippingport PRW plant cannot compete economically with any conventional plants fuelled with coal at a reasonable price.

(2) If the capital investment for nuclear-power plants could be brought down below about $500/kWh even a non-regenerative reactor with fuel costs as high as 7 mills/kWh, or 0·56 pence/kWh, could do good service in single areas where owing to transport difficulties the coal price exceeds 30 dollars a ton.

(3) Competition with modern efficient steam plants in areas of cheap coal prices will be possible for power converters with fuel costs as low as 1 mill/kWh—or even better for breeders with negligible fuel costs—provided that the investment costs do not exceed about $280/kW.

(4) Electric power from nuclear fuels cannot be expected to become cheaper than that of amortized hydro plants.

(5) Definite advantages of nuclear plants over all power stations using wind, water, or solar energy are: (*a*) the possibility of operating with greater load‑factor and (*b*) the avoidance of additional costs and losses caused by energy-storage and long-distance transmission.

The date of general use of nuclear power on a commercial scale will mainly depend on how long it will be until the technique of nuclear engineering is developed enough to bring the investment costs down to a level comparable to that of conventional plants, which lies between about 120 dollars and 180 dollars per kW capacity. In the studies made by American private firms estimates of investment costs are given as low as $110/kW for a 554-MW sodium-cooled reactor plant (Union Electric-Monsanto Chemical Team Report, in the *Electrical World,* June 29, 1953, pp. 85-87). This is in striking contrast to the data given in the official Report on Major Activities in the Atomic Energy Programmes of the U.S. AEC, January-July, 1954. The figures quoted for the power plants built within the five years' programme of the A.E.C. are:

PROJECT	TOTAL COST	$ PER kW
PWR	$85 million	1417
Fast Breeder	$40 million	2666

It would be misleading, however, to judge the economic aspects of nuclear power from the 1954 status of the experimental projects, with their heavy burden of developmental costs. An appreciable reduction of costs can be expected from the next set of plants, to be built in the near future. As a result of the recent U.S. Atomic Energy Act of 1954 a number of new projects were drafted by private firms on a more commercial basis, and the quotations of costs are accordingly lower.

Apart from a dozen smaller nuclear plants to be completed in the U.S.A. by 1960, two larger projects deserve particular interest. One of the most advanced designs is the large Fast-neutron Breeder Reactor Power Plant described by Cisler, Griswold, and Campbell (1956), which is an improved full-scale model of the Experimental Breeder Reactor EBR–2 shown in Fig. 91. Using fast neutrons and highly enriched uranium, it does not need any moderator, and accordingly its enormous heat output of 300,000 kW, or 1,000,000,000 B.Th.U., per hour is produced in a cylindrical core of only 30·5 in. diameter and 30·5 in. length. A very efficient cooling is necessary, which is performed by liquid sodium transferring the heat to a secondary NaK cooling circuit, which in its turn transfers the heat to the boiler and superheater system. The plant capacity will be 100 MW = 100,000 kW, and the costs of the plant, including electric equipment, but without charge for land and fuel inventory, are estimated at 54,572,000 dollars. Adding the expenses for the fuel inventory of 450 kilograms of U 235, the total costs will be in the neighbourhood of 700 dollars per kW installed capacity. This is considerably less than the costs of the U.S.A.E.C. projects of 1954 quoted above. With an annual consumption of 92 kg U 235, a plutonium generation of 109 kg Pu 239, along with the power production, can be expected, so that the net fuel costs will be quite insignificant.

Another, more conservative, design is a still larger plant described by Milne and Moore (1956), financed by the Consolidated Edison Company of New York, and proposed for erection at Indian Point, New York State. The reactor is a thorium thermal converter fuelled initially with highly enriched U 235, and later on with U 233, into which the thorium of the breeder blanket is converted. Moderator and coolant is pressurized light water. The dilemma of either operating at high temperatures, and accordingly high pressure in a relatively large vessel, or at lower temperatures, and accordingly low thermal efficiency of the steam plant, can be solved by a compromise proposed in one of the alternatives of the Indian Point project. This compromise means limiting the rôle of the reactor to the generation of saturated steam, which is subsequently superheated in a separate oil-fired superheater. According to this alternative the gross electric output would be 230 MW, and the total cost of the plant 55 million dollars, or 239 dollars per kW. The fuel costs, including the oil burned in the superheater, would, of course, be considerably higher than in a fully nuclear-fuelled plant. According to the other alternative of the project, using the pressurized water reactor as the sole heat-source, the gross electric power would be 140 MW, with total plant costs of 45 million dollars, or 322 dollars per kW. Both estimates represent a radical reduction of the first costs compared with the 1954 quotations of the U.S. Atomic Energy Commission.

Atomic power plants are on the march; they will cover quite a considerable percentage of electricity production in twenty or thirty years, probably with the effect of making obsolete some of the more expensive projects for obtaining power from recurring sources, like tidal, wind, and solar plants.

Life Expectancy of Nuclear Fuels

Viewed from a wider perspective, the question whether or not nuclear-power production can compete economically with that from present conventional plants is less important than that of the amount and life expectancy of fuel reserves. Once our oil and coal reserves are depleted we have to produce power from other sources at any price, and even the very expensive nuclear plants like the PWR at Shippingport might compete with the still more expensive production methods from recurring sources, as, for instance, solar power stations in North Africa supplying electric power over a 1200-miles transmission line to England. The significance of the new tool of atomic energy depends, therefore, primarily on the life expectancy of nuclear fuel reserves.

When the idea of boosting the world's energy resources by nuclear fuels cropped up first during World War II it was uncertain whether the global reserves of uranium would suffice to secure a supply which might last longer than that from fossil fuels. Certainly uranium does not belong to the rare elements; the average content in the earth's crust is of the order of magnitude of one part per million, which is, of course, quite a lot. The energy stored in the uranium impurity contained in each block of granite is larger than that of an equally heavy block of bituminous coal. But unfortunately nearly all the uranium content of the earth's crust occurs in such a diluted form that economic recovery seems to be scarcely possible. The amount of minable uranium is only a tiny fraction of the total occurrence, and, again, only 1/140 of the natural uranium is the fissionable U 235 with which nuclear chain reactions can be effected. For these reasons the first surveys of nuclear-energy reserves led to the conclusion that a supply could be expected which certainly might be larger than that of oil, but probably smaller than that of coal.

In the meantime the prospects of gaining considerably more power from uranium than from fossil fuels have been greatly improved. The extensive search for new sources was fairly successful, and within the last few years several new concentrated deposits have been discovered in Witwatersrand, South Africa; Rum Jungle, Northern Australia; and Radium-Hill, Southern Australia. These new finds will perhaps make a considerable contribution to the world's uranium production, in addition to what is being mined already in the classical deposits in Joachimstal, Erzgebirge, Czechoslovakia, and in the Belgian Congo, in Canada, and in Colorado.

Still more significant than the new finds of uranium ores is the fact that breeding has been accomplished. The transition from a mere converter reactor with a regeneration ratio of 0·8 to a full breeder with a regeneration ratio slightly higher than unity is not so much important in decreasing power costs as in prolonging the life expectancy of fuel reserves. It has been mentioned in the preceding sections that fuel costs are about 1 mill/kWh in a regenerative reactor, and negligible in a breeder reactor. Thus the aggregate production costs may be lowered by breeding from, say, 7 mills/kWh to 6 mills/kWh. On the other hand, the transition from 0·8 to 1·0 of the regeneration ratio has had the consequence that, instead of merely five times the natural content of fissionable material in uranium (see p. 341), 140 times as much can be utilized, which means that the transition from converter to breeder reactor might prolong the life expectancy of fuel reserves by a factor of 28.

A rough estimate of nuclear-energy reserves can be based on the following plausible assumptions: (*a*) The global amount of metallic uranium recoverable at competitive costs by present methods is about 25 million tons, (*b*) the whole quantity of U 238 can be converted into fissionable plutonium 239 by breeding, (*c*) the heat content of U 235 is about 24 million kWh per kilogramme. Multiplying $2 \cdot 5 \times 10^7$ tons $= 2 \cdot 5 \times 10^{10}$ kg with $2 \cdot 4 \times 10^7$ kWh/kg gives a total energy of 6×10^{17}/kWh, or about 2000 Q, which is nearly a hundred times more than Putnam's estimate of coal reserves (21 Q after deduction of the losses involved by hydrogenation—see p. 212). The date of depletion of the energy capital of our earth is therefore prolonged by at least a millennium.

A further still longer respite will be granted as soon as we succeed in developing economic methods for extracting uranium from the vast amount of existing low-grade deposits, as, for instance, igneous rocks, particularly granites. The possibility of securing long-range supplies of uranium and thorium from igneous rocks was indicated by Harrison Brown and L. T. Silver in their report at the Atoms for Peace Conference (Geneva, 1955). While the overall concentration of uranium and thorium in composite rocks of granite is only a few parts per million, over 90 per cent of these elements is concentrated in certain portions of the rock—as, for instance, the mineral phases sphene and zircon, which make up less than 1 per cent of the total weight of the rock. With increasing demand for fissionable material, a reasonably economic method of obtaining uranium and thorium from granite might be found by first separating from the rocks the mineral phases containing higher concentrations of U and Th. This might be done by the following two steps: (i) Quarry the rock and crush and pulverize it to grain size; (ii) leach with dilute acid (HCl or HNO_3). The rich mineral phases are dissolved by the process, and an extraction is gained containing concentrations of uranium and thorium which are twenty to a hundred times higher than in the original rock. Both elements may then be extracted from the enriched fraction by conventional chemical methods.

Considering that in the case of a serious demand the whole process from quarrying down to the final separation of elementary uranium and thorium will be almost fully mechanized, the problem of economy boils down to determining the ratio between the total energy requirements for extracting and processing a given quantity of uranium and the energy to be gained from the same quantity. According to Brown and Silver, the requirements would lie in the range of 19 to 37 kWh per ton of rock processed. The uranium content of the Essonville granite is 2·7 parts per million (in some European granites 4 parts per million, and even more), so that, taking into account all losses, at least one gramme of uranium with a heat content of 24,000 kWh could be obtained from a ton of rock. An atomic power plant with a breeder reactor and an overall efficiency of 25 per cent could therefore produce 6000 kWh electric energy from a quantity of fuel obtainable with an expenditure of 37 kWh or less. As soon as the necessity arises, therefore, for tapping other fuel sources than the rich ores, there would be a possibility of a considerable net gain of energy from igneous rocks. The reserves of nuclear fuel hidden there are sufficient for an indefinitely large number of millennia.

Thus the consequences of Hahn's great discovery will free us from the fear that our energy resources will run out within quite a short period of history. But another grave danger rises on the horizon: the radioactivity of the fission products, which, if let loose, might kill all life on earth. If we want to live safely in an atomic-powered world we must not only teach our engineers and workers to do their job with strict exactness, but also see to it that a better sense of human responsibility and of the need for appreciative co-operation (with definite abolishment of wars) is spread among the coming generations.

More recently the plans for European nuclear power stations have been extended. According to an address given by **Sir John Cockcroft** on September 12, 1957, the Euratom countries have prepared a target plan to install 15,000 MW of nuclear power stations by 1967. The United Kingdom Government has decided to build nuclear power stations with a capacity of 5000—6000 MW by 1965. Contracts have been placed for nuclear plants at Bradwell and Berkeley with net outputs of 300 MW and 275 MW respectively, a 360-MW station at Hunterston, in Scotland, and a 500-MW station at Hinckley Point, in Somerset.

The plans for nuclear power in the Soviet Union have been stepped up lately, and amount now to a total capacity of 2000 MW by 1960.

CHAPTER XVI

Thermo-nuclear Reactions

I T was explained in Chapter XIV that nuclear energy can be released either by the formation of middle-weight nuclei from light ones or by the disintegration of the heaviest ones. If the disintegration occurs in the form of a fission releasing neutrons a nuclear chain reaction may ensue, which, as explained in the previous chapter, can be used for power production.

We shall deal in this chapter with the other kind of processes— the *fusion reactions*. The synthesis of helium from two protons and two neutrons was mentioned in Chapter XIV as an example of a theoretically possible fusion reaction, which, however, does not occur in this simple, direct form. Vagabond neutrons do not wander about freely in sufficient quantities to make an accidental meeting of two protons and two neutrons probable enough. Free neutrons have a rather short life; either they are captured by nuclei in traversing matter, or they decay when moving in empty space like beta-emitting radioactive nuclei, disintegrating with a half-life of about thirteen minutes into a proton and an electron. We cannot expect, therefore, to obtain energy on a commercial scale by composing the helium nucleus directly from its nucleons.

The d-d Reaction

Some more binding energy is available as a result of the general tendency of lighter nuclei to form larger clusters, and particularly of the trend of deuterons to unite into a helium nucleus. It was explained in Chapter XIV how the strength of this tendency can be measured by combining Einstein's law with the observed mass defects. The trend to unite into larger nuclei cannot, however, be satisfied without overcoming first the serious obstacle of the repelling electrical forces. All atomic processes and nuclear reactions can be considered as a permanent struggle between the electrical forces (also called ' Coulomb forces ') with which the positively charged protons repel each other and the proper nuclear forces, which are attracting forces between both kinds of nucleons. At distances smaller than 10^{-11} cm the nuclear forces are much stronger than the electric repulsion. As soon, therefore, as two deuterons get close enough together the strong nuclear force will draw them still more tightly together, and this process occurs with such vehemence that the deuterons, bouncing together, loose one of the neutrons, which is emitted with high speed. This so-called *d-d* reaction, which has been investigated very thoroughly, is given by the formula:

$$_1^2 H + {}_1 H = {}_2^3 He + {}_0^1 n + 3 \cdot 3 \text{ Mev}$$

or, abbreviated, $\quad d + d = {}_2^3 He + n + 3 \cdot 3 \text{ Mev} \quad \dots \dots \dots (59)$

Expressed in words: Two colliding deuterons unite in forming the nucleus of the helium isotope with mass 3 while a neutron is expelled. The energy released is $3 \cdot 3$ Mev per synthesized helium atom, which makes 22 million kWh per kilogramme of 'burned' deuterium. With a price of about 220 dollars per kg of D_2O, and correspondingly 1100 dollars per kg of pure D_2 gas, we obtain fuel costs of $0 \cdot 05$ mills per kWh heat, and, assuming a 20-per-cent plant efficiency, $0 \cdot 25$ mills per kWh electric power, in a plant which would burn deuterium into helium and would use the heat for generating electricity with a plant efficiency of 20 per cent.

Considering the fact that enormous quantities of deuterium are stored in all the oceans of the world, one might be tempted to feel reassured about future energy supply in the belief that nuclear plants using fusion reactions could produce electricity from practically inexhaustible sources, and do it at costs quite competitive with conventional thermal plants. As a matter of fact, some scheme of a similar kind is likely to become the definite solution of the world's energy problem in a rather distant future. But to-day we cannot so far see any possibility of realizing power production from the heat of fusion reactions.

The reason preventing us from taking the last decisive step towards secure everlasting energy supply will be understood by comparing the ways in which we can handle fission reactions and fusion reactions. In both cases any physicist equipped with the proper tools of a modern atomic laboratory will be able to produce fission reactions and fusion reactions any time he likes. But his experiments will be restricted to small-scale laboratory tests. By means of highly sensitive devices like the Wilson Cloud Chamber, or the Geiger Counters, or the modern Crystal Counters triggered by single high-speed nuclei, he can demonstrate that reactions like the fission of the U 235 nucleus or the fusion of two deuterons into a helium nucleus do actually occur, and he can even determine the energy of single particles emitted during the process. But, as explained in Chapter XV, no measurable amount of heat can be produced as long as the number of atoms participating in the reaction remains so incommensurably small compared to the many trillions of carbon atoms reacting per second in the flame of any heating device.

In the case of the uranium fission the step from laboratory reactions with infinitesmally small energy output to proper energy production has been done by effecting nuclear chain reactions. Will it be possible to start chain reactions also in the case of nuclear fusion processes like the *d-d* reaction given by equation (59)? The answer is: Yes, it is basically possible; there is no fundamental obstacle. As a matter of fact, fusion chain reactions play the decisive rôle in the energy household of the universe, and even man-made reactions of this kind have been accomplished already. But they can occur only under peculiar conditions which do not permit their use for orderly power production in the present state of our technical knowledge.

The difference between the cases of fission and fusion reactions will be understood at once. The fission of U 235 and Pu 239 produces, among others, two or three neutrons which under proper conditions cause more fissions, and thus start a chain reaction. Curiously enough, it is just the slow neutrons which are most suited to split fissionable nuclei like U 235 and Pu 239, and therefore the chain reactions might proceed even at very low temperatures in a particularly well-cooled uranium pile. If desired for any reason—a very unlikely thing to happen—heat might be produced in an alcohol-cooled reactor working at temperatures below freezing-point.

In some fusion processes, like the *d-d* reaction, for instance, neutrons are also generated, but they do not contribute directly to causing other pairs of deuterons to react. The fusion between two deuterons occurs when they are bumping against each other with an energy sufficient to overcome the repelling electric force. The positively charged nuclei are shielded from each other by what is called their *potential barrier*, and a collision between two nuclei will result in a reaction only when their kinetic energy is great enough to penetrate the potential barrier. The situation is analogous to that of ordinary chemical reactions. A mixture of oxygen and hydrogen can exist stably at room temperature for a long time. It is only when it is ignited by a spark that the energy of the molecules near the ignition spot is raised beyond the threshold of the reaction $2H_2 + O_2 = 2H_2O$, and then the energy of this process heats neighbouring parts of the gas, lifting the kinetic energy of the molecules over the threshold again, so that the chain reaction is released and an explosion occurs.

Pure gaseous deuterium consisting of D_2 molecules may be regarded as a kind of super-explosive in which the *d-d* reaction might take place under suitable conditions. However, the energy release as well as the threshold energy required for ignition is immensely greater than with chemical explosives. The threshold is of the order of magnitude of 0·1 Mev, which is certainly less than the 3·3 Mev energy gain. But it would be a mistake to believe that the ample surplus of energy gain over the threshold for initiating a new *d-d* process would suffice to release a chain reaction. For the neutron emitted in the *d-d* process with high speed does not itself start a new reaction (as it does in fissionable fuel), but can contribute to it only by acting as a kind of golf-club, which, when hitting a deuteron, gives it a kick so that the latter, if it hits in its turn another deuteron, penetrates the potential barrier and fuses into it according to equation (59). There are two if's in this statement, and it must be borne in mind what has been said in Chapter XIV about the tiny dimensions of the nuclei and the vast empty space between them. The neutron coming fresh from the *d-d* reaction with a kinetic energy which is about three-quarters of the total energy gain of 3·3 Mev will pass and perhaps miss a million or more deuterons in the gas before it happens to collide with one of them. Then it will give it a kick which under most favourable conditions transfers three-quarters of the kinetic energy to the deuteron. This amount would, of course, suffice for the deuteron to penetrate the potential barrier and fuse into another deuteron *if* a collision were to occur before its energy is dissipated on its way through the gas. But let us remember that collisions close enough to bring the attracting nuclear forces into play are very infrequent. On the other hand, the electric forces of the Coulomb field of the electrons and nuclei of the matter traversed have a much wider range than the nuclear attractions, and thus our kicked deuteron, which moved quite swiftly immediately after receiving its impulse from the neutron, will gradually lose its energy in the jungle of electric fields. By its electrical interaction with the many nuclei and electrons which it passes it will set them into motion, and thus transfer a part of its kinetic energy to each of them. When at last a real close collision occurs the deuteron may have lost so much of its initial energy that it is too fatigued to penetrate the potential barrier of its collision partner.

This explains why no self-sustaining chain reaction occurs in pure deuterium gas under terrestrial conditions, although the energy released by a single *d-d* process is more than thirty times larger than the energy threshold of this process. What is required to maintain a nuclear chain reaction of *d-d* processes is a very large number of high-speed deuterons which, by starting many single fusion reactions at a time, may raise the temperature of the gas to the necessary degree to make further high-energy collisions frequent enough. Nuclear chain reactions in which the means of maintaining the chain are not neutrons, as in the fission processes, but fast collisions due to high temperature, are called *thermo-nuclear reactions*. The necessary temperatures of the order of 10^7 degrees centigrade exist in the interior of the stars.

Energy Production of Stars: The p-p Reaction and the Carbon Cycle

Before the advent of nuclear physics the source of the energy of the stars was a riddle for astronomers. Neither gravitational energy of the contracting great cosmic bodies nor conventional chemical processes could account for the prodigal heat output which has gone on for at least two or three thousand million years. Although our sun belongs to the stars with relatively small energy output, its heat radiation is so great that an anthracite ball of the size of the earth would have to be burned every five days to produce the necessary amount of heat by combustion.

As early as 1928 **Atkinson** and **Houtermans** suggested that thermo-nuclear reactions between light nuclei might be the source of the energy of the stars and the sun, and thus indirectly also the ultimate source of all life on the earth. It is believed to-day that an assumption of this kind actually furnishes the solution of the old riddle of the origin of stellar radiation. Although, however, the temperature in the interior of the stars, which is estimated to be about 2×10^7 degrees on the average, would be well suited to maintain the *d-d* reaction in a stellar core of deuterium gas, it is not this particular reaction which supplies the energy of the stars. The reason is that the concentration of deuterium in stars is much too small to make collisions between two deuterons frequent enough. The main constituents of the interior of the stars are light hydrogen, helium, small amounts of carbon, very small traces of lithium and beryllium, and a small admixture of heavier components, the so-called Russell mixture, which contains the most abundant elements found spectroscopically in stellar atmospheres.

CONSTITUTION OF THE RUSSELL MIXTURE

Oxygen	50 per cent
Sodium + magnesium	25 per cent
Silicon	6 per cent
Potassium + calcium	6 per cent
Iron	13 per cent

Temperatures of some 10^7 degrees do not suffice to maintain thermo-nuclear reactions between heavier nuclei because the potential barrier of their higher charges prevents them from fusion. Therefore thermo-nuclear reactions in stars can be expected to occur only between protons and protons or between protons and light nuclei, as, for instance, lithium or carbon.

According to our present knowledge, there are two processes which supply the energy of the stars. One is the so-called *p-p* reaction given by the formula

$$_1^1H + \; _1^1H = \; _2^2He \rightarrow \; _1^2H + e^+ + 1.5 \text{ Mev} \quad\quad\quad (60)$$

In words: Two colliding protons unite in forming a transitory unstable helium nucleus with mass 2 which immediately disintegrates into a deuteron and a positron. (The positron is the positive counterpart of the electron. In contrast to the latter, it does not exist in a stable form, but dematerializes by colliding with an electron and radiating away both masses in the form of a short-wave gamma quantum.) Leaving out the transitory helium nucleus, the reaction can be shortly written

$$p + p \rightarrow d + e^+ + 1.5 \text{ Mev} \quad\quad\quad (61)$$

The energy released in this process is about 1.5 Mev per synthesized deuteron. **H. A. Bethe** and **C. L. Critchfield** were the first to draw attention to this process, and it is assumed to-day that most of the stars smaller than the sun draw their life energy from the *p-p* reaction.

At about the same time **Weizsäcker** (1938) in Germany and **H. A. Bethe** (1939) in the U.S.A. found that another process, the so-called *carbon cycle*, might account for the energy production of the brighter stars. The process may be best explained by reference to Fig. 95.

The wavy lines represent gamma quantums emitted by some of the steps of the process. The cycle starts with the reaction shown in the bottom centre of the figure, where a $^{12}_{6}C$ nucleus (carbon) is hit by a proton p. The collision results in the emission of an electromagnetic wave (a so-called gamma quantum) and the capture of the proton into the carbon nucleus, which is transmuted thereby into a nitrogen nucleus $^{13}_{7}N$ (step (a) of Fig. 95). This latter is, however, not stable, but suffers a beta$^+$

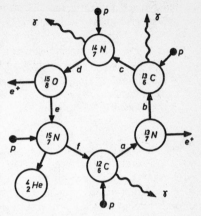

FIG. 95. CARBON CYCLE

decay which transmutes it into $^{13}_{6}C$ (step (b)). Another collision with a proton occurs, releasing a gamma quantum again, and transmutes the $^{13}_{6}C$ into a nitrogen nucleus $^{14}_{7}N$ (step (c) leading to the top centre of the figure). As soon as this is hit by another proton it is converted into a $^{15}_{8}O$ nucleus, which is a beta$^+$ emitter and decays into $^{15}_{7}N$. One might expect that a collision of $^{15}_{7}N$ with a proton would result in the formation of $^{16}_{8}O$, which is a very stable nucleus. As a matter of fact, such reactions do occur occasionally, but in the majority of such collisions the energy of the impact together with the binding energy of the nuclear forces is so strong that it breaks the newly built compound nucleus into two parts, which, as shown at the bottom left of Fig. 95, are a $^{4}_{2}He$ nucleus and a $^{12}_{6}C$ nucleus. Thus at the end of the cycle the carbon nucleus with which the chain started is restored again; no carbon is used up. The net result is that four protons have been converted into one helium nucleus and two positrons. The latter will react with the many electrons swarming about in the highly ionized stellar matter and radiate the mass away in the form of gamma quantums, thus making a contribution to the energy production of the process, which is a total of about 23 Mev per generated helium nucleus. (It may be added for the physicist that the two acts of beta$^+$ emission—steps (b) and (e) of the cycle—are accompanied by the emission of neutrinos which seem to have scarcely any interaction with matter at all, and which therefore escape from the star into the surrounding space and carry with them a certain amount of kinetic energy.

This part of the process was first studied by **Gamow** and **Schoenberg** (1941), who called it the *Urca process*, because of the obvious similarity between the traceless disappearance of energy stolen by neutrinos and the similar traceless disappearance of the gamblers' money in the crowded playrooms of the famous Casino da Urca in Rio de Janeiro.)

Both processes, the $p–p$ reaction and the carbon cycle, are continuously going on in the stars, and supply the vast amount of energy which they spend so generously with their radiation. In view of this fact one might ask: Why, then, don't we use these processes for power production from the practically inexhaustible hydrogen sources of the earth?

The answer is that the $p–p$ reaction, as well as the carbon cycle, can proceed as self-sustaining thermo-nuclear chain reactions only in bodies of the size of the fixed stars, because the rate of energy production is so poor that bodies of terrestrial dimensions could not be kept at sufficiently high temperatures. On account of the tiny size of the nuclei the likelihood of nuclear collisions leading to a reaction is so small that it takes on the average about six million years for a single carbon nucleus to perform the whole cycle shown in Fig. 95. Accordingly, the rate of energy production is of the order of magnitude of one or two ergs per gramme per second in the interior of the stars, which is about 1 per cent of the heat production by metabolism in the human body. The reason why, in spite of the slow rate of energy production, the high temperature of millions of degrees can be maintained lies in the huge size of the stars. The production of energy goes on in the interior of the stars while the radiation is emitted from the surface. Since energy is expended only by radiation, the ratio between income and expenditure of energy will be proportional to the ratio between volume and surface of the star—that is to say, proportional to its radius. Hence the core of a body with a diameter of about a million miles can be kept hot enough to maintain a thermo-nuclear reaction like the $p–p$ process, while a body of earthly dimensions cannot.

The Hydrogen Bomb

In trying to imitate the power production of the cosmos human efforts are handicapped by the small size of the bodies which we can handle, but, on the other hand, we have two distinct advantages over Mother Nature. We can by means of a plutonium bomb as a detonator achieve higher temperatures than those existing in the interior of the stars, and we can by assembling round the detonator a suitably chosen thermo-nuclear explosive start much faster reactions than the p–p process or the carbon cycle. With reactions yielding many million calories, instead of merely two ergs per gramme per second, it has been possible to achieve thermo-nuclear reactions in the hydrogen bombs exploded both in Russia and at the U.S. test sites in the Pacific.

When during World War II the idea of using a fission bomb as a detonator for a thermo-nuclear bomb was first conceived it was thought that by surrounding a plutonium bomb with a mantle of a deuterium compound a d–d reaction might be started when the plutonium detonates. In the meantime it has been found that there are other nuclear processes which yield more energy per reaction, and have, in addition, a lower energy threshold for initiating the reaction. A widely discussed case is the so-called d–t reaction which follows a close collision of a deuteron and a *triton*. The latter is the nucleus $_1^3 H$ of superheavy hydrogen, also called tritium, which is an isotope of hydrogen with the mass number 3. It is a radioactive substance, decaying with a half-life of about twelve years. On account of its short life it does not exist in nature, but it can be made artificially in the great accelerating machines—and much better still in the reactors—for instance, by the process:

$$_3^6 Li + _0^1 n = _1^3 H + _2^4 He + 4 \cdot 7 \text{ Mev} \dots\dots\dots\dots(62)$$

A mixture of deuterium and tritium gas is a good thermo-nuclear explosive ready for the reaction:

$$_1^3 H + _1^2 H = _2^4 He + _0^1 n + 17 \cdot 6 \text{ Mev} \dots\dots\dots\dots(63)$$

As shown in Fig. 96, the rate of energy production is about a hundred times greater in the d–t reaction than in the d–d process. In both cases the rate of production increases with a very high power of the temperature. It can be seen from Fig. 96 that the energy production of the d–t reaction is less than 1 cal/g/sec at a temperature of 400,000 degrees, but amounts to 10^{10} cal/g/sec at a temperature of 2,000,000 degrees.

A tritium-deuterium mixture would therefore be the ideal substance for making hydrogen bombs if tritium were available in sufficient quantities This, however, is not the case, because the output of artificially made tritium is quite small. The constructors of the thermo-nuclear bomb were therefore faced by the problem of how to obtain the necessary amount of tritons for the reaction. No data have been released, of course, about the material used in the bomb, but it has been guessed as a theoretically possible solution of the problem that the mantle surrounding the plutonium bomb (which acts as the detonator) contains 'light heavy' lithium hydride ${}^{6}_{3}$LiD, to which some tritium is admixed. 'Light heavy' means that the lithium hydride, whose conventional chemical formula is LiH, is not composed of natural lithium (which is a mixture of 92·7 per cent ${}^{7}_{3}$Li and 7·3 per cent ${}^{6}_{3}$Li) and natural hydrogen (a mixture of 99·985 per cent ${}^{1}_{1}$H and 0·015 per cent D), but of light lithium ${}^{6}_{3}$Li and heavy hydrogen D. As soon as the plutonium detonates, the temperature of the core of the bomb is raised to some hundred million degrees, and at the same time a

vast number of neutrons released by the atomic explosion penetrate the LiD blanket. Collision of some of the neutrons with Li nuclei is followed by the reaction (63), which, using the abbreviations ${}^{2}_{1}$H $= d$, ${}^{3}_{1}$H $= t$, and ${}^{4}_{2}$He $= \alpha$, may be written shortly

$${}^{6}_{3}\mathrm{Li} + n = \alpha + t + 4\cdot7 \ \mathrm{Mev} \dots (63a)$$

This reaction produces tritons and additional heat, on account of the energy release of 4·7 Mev per each single process. The high velocity with which the tritons and alphas created in the process (63a) are flying apart will enable a fraction of the tritons to collide with deuterons of the LiD blanket vehemently enough to react according to equation (63)— or, written in the simplified form, $d + t = \alpha + n + 17\cdot6 \ \dots(64)$

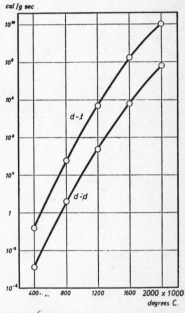

FIG. 96. RATE OF ENERGY PRODUCTION IN THE d-t REACTION AND THE d-d REACTION PLOTTED AS A FUNCTION OF TEMPERATURE

so that more neutrons are produced again. In this way the detonating plutonium core of the bomb might start a cycle consisting of the two alternating reactions (63a) and (64), the first of them consuming neutrons and producing tritons, the second consuming tritons and reproducing neutrons. Both together make helium from lithium and hydrogen, releasing thereby a considerable amount of energy, which is approximately 74 million kWh per kilogramme of LiD.

This is one of the theoretical possibilities for realizing explosive thermo-nuclear reactions, but it may be that the real hydrogen bombs made by the Americans and Russians use other reactions which are better suited. One fact seems to be certain, however, that the ingredients of the bomb have to be found among the isotopes of the elements with low nuclear charge, probably hydrogen and lithium.

The progress—if it may be called so—made by the introduction of thermo-nuclear reactions into the technique of atomic weapons consists in the fact that no limit other than the carrying capacity of the means for delivery is set to the size of the bomb. In the case of the uranium and plutonium bomb the size of the bomb is limited by the critical mass: a block of pure U 235 or Pu 239 explodes by itself as soon as its weight surpasses a certain critical value of just a few pounds. A thermo-nuclear explosive, however, like LiD or a deuterium-tritium mixture, might be as large as our whole globe without exploding before a part of it is heated to a temperature of many million degrees. This is the reason why the destructive power of the Eniwetok bomb of March 1954 was about seven hundred times that of the Hiroshima bomb of 1945.

Although it might appear regrettable that the most spectacular scientific progress of our time has so far resulted only in a vast increase of the means for mass annihilation, the net effect on our civilization will, after all, be quite beneficial. The ' balance of terror ' which has been established by the ability of both of the two Great Powers to strike annihilating retaliatory blows has not only prevented the leaders from taking hasty steps in critical situations, but will, moreover, lead finally to the definite abolition of the primitive pre-atomic barbarism of settling international disputes by war. The transition from the present state of working against each other to a new system of intelligent co-operation will be a greater benefit for mankind than anything that can be expected from peaceful uses of atomic energy.

Controlled Thermo-nuclear Reactions

In the case of the fission reactions the first step was a controlled chain reaction (achieved by Fermi and co-workers in Chicago on December 2, 1942), while more than two and a half years later the test explosion of an A-bomb at Alamogordo on July 16, 1945, was the second step, consisting of a most powerful explosive chain reaction which proved the practicability of the new tool.

In the case of the fusion reactions the development is reversed. Thermo-nuclear explosions have been made in test experiments which demonstrated their tremendous force. The corresponding step to a controlled fusion reaction, analogous to the well-controlled fission reactions proceeding in the uranium piles, might be the greatest achievement of human science and technology, because it would secure for mankind practically inexhaustible sources of energy. The total reserves of deuterium contained in the oceans are about 2×10^{16} kg, and with the energy release of the $d-d$ reaction amounting to 22 GWh per kilogramme of deuterium we obtain 4.4×10^{23} kWh, or about $1.5 \cdot 10^9$ Q, as the energy reserve stored in heavy hydrogen. This is more than a hundred million times the energy of the global coal reserves, and would suffice to cover our present energy demand for fifteen thousand million years.

The challenge to take this last and decisive step in the field of power production has been answered by the systematic efforts of competent working groups in the U.S.A., in Great Britain, and in the Soviet Union. In the early stages of the development of the hydrogen bomb a long-range research programme aimed at achieving controlled fusion for peace-time uses was started by the U.S. Atomic Energy Commission under the code name Project Sherwood. In the course of the extension of this programme experimental and theoretical work on controlled nuclear fusion is being carried out at five places in the United States—Los Alamos, New Mexico; Livermore, California; Princeton, New Jersey; Oak Ridge, Tennessee; and New York University. The particular code name given to the branch of Sherwood working at Princeton is Project Matterhorn. Contributions to the entire enterprise have been made by some of the most prominent nuclear physicists in the States—for example, **Enrico Fermi, Edward Teller, Lyman Spitzer, H. W. Smyth, James L. Tuck,** and **William Brobeck.**

In Great Britain the same problem is being tackled by a well-equipped group of scientists at the Atomic Energy Research Establishment at Harwell, working under the direction of **Sir John Cockcroft,** while in the Soviet Union research work on this subject is being carried out under the direction of Academician **L. A. Artsimovich** at the Academy of Science of the U.S.S.R. The Academicians **Igor Kurchatov** and **M. A. Leontovich** appear to be playing a leading rôle in the Russian thermo-nuclear project.

Up to 1955 strict silence was kept, not only about the details of the work on controlled fusion, but even as to the fact that extended research in this field was being carried out under the auspices of the American, British, and Russian Governments. The first to break this silence, by means of a widely discussed statement, was the Indian physicist **Homi L. Babha,** Chairman of the Atoms for Peace Conference held in Geneva in August 1955, who in his Presidential Address expressed the hope that " a method will be found of liberating fusion energy in a controlled manner within the next two decades."

As a consequence of Babha's statement, the question was raised in some quarters whether it would be wise to continue to spend innumerable millions for the development and construction of atomic-power plants if fission power were soon to be outdated by fusion power. Still, in spite of the economic significance of the topic, the veil of secrecy surrounding Project Sherwood and its British counterpart was not lifted.[1] Eight months later, however, a worldwide sensation was caused in scientific circles when the first details of Russian thermo-nuclear experiments were revealed in Mr Igor Kurchatov's lecture at Harwell, England, which was delivered on April 26, 1956, with the permission of the Soviet Government. The presence of Prime Minister Bulganin and Party Secretary Khrushchev underlined the official character of Kurchatov's lecture, which will be dealt with in greater detail below.

Although, even with all the progress made in Russia, the fusion reactor is not yet just round the corner, it seems obvious that Kurchatov gave quite a competent account of experiments basic to eventual realization of controlled thermo-nuclear power. With little or no lag in Eastern development of fusion devices, and with thermonuclear weapons of almost equal strength in the possession of both the U.S.A. and U.S.S.R., the mutual playing at hide-and-seek in the field of the peaceful applications of fusion processes seems to have become obsolete. Accordingly, a strong demand for declassification of papers on controlled fusion reactions has been made by leading American and British scientists, who argue that secrecy in this field,

[1] When the National Industrial Conference Board decided to include a report on controlled hermo nuclear reactions in the programme of its October 1955 Conference in New York, none of the real experts in this field was ready to accept an invitation to lecture on this important matter. Finally, the Board, looking for scientists outside the United States, chose the author of this book as their reporter on Non-weapon Potentials of The-mo-nuclear Reactions. Being in the lucky position of not having access to classified work, I was free to speak and to explain to my audience at the Conference the grave obstacles which in my opinion will ~ot be overcome soon enough to make useless the fission power plants which will shortly be built. No comment on my rather sceptical views was made by any of the well-informed members of Project Sherwood, but Mr Will am Laurence, science expert of the *New York Times*, ventured to lay a wager that the first fusion reactor would be working by 1960.

instead of contributing to the security of the West, is only hampering further progress.

Complete success has not yet been achieved by these efforts to restore freedom in the international communication of scientific knowledge. Yet a number of papers on the formerly taboo subject of controlled fusion reactions have appeared since Babha's and Kurchatov's speeches, and from these papers quantitative data about the almost insurmountable difficulties of realizing the fusion reactor can be gained. Here is a list of the authors; the full sources are given in the bibliography at the end of this book:

E. Sänger (1955); R. F. Post (1956); Lyman Spitzer (1956); Igor Kurchatov (1956); Ralph E. Lapp (1956); F. Winterberg (1956); W. B. Thompson (1957); J. D. Lawson (1957); R. S. Pease (1957); J. E. Allen (1957); R. J. Taylor (1957); R. Carruthers and P. A. Davenport (1957).

What are the real obstacles against making a fusion reactor? Two basically different approaches can be considered—electrically excited fusion reactions and self-sustaining thermo-nuclear reactions. Fusion reactions of the first kind were already accomplished a quarter of a century ago, when **Cockcroft** and **Walton** bombarded a lithium target with electrically accelerated protons of 0·7 Mev energy, and observed what later on proved to be the reaction

$$\textstyle{7 \atop 3} \text{Li} + {1 \atop 1} \text{H} \rightarrow {4 \atop 2} \text{He} + {4 \atop 2} \text{He} + 17\text{·}3 \text{ Mev} \dots\dots\dots\dots(65)$$

Here we have a fusion process in which every single nuclear reaction yields a considerable net gain of energy (17·3 Mev output with only 0·7 Mev expenditure), while, on the other hand, the process taken as a whole is an enormous waste of energy. The reason is that the entire bombarding beam of protons must be accelerated to 0·7 Mev energy, and that only one among many million protons of the beam happens to start the reaction (65) by properly hitting a nucleus, while all the others miss the tiny nuclei and lose their energy by the so-called Coulomb interactions with the electric fields of the atoms and electrons of the target material. No way is visible to remedy this disproportion between hits and misses of charged nuclei. It is felt, therefore, that the simple old method of bombarding a target—though certainly producing single fusion reactions—will never be sufficiently improved to achieve an overall gain of energy for useful power production.

The other method is that of thermo-nuclear reactions like those occurring in the stars and in the hydrogen bomb—that is, to heat deuterium gas or deuterium-tritium mixture to a temperature high enough for the thermal velocity of the atomic nuclei to suffice to overcome the repelling forces of the potential barrier. In this case, too, the number of hits—that is, of real nuclear collision leading to a fusion reaction—will be small compared with the number of Coulomb interactions with the electrical fields of the electrons and nuclei. While, however, electrically accelerated particles striking a cold target will lose their energy by the Coulomb interactions, the case is quite different in a sufficiently hot gas, where the velocity of the thermal agitation of all the nuclei is on the average high enough to start a fusion reaction as soon as a nuclear collision occurs.

It may be mentioned that the expression ' sufficiently hot ' means here temperatures of a hundred million degrees or even more. At these temperatures the gases are fully ionized. This means that the nuclei and electrons are whirling round in their disordered thermal agitation without regard to their mutual bonds, which at lower temperatures keep a certain fixed number (equal to the ' atomic number ' of the element) properly circulating round the nucleus. This fully ionized state of matter, which is realized on earth in the electric arc and in electric discharge tubes, has been termed ' plasma.' The theory of plasma physics is playing an important rôle in astrophysics, because all the fixed stars, and therefore the overwhelming majority of all matter in the universe, are kept in the plasma state. Nowadays plasma physics belongs to the tools of thermo-nuclear research, and the recent progress in astrophysics has been very helpful in tackling the problem of the fusion reactor.

The successful realization of the hydrogen bomb proved the possibility of starting self-sustaining energy-producing thermonuclear reactions. The following conditions must be fulfilled in order to achieve the transition from the explosive to the controlled fusion reaction :

(*A*) The nuclear fuel must be heated to a temperature of about a hundred million degrees without using an A-bomb as a detonator.

(*B*) The system must be made large enough to keep the reaction self-sustaining.

(*C*) The system must be made small enough to keep the power output within reasonable limits.

(*D*) The container vessel must be protected from destruction by the enormous temperature and pressure of the nuclear fuel.

The present state of thermo-nuclear research is characterized by the fact that some progress has been made towards the fulfilment of conditions (*A*) and (*D*), while no way is visible yet of finding a compromise between the antagonistic conditions (*B*) and (*C*).

Since the burning temperature of the nuclear fuel (at least 46 million degrees K for the tritium-deuterium mixture and 400 million degrees K for pure deuterium) is many thousand times higher than the melting-point of any existing substance, no direct contact between fuel and container vessel is permissible. It was suggested, therefore, some time ago that the reacting plasma should be contained in what **Edward Teller** called a ' magnetic bottle '—that is, a strong magnetic field surrounding the nuclear fuel. Since all components of the plasma, electrons and nuclei, are electrically charged, the electro-dynamic forces of the field would cause a curvature of the paths of all the particles trying to penetrate the field, and thus prevent them from doing so. Instead of dashing right through the walls of the magnetic bottle, the particles are kept running round like the squirrel on the rotating wheel.

The idea of containing the fuel by a magnetic bottle can be combined with a method of electrically heating it by using the so-called ' pinch effect ' of strong electric discharges. According to elementary laws of electrodynamics, two parallel electric circuits carrying currents in the same direction attract each other with a force proportional to the product of the two currents. This force is acting also between each pair of parallel filaments of a current flowing through a conductor, and causes a lateral pressure. In extreme cases—when, for instance, in a short circuit the current is raised to some 100,000 amps—the pressure is strong enough to pinch and deform the conductor. Strong copper bus-bars of one inch by three inches cross-section have been crumpled sometimes by the tremendous pinching force of short-circuit high currents.

If, therefore, an electric discharge of several hundred thousand amperes is sent through a plasma the pinch effect will concentrate the gaseous conductor to a small plasma column which, like any other electric current, is surrounded by a circular magnetic field. Since the magnetic force at the circumference of the plasma column is proportional to the current, and inversely proportional to the diameter of the column, the field around the thinly pinched plasma will be strong enough to exert considerable pressure, and thus to act as a magnetic bottle. Moreover, the high current density will produce sufficient heat to attain temperatures of the order of a million degrees in the plasma, so that the occurrence of fusion reactions might be considered possible.

This programme of using the pinch effect—contemplated, and probably also tried, by the American group within Project Sherwood —has been carried out by the Russians, according to Kurchatov's report. Their experiments consisted essentially of observing what happens when pulses of high current discharges are sent through gases of low atomic weights. Straight cylindrical glass discharge tubes filled with hydrogen, deuterium, or other light gases were used, with diameters ranging from five to sixty centimetres, while the length of the discharge gap varied from several centimetres up to two metres. Banks of high-voltage condensers with a total capacity of several hundred microfarads, charged to voltages up to fifty kilovolts, were used to produce short pulses of high current discharges. The peak current varied from 100,000 to 2,000,000 amperes, and the maximum instantaneous power released in the plasma amounted to 40 million kilowatts, or about fifteen times the power capacity of Grand Coulee. On the other hand, the duration of the discharge is only several microseconds, so that the energy released in a single pulsed discharge is of the order of one kilowatthour.

The sequence of events during the build-up and subsequent decrease of the discharge was studied by means of oscillograms of current and voltage, and also by taking pictures of the process with ultra-high-speed moving-film cameras (up to two million frames per second). During the build-up period, while the current rises from zero to its peak value, the well-known skin-effect causes the current to start flowing along the walls of the cylindrical discharge vessel, so that the current-carrying part of the plasma is a hollow tube. When, a few microseconds later, the current approaches its peak value and the rate of growth decreases, then the pinch effect exceeds the skin effect, so that the current-carrying plasma tube detaches itself from the walls and contracts quickly until it is pinched to a thin column. The plasma inside the rapidly contracting column is strongly pressurized and heated by this process, so that the contracting layers rebound under the elastic forces of the compressed plasma, and expand again. This change of the geometry of the discharge is accompanied by a pronounced kink in the oscillograms, showing the increase of the current. Two to three consecutive kinks have been observed in a single discharge.

So far the observations confirmed what classical electrodynamics might have predicted about high current discharges through a plasma. The most interesting result of these investigations was, however, the discovery that at sufficiently high currents the discharge in deuterium becomes a source of neutrons. This is of importance

because—as stated earlier in this chapter, under equation (59)—
the *d-d* reaction produces neutrons[1]:

$$d + d = {}_2^3 He + n + 3.3\ Mev \dots\dots\dots\dots\dots(59)$$

At the early stages of investigation it seemed, therefore, as if the
expectation of achieving thermo-nuclear reactions in a high-current
pinched discharge had been fulfilled. Later on it was found, however, that neutrons can be observed at comparatively small discharge
currents which do not heat the plasma to a temperature necessary
for thermal excitation of the fusion reaction (59). Closer investigation showed that the neutrons were always emitted at the moment
when the plasma, after having rebounded, was subjected to its second
contraction. At the same time X-rays were also emitted, and the
energy of the X-ray quanta was found to correspond to an excitation
voltage of 300–400 kilovolts, which is about eight times the voltage
applied to the discharge tube.

This shows that a million amperes discharge in a pinched plasma
is a rather complex phenomenon whose details are not yet fully
understood. It looks as if the rapidly contracting plasma column
(with a radial velocity of hundreds of kilometres per second),
accompanied by a corresponding lateral displacement of the surrounding magnetic field, would act like a high-tension transformer
producing a secondary voltage which is a multiple of the primary
one. If the observed neutrons originate actually from the *d-d*
reaction (59) the fusion might perhaps be caused by the electric
acceleration of the deuterons in the 400-kilovolt field.

The overall result of the Russian experiments is, therefore:

(1) The magnetic bottle works all right in keeping the hot
     plasma clear of the vessel walls.

(2) Fusion reactions seem to be produceable electrically with
     a primary voltage less than a tenth of that used in the
     classical 1932 experiments of Cockcroft and Walton.

These interesting accomplishments are, however, far from realizing a power-producing fusion reactor, because the energy released
from the fusion of the deuterons during a discharge pulse is certainly
far less than a millionth of the necessary energy input, which is about
$\frac{1}{6}$ kWh.

---

[1] It should be noted that the fusion of deuterons can also lead to the formation
of tritium by means of the reaction:
$$d + d \rightarrow t + p + 4\ Mev \dots\dots\dots\dots\dots\dots\dots(66)$$
Both processes (59) and (66) occur with roughly equal probability.

What has been achieved by the analogous American and British experiments has not yet been revealed, because, more than a year after Kurchatov's instructive lecture, the thermo-nuclear work for peaceful energy production is still strictly classified. With the exception of the above-quoted papers on the theory of thermo-nuclear reactions and of plasma physics, nothing seems to have been released officially.[1]

A numerical evaluation of the theoretical data about plasma and fusion shows the tremendous difficulties still to be encountered. The plasma temperature of about a million degrees obtained by the Russians may be remarkable as the highest temperature made artificially outside A-bombs or H-bombs. But it is still far too low to achieve self-sustaining thermo-nuclear reactions. Will it be possible to increase the temperature up to the 40 or 400 million degrees necessary for the *d-t* and *d-d* reactions respectively? The theoretical predictions are rather discouraging. As shown by Pease (1957), the current of about two million amperes recorded in the Russian experiments appears to be a kind of upper limit of currents obtainable in pinched plasma discharges. Therefore, even by increasing the applied voltage and the capacity of the condensers, no further increase of the current can be expected unless the theory proves to be wrong.

The temperature obtained in the plasma is proportional to the square of the current, and inversely proportional to the plasma density. With a given ceiling of the current, further increase of the temperature could be achieved only by decreasing the plasma density—that is, by decreasing the initial pressure of the gas in the discharge tube. This, however, entails other difficulties. A thermo-nuclear chain reaction with positive energy balance can proceed only when a sufficient fraction of the energy gained in the fusion process is supplied by collisions to the rest of the fuel, thus holding its temperature at the necessary level. To fulfil this condition the fast-moving reaction products must be prevented from leaving the plasma before they have delivered a part of their energy to other particles of the fuel. This may be achieved in the case of the charged reaction products—for instance, with the $\frac{3}{2}$ He produced in the reaction (59)—by winding up their paths in a strong magnetic field. But, apart from that, sufficient time for collisions must be left to the high-speed particles issued from a fusion reaction. The poor energy out-

[1] The only information obtainable seems to be that the gadget used in the laboratories of Project Sherwood was christened " Stellatron," as an engine with a medium work.ng at stellar temperatures.

put of the experiments reported by Kurchatov is due partly to insufficient temperature and partly to the short duration of the single discharge pulse. If we want to obtain power from a device working with pulses of pinched discharges the strong effort to start the discharge (about one-sixth of a kilowatthour per pulse) must be rewarded by an appropriate energy output per pulse.

What 'appropriate' means in this connexion can be found by realizing what a fusion power reactor operating with pulsed discharges would be like. The energy of the *d-d* or *d-t* reaction, being primarily kinetic energy of neutrons, alphas, and other nuclei flying apart, is dissipated into heat, and can be utilized by surrounding the discharge vessel with a boiler or a heat exchanger, which in its turn feeds steam or hot air into a conventional power station. Let E designate the electric power input per single pulse and RE the energy released from the fusion reaction per pulse. Since the primary input E is also dissipated into heat, and finally fed into the boiler, the total heat supply per pulse is

$$RE + E = E(R + 1) \quad\quad\quad\quad (67)$$

With an overall efficiency η of the plant, the electric output per pulse will be

$$\eta \, E(R + 1) \quad\quad\quad\quad (68)$$

from which the expenditure E for starting the discharge must be deducted. This gives a useful output of

$$E \text{ useful} = \eta \, E(R + 1) - E \quad\quad\quad\quad (69)$$

which is positive when

$$\eta \, (R + 1) \geqq 1, \text{ or } R \geqq \frac{1}{\eta} - 1 \quad\quad\quad\quad (70)$$

With $\eta = 0\cdot 2 = 20$ per cent, we find $R \geqq 4$.

This figure can be used to make estimates as to how long the pinched state of the plasma must be upheld in order to secure a positive power output. The following assumptions concerning the dimensions and the electrical data of a pulsed discharge *d-d* reactor are probably not inconsistent in themselves, and are in accordance with data given by Post (1956):

Volume of pinched plasma	20 cm^3
Temperature of pinched plasma	$4 \times 10^8 \text{ deg.K}$
Number of deuterons per cm^3 in the plasma ...	10^{17}
Reaction power density (heat)	20 kW/cm^3
Electrical energy input per pulse	$\frac{1}{6} \text{ kWh} = 600 \text{ kWsec}$

Taking a minimum relative gain of $R = 4$, as calculated above, we find that the heat released during a discharge should be at least $4 \times 600 = 2400$ kWsec, and therefore, with a reaction power output of $20 \times 20 = 400$ kW during the discharge, the duration of a single discharge should be at least six seconds.

Even allowing for overstatements or understatements made in the foregoing assumptions, the result is that to obtain useful power each single discharge would have to last for a period of the order of a second or more. That appears to be quite impossible, however, according to our present knowledge, because a pinched discharge is a transient phenomenon, like a spark or a lightning-flash, and cannot be upheld for a longer period, as can an electric arc. As explained by **Kruskal** and **Schwarzchild** (1954), there is an inherent instability in the pinched discharge. Any kink or lateral displacement of the pinched column will grow rapidly and cause a disruption of the column. As a matter of fact, the discharges photographed in the Russian experiments lasted only for several microseconds.

In two respects, therefore, a gap of several orders of magnitude is open between the present achievements of pulsed discharge reactors and the necessary conditions for realization of fusion power. The burning temperature of the deuterium plasma should be 400 million instead of one million degrees, and the burning period of a single pulse should be ten or a hundred thousand times longer than has been obtained so far.

Lacking authentic information about the accomplishments of Project Sherwood, this is all that can be said about the pulsed discharge reactor, which appears still to be the most promising method of realizing controlled thermo-nuclear reactions. It is reasonable to expect that it will take some time to close the wide gap between present performance and minimum conditions for success. There is no need, therefore, to worry about the capital invested in fission reactors. They will serve their purpose and refund a good deal of the investments before the final solution of the power problem in the form of the fusion reactor is ripe for practical use.

The situation with regard to our knowledge about inexhaustible energy resources existing in the form of global deuterium reserves is therefore similar to that of a man who discovers a locked treasure-chest containing immense riches which he cannot open because he has not found the key. Still, we have the fissionable fuels, which we shall soon be able to utilize fully. They will cover the power demand for several centuries, or even millennia—certainly for a much longer period than most geologists expected when at the early stages of nuclear power development an extensive world-wide search for uranium was started. Considering the ever-increasing rate of progress of men's ability to investigate and to make use of the forces of Nature, it does not seem too optimistic to expect that a method of operating controlled thermo-nuclear reactions could be found before the global uranium and thorium reserves are depleted.

Summary

The general teachings of this book can be summarized briefly as follows:

1. The world consumption of energy, which during the last century increased by about 2 per cent annually, will continue to increase at a still higher rate in the foreseeable future.

2. While the bulk of global energy production was derived from recurrent sources like wood up to the beginning of the machine age, in the early years of the nineteenth century, we are obtaining to-day more than three-quarters of the total consumption from fossil fuels. This situation, which is a characteristic feature of twentieth-century economy, will, however, be only a short episode in human history, because in the near future a shortage of natural petroleum can be expected, and at some time in the twenty-first or twenty-second century coal reserves will be depleted unless a great deal of the global energy demand is covered from other sources.

3. Energy production from recurrent sources like plant fuel, water- and wind-power, tidal and solar energy, is theoretically possible, but—except for water-power in particularly favoured countries—only at costs which are many times higher than present production costs.

4. Power production from nuclear fuels will be possible at costs comparable to those of conventional thermal plants, but not as cheaply as from amortized hydro-plants. It can be anticipated that a growing percentage of the world's electrical power will be produced from atomic energy as early as the sixties and seventies of this century.

5. One of the consequences of this development will be the danger of radioactive contamination. To whole nations it is a menace of a hitherto unknown kind, and particularly insidious because the effects of perilous irradiation may make themselves felt only after a dose leading to serious consequences has already been applied. With large quantities of radioactive fission products available in all countries it would be suicidal to wage wars, and, indeed, the utmost conscientiousness and scrupulous exactitude in operating nuclear power plants and handling the fission products will be necessary to protect the population from grave perils.

6. Nuclear fuel will be a substitute for conventional fuels in large thermal power plants, and also in ship and large-aircraft propulsion. It is neither a substitute for liquid fuels in motor-cars nor one for electricity. With the exception of marine and air transport, the use of atomic power will be made only through the medium of electricity, and therefore increased use of atomic energy will necessarily entail increased consumption of electric power.

7. In a similar way, as a few centuries ago the increasing use of coal helped to save British forests from depletion, a step-by-step increase in the use of nuclear fuel may help to save the coal-mines from depletion. As explained in Chapter VIII, coal is a very important raw material as the basis for making plastics and other high-polymer substances, as well as a great variety of other valuable goods. Saving coal means, therefore, saving raw material of vital importance for the chemical industry.

8. Atomic energy will also help to lengthen the supply of liquid fuels by restricting as far as possible the use of motor petrol and Diesel oil to road-traffic vehicles, while stationary plants and large ship's engines will be fed increasingly from nuclear fuels. It may be expected, too, that coal will be used in increasing measure for making synthetic liquid fuel. Even then the ultimate exhaustion of fossil reserves will cause a shortage of liquid fuels, and, following the necessary electrification of railways, an electrification of road traffic with overhead lines on the main roads can be expected. This lies, however, in such a distant future that it is too early to-day to make reliable forecasts. What can be predicted with a good degree of certainty is the rising use of electricity from hydro-electric and atomic plants within the next decades.

Bibliography

ENERGY, GENERAL

AYRES, E., AND SCARLOTT, C.: *Energy Sources—the Wealth of the World* (McGraw-Hill, New York, Toronto, London, 1952).

FIFTH WORLD POWER CONFERENCE, VIENNA, 1956 (Reports published 1956).

GUYOL, N. B.: *Energy Resources of the World* (Department of State Publication 3428, U.S. Government Printing Office, Washington, D.C., June 1949).

PUTNAM, P. C.: *Energy in the Future* (Van Nostrand, New York, 1953).

RUHEMANN, MARTIN: *Power* (Sigma Books, London, 1946).

Statistical Year Book of the World Power Conference.

MISCELLANEOUS

DANIELS, FARRINGTON, AND DUFFIE, J. A.: *Solar Energy Research* (University of Wisconsin Press, Madison, 1955).

DAVIES, S. J.: *Heat Pumps and Thermal Compressors* (Constable, London, 1950).

DEAN, GORDON: *Report on the Atom* (Eyre and Spottiswoode, London, 1954).

GAMOW, G., AND CRITCHFIELD, C. L.: *Theory of Atomic Nucleus and Nuclear Energy Sources.* (Clarendon Press, Oxford, 1949).

GLASSTONE, SAMUEL: *Sourcebook on Atomic Energy.* (Macmillan, London, 1951).

HALE, WILLIAM J.: *Chemistry Triumphant* (Baillière, London, 1933).

HOGBEN, LANCELOT: *Science for the Citizen* (Allen and Unwin, London, 1951).

INTERNATIONAL CONFERENCE ON THE PEACEFUL USES OF ATOMIC ENERGY, GENEVA, 1955 (Reports published 1956).

JAY, K. E. B.: *Britain's Atomic Factories. The Story of Atomic Energy Production in Britain* (H.M.S.O., London, 1954).

LILIENTHAL, DAVID E.: *T.V.A. Democracy on the March* (Harper, New York, 1944).

LUCAS, S. H.: *Modern Ideas of the Atom* (Harrap, London, 1947).

PALEY, W. S.: *Resources for Freedom. A Report to the President.* (5 vols., U.S. Government Printing Office, Washington, D.C., 1952).

PUTNAM, P. C.: *Power from the Wind* (Van Nostrand, New York, 1948).

ROWLANDS, DOROTHY HOWARD: *Coal: Yesterday, To-day, and To-Morrow* (Harrap, London, 1949).

SHELL: *Petroleum Hand Book.*

SMYTH, HENRY DE WOLF: *Atomic Energy for Military Purposes: Official Report* (Reprinted by H.M.S.O., London, 1945).

UNITED NATIONS: *Proceedings of the United Nations Scientific Conference on the Conservation and Utilization of Resources* (United Nations Department of Economic Affairs, New York, 1950).

WORLD SYMPOSIUM ON SOLAR ENERGY, PHŒNIX, 1955 (Proceedings published 1956).

SOURCES CITED

ABBOT, C. G.: " Solar Radiation as a Power Source," in the *Smithsonian Institution Annual Report,* 1943 (Publication No. 3741, p. 99 (1944)).

ACKERET, J., AND KELLER, C.: " Hot-air Power Plant," in *The Engineer,* Vol. 169, No. 4397, p. 373 (1940).

ALLEN, J. E.: "An Elementary Theory of the Transient Pinched Discharge," in *Proc. Phys. Soc.,* Sec. B, Vol. 70/1, p. 24 (January 1957).

ATKINSON, R., AND HOUTERMANS, F.: " Zur Frage der Aufbaumöglichkeit der Elemente in Sternen," *Zeitschrift für Physik,* Vol. 54, p. 656 (1929).

AYRES, E.: " Power from the Sun," in the *Scientific American,* Vol. 183, p. 16 (August 1950).

BAILEY, F. M: " Exploration of the Tsangpo," in *Journal of the British Geographical Society,* 1914, p. 341.

BETHE, H. A.: " Energy Production in Stars," in the *Physical Review,* Vol. 55, p. 434 (1939).

BETHE, H. A., AND CRITCHFIELD, C. L.: " The Formation of Deuterons by Proton Combination," in the *Physical Review,* Vol. 54, pp. 248 and 862 (1938).

BROWN, H., AND SILVER, L. T.: " The Possibilities of Securing Longrange Supplies of Uranium, Thorium, and other Substances from Igneous Rocks," in Geneva Reports, Vol. 8, pp. 129–132.

CARRUTHERS, R., AND DAVENPORT, P. A.: " Observations of the Instability of Constricted Gaseous Discharges," in *Proc. Phys. Soc.,* Sec. B, Vol. 70/1, p. 49 (January 1957).

CISLER, W. L., GRISWOLD, A. S., AND CAMPBELL, F. D.: *A Large Fast-Neutron Breeder Reactor Power Plant* (Report 270 J/18 of the Fifth World Power Conference, Vienna, 1956).

CRICHTON, A. B.: " How much Coal do we Really Have? The Need for an Up-to-date Survey," in the *American Institute of Mechanical Engineers Technical Publication,* 2428 (1948).

FIELDNER, A. C.: " United States Fuel Reserves; Solid Fuels and their Suitability for Production of Liquid and Gaseous Fuels," in the *Oil and Gas Journal,* Vol. 47, p. 138 (March 17, 1949).

EICHELBERG, G., AND PFLAUM, W.: "Highly Supercharged M.A.N. Machinery," in *The Motor Ship* (April 1952).

GIBRAT, R., AND AUROY, F.: *Problémes posés par l'utilisation des marées,* Report III H/22, Fifth World Power Conference, Vienna, 1956.

HAHN, O., AND STRASSMAN, F.: " Über den Nachweis und das Verhalten der bei der Bestrahlung des Urans mittels Neutronen entstehenden Erdalkali-Metalle," in *Die Naturwissenschaften*, Vol. 27, p. 11 (6 I 1939).

HEADLAND, H.: " Tidal Power and the Severn Barrage," in *Proceedings of the Institution of Electrical Engineers*, Vol. 96, Part 2, p. 427 (June 1949).

HINTON, SIR CHRISTOPHER: " Nuclear Reactors and Power Production," in *Atomics*, Vol. 5, No. 4, p. 115; No. 5, p. 147; No. 6, p. 174 (April-June 1954).

HOTTEL, H. C., AND WOERTZ, B. B.: " The Performance of Flat-plate Solar-heat Collectors," in *Transactions of the American Society of Mechanical Engineers*, Vol. 64, p. 91 (February 1942).

HUBBERT, M. K.: " Energy from Fossil Fuels," in *Science*, Vol. 109, p. 103 (February 4, 1949).

KELLER, C.: " Closed-cycle Gas Turbine, Escher-Wyss-AK Development, 1945–50," in *Transactions of the American Society of Mechanical Engineers*, pp. 835–850 (August 1950).

KRUSKAL, M., AND SCHWARZCHILD, M.: *Proc. Roy. Soc.* (London), A. 223 (1954).

KURCHATOV, I.: " Russian Thermo-nuclear Experiments," in *Nucleonics*, Vol. 14, No. 6, pp. 36–43 (June 1956).

LANE, J. A.: " Growth Potential of U.S. Nuclear Power Industry," in *Nucleonics*, Vol. 12, No. 6, pp. 12–17 (June 1954).

LAPP, R. E.: *Atoms and People* (Harper, New York, 1956).

LAWSON, J. D.: " Some Criteria for a Power-producing Thermo-nuclear Reactor," in *Proc. Phys. Soc.*, Sec. B, Vol. 70/1, p. 6 (January 1957).

LESSING, L. P.: " The Gas Turbine," in the *Scientific American*, Vol. 189, No. 5, p. 65 (November 1953).

MILES, F. T., AND WILLIAMS, CLARKE: *Liquid Metal Fuel Reactor* (Report 8/P/494 of the International Conference on the Peaceful Uses of Atomic Energy, Geneva, 1955).

MILNE, G. R., AND MOORE, W. T.: *The Proposed Consolidated Edison Company of New York Nuclear Power Plant* (Report 269 J/17 of the Fifth World Power Conference, Vienna, 1956).

PEASE, R. S.: " Equilibrium Characteristics of a Pinched Gas Discharge cooled by Bremsstrahlung Radiation," in *Proc. Phys. Soc.*, Sec. B. Vol. 70/1, p. 11 (January 1957).

POST, R. F.: " Controlled Fusion Research—An Application of the Physics of High-temperature Plasma," in *Review of Modern Physics*, Vol. 28, No. 3, pp. 338–362 (June 1956).

QUIBY, H.: " Trials of an Aerodynamic Turbine," in *The Oil Engine* (November 1945).

SÄNGER, E.: " Stationäre Kernverbrennung in Raketen," in *Astronautica Acta*, Vol. 1, pp. 61–88 (1955).

SPITZER, LYMAN: *Physics of Fully Ionized Gases* (Interscience Publishers, Inc., New York, 1956).

TAYLOR, R. J.: " Hydromagnetic Instabilities of an Ideally Conducting Fluid," in *Proc. Phys. Soc.*, Sec. B, Vol. 70/1, p. 31 (January 1957).

TELKES, M.: " Review of Solar House Heating," in *Heating and Ventilating*, Vol. 46, p. 68 (September 1, 1949).

TELKES, M.: " Low-cost Solar-heated House," in *Heating and Ventilating*, Vol. 47, p. 72 (August 1950).

THIRRING, HANS: " Thermo-nuclear Power Reactors—Are They Feasible? " in *Nucleonics*, Vol. 13, No. 11, pp. 62–66 (November 1955).

THOMPSON, W. B.: " Thermo-nuclear Reaction Rates," in *Proc. Phys. Soc.*, Sec. B, Vol. 70/1, p. 1 (January 1957).

TROMBE, FELIX: " Le four solaire, le plus grand du monde," in *Atomes*, Vol. 8, No. 91, pp. 327–332. (October 1953).

UNION ELECTRIC-MONSANTO CHEMICAL TEAM REPORTS: " Progress on Nuclear Power Plants," in *Electrical World*, Vol. 139, No. 26, pp. 85–87 (June 29, 1953).

U.S.A.E.C.: *Major Activities in the Atomic Energy Programs* (January–July 1954).

WARD, F. KINGDON: *The Riddle of the Tsangpo Gorges* (E. Arnold, London, 1926).

WEISSÄCKER, C. V.: Über Elementverwandlungen im Inneren der Sterne," in *Physikalische Zeitschrift*, Vol. 39, p. 633 (1938).

WHITE BOOK: *Programme of Nuclear Power* (Cmd. 9389) (H.M.S.O., London, 1955).

WINTERBERG, F.: " Thermonukleare Reaktionen in technischer Sicht," in *Atomkernenergie*, Vol. 1, No. 6, pp. 199–201 (June 1956).

ZINN, W. H.: " Basic Problems in Central Station Nuclear Power," in *Nucleonics*, Vol. 10, No. 9, p. 8 (September 1952).

Conversion Tables of Units

Symbols for Decimal Multiples

Prefix	Symbol	Multiple
micro	μ	10^{-6}
milli	m	10^{-3}
centi	c	10^{-2}
deci	d	10^{-1}
hecto	h	10^{2}
kilo	k	10^{3}
Mega	M	10^{6}
Giga	G	10^{9}
Tera	T	10^{12}

Length

mm = millimetre; cm = centimetre; dm = decimetre; m = metre;
km = kilometre

1 cm = 0·3937 in;	1 in = 2·540 cm
1 m = 3·281 ft;	1 ft = 30·48 cm = 0·3048 m
1 m = 1·0936 yd;	1 yd = 0·9144 m
1 km = 0·62140 miles;	1 mile = 1·609 km
1 km = 0·5396 nautical miles;	1 naut. mile = 6080 ft = 1·853 km

1 micron = 0·001 mm = 10^{-4} cm = 10^{-6} m
1 Ångstrom = 10^{-8} cm
1 X.E. = 10^{-11} cm

Units of Length used in Astronomy

1 astronomical unit (average distance of the earth from the sun) =
   0·93003 × 10^{8} miles = 1·4964 × 10^{8} km
1 light-year = 5·88 × 10^{12} miles = 0·9461 × 10^{13} km
1 parsec = 3·26 light-years = 1·917 × 10^{13} miles = 3·084 × 10^{13} km

Area

cm² = square centimetre; m² = square metre, etc.

1 cm² = 0·155 sq in;	1 sq in = 6·452 cm²
1 m² = 10·764 sq ft;	1 sq ft = 0·09290 m²
1 hectare = 10^{4} m² = 2·471 acres;	1 acre = 0·4047 hectare

1 km^2 = 247·1 acres;   1 acre = 0·004047 km^2
1 km^2 = 0·3861 sq. miles;   1 sq mile = 640 acres = 2·59 km^2

Unit of Area used for Nuclear Cross-sections:
1 barn = 10^{-24} cm^2

Volume

cm^3 = cubic centimetre; m^3 = cubic metre, etc.

1 cm^3 = 0·0610 cu. in;   1 cu in = 16·387 cm^3
1 litre = 10^{3}cm^3 = 0·220 imp gal;   1 imp gal = 4·546 litres
1 litre = 0·2643 U.S. gal;   1 U.S. gal = 3·785 litres
1 m^3 = 35·314 cu ft;   1 cu ft = 0·028317 m^3
1 m^3 = 1·308 cu yd;   1 cu yd = 0·7646 m^3
1 m^3 = 6·292 barrels;   1 barrel = 42 U.S. gal = 0·159 m^3
        = 159 litres
1 m^3 = 8·107 × 10^{-4} acre-ft;   1 acre-ft = 4·356 × 10^4 cu ft
        = 1233·5 m^3

Mass and Weight

g = gramme; mg = milligramme; kg = kilogramme

1 g = 0·03527 ounces avdp.;   1 oz avdp. = 28·35 g
1 kg = 2·205 lb avdp.   1 lb = 16 oz = 0·4536 kg
1 metric ton = 1000 kg = 0·9842 long tons = 1·1032 short tons
        1 long ton = 1·016 metric tons

MU (mass unit) = Unit of Atomic Weights
1 MU = 1·66 × 10^{-24}g;   1 g = 6·025 × 10^{23} MU

Pressure

at = technical atmosphere = kg per cm^2
1 at = 14·223 lb per sq in;   1 psi = 0·070307 at

Work, Energy, and Heat

kgm = kilogramme-metre; ft–lb = foot-pound; j = joule; kj = kilojoule;
Ws = Watt-second; kWh = kilowatt-hour; H.P.h = horse-power-hour; cal
= calorie; kcal = kilocalorie; B.Th.U. = British Thermal Unit;
ev = electron-volt; Mev = Megaelectron-volt

1 kgm = 7·233 ft–lb;   1 ft–lb = 0·1383 kgm
1 kgm = 9·8037 j;   1 kj = 10^{10} ergs = 102 kgm
1 kWh = 3·67 × 10^5 kgm = 2·656 × 10^6 ft–lb
1 kcal = 3·968 B.Th.U.;   1 B.Th.U. = 0·252 kcal
1 ft–lb = 1·285 × 10^{-3} B.Th.U.;   1 B.Th.U. = 778·3 ft–lb
1 kgm = 2·342 cal;   1 cal = 0·427 kgm
1 kgm = 9·294 × 10^{-3} B.Th.U.;   1 B.Th.U. = 107·6 kgm
1 j = 0·239 cal;   1 cal = 4·186 j

1 kj = 1 kWs = 0·948 B.Th.U.;   1 B.Th.U. = 1·055 kj
1 kWh = 860 kcal;   1 kcal = 1·163 × 10^{-3} kWh
1 kWh = 3413 B.Th.U.;   1 B.Th.U. = 2·93 × 10^{-4} kWh
1 therm = 10^5 B.Th.U. = 2·52 × 10^4 kcal
1 Q = 10^{18} B.Th.U. = 2·52 × 10^{17} kcal = 2·93 × 10^{14} kWh = 293,000 TWh
1 Mev = 1·6 × 10^{-6} ergs = 1·6 × 10^{-13} j = 1·52 × 10^{-16} B.Th.U. = 3·82 × 10^{-17} kcal = 4·45 × 10^{-20} kWh

Heat Value

1 kcal/kg = 1·8 B.Th.U./lb;   1 B.Th.U./lb = 0·556 kcal/kg
1 kcal/m^3 = 0·1124 B.Th.U./cu ft;1 B.Th.U./cu ft = 8·9 kcal/m^3
1 kcal/litre = 15 B.Th.U./U.S. gal;1 B.Th.U./U.S. gal = 0·0666 kcal/litre

Power

kW = kilowatt; H.P. = horse-power. kcal/s = kilocalorie per second
1 kW = 1·341 H.P.;   1 H.P. = 0·7455 kW
1 kW = 0·239 kcal/s;   1 kcal/s = 4·186 kW
1 kW = 0·948 B.Th.U./s;   1 B.Th.U./s = 1·055 kW
1 H.P. = 0·178 kcal/s;   1 kcal/s = 5·615 H.P.
1 HP = 0·7067 B.Th.U./s;   1 B.Th.U./s = 1·415 H.P.

Electric Charge

1 Coulomb = 1 Ampere-sec = 3 × 10^9 electrostatic units

APPENDIX II
Physical Constants

Gravitational acceleration: g = 9·806 m/s^2 = 32 ft 2 in per sec^2
Mechanical equivalent of heat: 427 kgm/kcal = 778·3 ft–lb/B.Th.U.
Gas constant for one mol: R = 8·31 joule/grad
Faraday Constant (electric charge transported by one gramme-atom of univalent ions): F = 96494 Coulombs
Velocity of light: c = 2·9978 × 10^{10} cm/·s
Planck Constant: h = 6·626 × 10^{-27} ergsec
Loschmidt (or Avogadro) Number: L = 6·025 × 10^{23}
Boltzmann Constant: k = R/L = 1·38 × 10^{-16} erg/grad
Mass of proton: m_H = 1·67 × 10^{-24} g
Mass of electron: m_0 = 0·9107 × 10^{-27} g
Charge of electron: e = 4·80 × 10^{-10} electrostatic units
    = 1·6 × 10^{-19} Coulombs
Rest energy of electron: m_0c^2 = 0·819 × 10^{-6} ergs = 0·512 Mev

Index

(Personal names are set in italic type)

621.4
T447

94689